# The Second Passive Solar Catalog



David A. Bainbridge

The Passive Solar Institute
P.O. Box 722
Davis, CA 95616



ISBN 0-933490-02-X

## a) How to Use This Book

This book, the second in the series of Passive Solar Catalogs, emphasizes passive solar design for retrofit applications. Passive solar retrofitting can make your existing building more economical to operate—and also more comfortable—both in winter and in summer.

The 1st and 2nd Catalogs are designed as a set and we recommend buying the First Catalog if you don't already have it, at $7.50 ppd. It includes much more detailed information and extensive references on The Foundations of Passive Heating and Cooling (Chapter 4), an introduction to Passive Solar Heating and Cooling (Chapter 5), twelve examples of Passive Design (Chapter 6), Principles and Practice of Passive Solar Water Heating (Chapter 7), Daylight Design for Natural Lighting (Chapter 8), and Planning for Solar Access in Subdivision Design (Chapter 9). It is a sound foundation for this second catalog.

This second Catalog can be used in three ways depending on your interest and background.

1st: If you are interested in Passive Solar retrofit, read Chapters 1, 2, 3, 4 and 5 and skim 6, 7, 8, 9 and 10. (Also read Chapters 3, 4, 5, 7, 8 of the First Catalog).

2nd: If you are alread familiar with Passive Design and are looking for specific information go to Chapter 6: Architectural Details, Chapter 8: Passive Components, and Chapter 9: Passive Consultants.

And finally, if you are more interested in the role Passive Solar Design will play in America's future read Chapters 1, 2, 3, 11, 12, 13, and 14 (also Chapter 12 of the 1st Catalog). We show why it is imperative for America to reduce her energy dependence and how passive solar retrofitting can help.

## Table of Contents



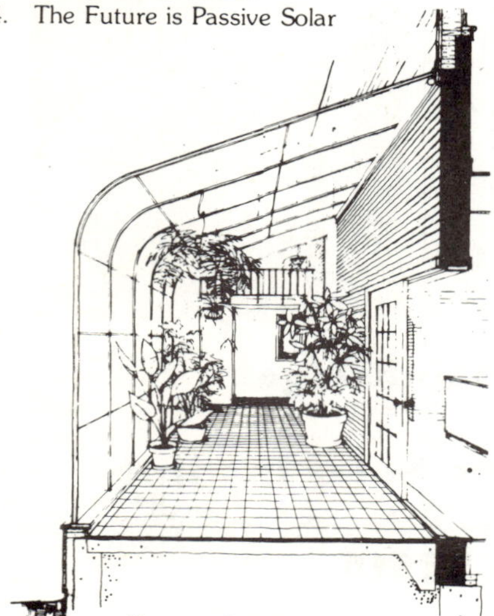

*Courtesy: Four Seasons Green houses*

## b) How This Book Came to Be

The Passive Solar Catalogs were conceived while I was working as a judge on the HUD Cycle 4 Passive House Design Review. Designs were overly complex, expensive, relied too heavily on mechanical equipment, and all too often would not work.

Many of the submittals ignored or hadn't understood the basics of passive design, climatically adapted building, and energy conservation. They had often also failed to design architectural details that would work and custom designed components that were already available off-the-shelf.

The First Catalog was designed to get the basic information out and serve as a vehicle for letting designers know what products and services were available. It has been an unqualified success with over 7,000 copies in print, enthusiastic reviews, and very favorable comments from readers.

The Second Catalog complements the First issue and tackles the more critical question of passive solar retrofit. As the final chapter of the 1st Catalog pointed out, we could save 3 quadrillion BTU/year by 1985 if we embarked on an ambitious retrofit program. It is hoped that this book will provide some of the stimulus needed to get this program established. Our survival as a free and economically strong country depends on it.

Your comments on the format and content of this Catalog would be appreciated. The lst Catalog was revised in later printings in response to comments from readers. We would particularly like to encourage you to let us know what your experience with components listed here is—do they work, ship on time, etc. We'll use your comments in the 3rd Catalog. Thank you.

With best wishes for your Solar Future,


David A. Bainbridge
President
Passive Solar Institute
P.O. Box 722
Davis, CA 95616

## c) Credits


WRITING: David Bainbridge
EDITING: David Bainbridge
PHOTOS: David Bainbridge and others
ILLUSTRATIONS: David Bainbridge and Greg Wolters
TYPING: Lola Brocksen and Priscilla Smith
TYPESETTING: Ronnee Ubeda, Connie Yee Michael Kearns, Jeff Frane
PMTS: Susan Johnson and Gary Robeck
PLATE MAKING: Johnny Print
PRINTING: Johnny Print
DISTRIBUTION: Book people


## d) Acknowledgements


Thanks to: All those who helped, including special thanks to:
Chet Mosley
Lola Brocksen
Instant Copy
Anchor Printing
Johnny Print
Aurora Typesetting
Betsy Wauchope
Mountain Photo
Net Energy
Ericksen Insul-shutter
Jeff Reiss
Virginia Thigpen
Ken Haggard
Phil Niles
The Rodale Press for permission to use materials on shutters and retrofitting that first appeared in their publications.
The California Energy Commission for permission to reprint some of the details from the Passive Handbook.
Greg Wolters
Joe Carter
Don Wright
Howard Smukler
Gil Souza
Jack Klug
Bob and Chris Sowers


## e) Dedication

To Victor and Aladar Olgyay for their many years of research and writing on climatically adapted building design. Their major contributions are listed below and are must reading for the passive solar designer.


Olgyay, Olgyay and Associates (1953) **Application of Climate Data to House Design,** Housing and Home Finance Agency, Washington, D.C.

Olgyay and Olgyay (1954) "Sol-air Orientation," **Architectural Forum,** March.

Olygyay and Olgyay (1975) **Solar Control and Shading Devices,** Princeton University Press, $7.50.

Victor Olgyay (1953) **Design with Climate,** Princeton University Press $25.00.

# Chapter 1. What is Passive Solar Design

Many of the most effective and workable solar applications for new and retrofit installations are those called *Passive* Solar Systems. Although there are many definitions of this term several points are common to most of them. The definition we like at PSI is that *Passive* solar systems are those that:

- Use energy from the sun and climate resources to provide heating and cooling.
- Do not rely on external sources of energy to function. (They will function even if the power goes off).
- Decrease the building's use of auxiliary energy.
- Combine collection and storage.
- Are simple and low in cost.
- Generally have a very long lifetime.
- Require little non-renewable energy to to build and maintain.

It is a just commentary on our society that *Active* systems, which won't work without fans, pumps, controls and power, are given the strong name and that Passive Systems which typically run by themselves, are given the weaker name. But then as Malcolm Wells puts it, "An SST, smashing its way through the sky by burning lakes of fuel is high technology. A bicycle, run by a carrrot-eating machine, and thereby operating at a far, far lower cost per passenger-mile is low technology."

Three examples will perhaps more clearly illustrate what passive solar design is:

## 1. Passive Space Heating

A south window is the simplest passive heater. Any cat can find the warmest spot in the house on a cold winter day—in the sun streaming in a south window.

A solar green house is one of the best passive heating systems.

## 2. Passive Space Cooling

Any dog can find the coolest spot on a hot summer day—under a big shade tree with a gentle breeze. It combines ventilation, shade, and evaporation to provide comfort-much like many passive solar systems.

## 3. Passive Solar Water Heater

A black can of water sitting in the sun is a simple hot water heater. Putting it in an insulated box with south glass will make it work better.

And that is really all there is to it! The basics of passive solar design can be understood by anyone. Applying them to your house, office, or farm is straight-forward, simple, and economical. Take a careful look through this book and decide on your first step toward energy self-sufficiency. It will prove interesting, fun, and very rewarding.

# Chapter 2: Why Passive Solar Makes Sense

Now that you have an idea of what Passive Solar is all about, let's take a look at why it deserves even greater attention that it has received to date. The following could and perhaps should be expanded and added to but I think these reasons will demonstrate why passive solar makes sense.

## 1. Heating and Cooling

Passive systems can provide both heating and cooling by using the sun and climate resources. This reduces the cost, energy demand, and increases the comfort level. In many areas of the U.S. and Canada they can provide full heating, full cooling, or both.

## 2. Simplicity

Passive systems utilize basic principles, common materials, and construction skills that are available locally. This simplicity puts passive systems within reach of every one, both here and abroad. The simple materials ensure that most passive systems will have very low "net" energy cost as well, which is also very important.

## 3. Reliability

Passive systems are inherently reliable as they have few or no moving parts. In addition, they will work even if the "power" goes off and can provide some security in an increasingly troubled world. The biggest advocates of solar green house in Vancouver, B.C. are the older British couples who went through the blitz in England and know what it means when the power goes off.

## 4. Reduce First Cost To Builder (and potentially for buyer)

A well designed passive house will often have a lower first cost than a standard house with air conditioning and central heating. This saving can be used for greater profit or as a pass through to reduce the sales price. The Federal and State tax credits are increasing these savings considerably.

## 5. Reduce Operating Cost For Buyer

An even bigger incentive is the potential for reducing utility bills for the buyer. This can be used to help the buyer get a bigger loan, because banks are beginning to consider expected utility bill in their evaluation of home loans. Savings are also very attractive for commercial applications, although they have received less attention to date.

## 6. Comfort

Passive buildings are extremely comfortable. This comfort results from more balanced surroundings, particularly the stable radiant temperatures and humidity.

## 7. Quiet

Passive structures are also very quiet due to added mass, added insulation, and double and triple glazing. The lack of noisy and irksome furnaces, compressors, fans, and ducts helps contribute to peace of mind and health.

## 8. Health

Passive solar heating and cooling systems are much healthier than regular mechanical systems for a variety of reasons. These include: lack of air pollutants associated with combustion, improved humidity regime, and better ion balance. This area has only recently been explored. Studies on indoor air pollution from existing stoves and furnaces are not encouraging. Levels often exceed E.P.A. long term exposure limits and as one researcher says, "It is thus essential that air pollution characterization of the indoor environment be undertaken in order to assess correctly the human exposures in epidemiological studies ..." Passive Solar heating with wind electric or active solar back up would be healthiest.

## 9. Ease of Mass Marketing

Market penetration will be very rapid due to the simplicity, low cost, and readily available materials and skills needed for passive systems. While this is particularly true for new buildings, it will also prove true for retrofitting existing structures.

The ease of mass marketing is primarily a result of three factors. First, the materials used are commonly low in cost and readily available. Second, the skills and tools needed to manufacture most passive solar components are readily available and low in cost. And finally, passive solar systems can in most cases be installed by local labor with tools and skills already in use.

All three of these factors are important. The availability of materials in the local market insures that few supply hang-ups will occur. Installation can occur rapidly without fear of delivery or manufacturing problems. The low materials cost reflects the low energy cost of most materials used and ensures that rapidly increasing energy prices will not disrupt the market.

The fact that skills and tools needed to manufacture passive solar components are readily available and low in cost is also very encouraging and suggests that very rapid market penetration can occur. For example, steel tanks for water storage in a passive solar house can be welded by almost any local welding shop. Thermal curtains and shutters, another vital component, can be manufactured by almost every drapery shop or seamstress, or cabinet and door company using standard techniques.

Finally, these passive components and passive solar systems can be installed by existing contractors and trades-people with their existing tools. No special equipment is needed nor is sophisticated training required. This ensures rapid delivery, fast penetration of the market, and good profitability for the installers.

## 10. More Jobs

An exact estimate of the employment impact of widespread use of passive solar systems is difficult to calculate. However, existing information does provide a rough comparison of employment impacts of a given investment in passive solar versus active solar or traditional sources used by public utilities. An investment of $100,000 will provide 10 jobs in passive solar, 5 jobs in active solar, or one job in a conventional power plant.

This figure can be used to generate an estimate of the projected employment impact of investment in passive solar systems in the United States from 1980-1985 based on the estimate of desired investment in solar systems. Assuming the estimate of desired investment in solar systems. Assuming the estimates made in Chapter 14 are reasonable (they are probably conservative) passive solar systems could provide 1,000,000 jobs per year between 1980-1985. Much of this work will be in retrofits and it will help balance out the loss in construction jobs from higher interest rates.

## 11. Local Jobs

These jobs are primarily local jobs which is very desirable. First, it can reverse local and national balance of payment problems. Money now going to distant heater and air conditioner manufacturers, and money flowing to the Mid-east for the oil to fire generating plants can now remain and be active in the local market place. The tremendous value of this decentralization should receive more emphasis and research than it has to date.

## 12. Regional Identity

One of the more subtle, positive impacts of widespread use of passive solar systems will be the re-establishment of regional architectural types. This diversity and identity have values which have only recently been realized. The universal house and office now applied indiscriminently in New York, Los Angeles, Winnipeg, Cairo, Delhi, and Tokyo will be replaced by more comfortable, efficient, and appropriate designs. When we get off a plane in the future it may not be necessary to consult a map to find out where we are.

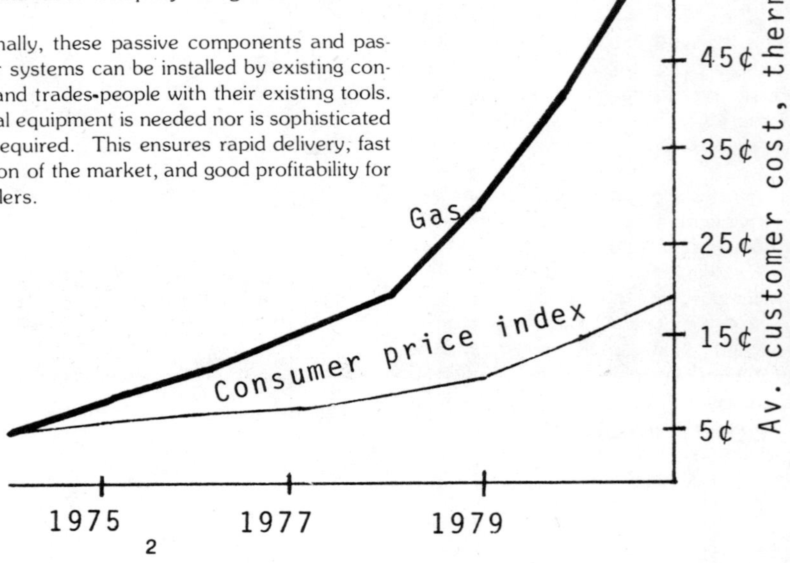

## 13. Peak Load Reduction

Although utilities in the past have, in most cases, studiously ignored passive solar systems as a load management tool, (rightly so according to the public utility balance sheet with profit based on total investment), they are increasingly interested in any methods which can help reduce peak power demands. As a result, we are now seeing passive systems promoted by the utilities. This will be to society's benefit as passive solar systems cost about 1/100 as much as adding generating capacity and will employ 6-10 times as many people for a given investment. Passive systems are particularly effective at reducing peak heating and cooling loads.



## 14. Environmental Impact

In addition to reducing peak load, passive solar systems have few detrimental environment impacts. The materials used are by and large renewable, non-toxic, recyclable, and labor intensive. This is in contrast to virtually all of the other methods of generating energy. When compared to the impact of coal, nuclear, wood, or hydro power there are "no" adverse impacts.

## 15. Saving to Society

The reduced energy use made possible by the passive systems also provides several dollar savings to society. It reduces the need for investment in additional power plants - freeing capital for better uses. Consider that the expected capital requirements for large central power plants was calculated at $1 trillion between 1976 and 1985. That money would easily build 10 million new passive solar houses and remodel almost all of the houses in America in the same period. Which is the better investment?

The use of passive solar systems has been "passive" long enough. The need for action is inescapable. Many passive systems are well understood and many more are being developed. The sun is shining. Let's get to work.

## 16. Risk

Perhaps the most important reason for favoring passive solar design is the low element of risk involved. Passive systems have low risk from: supply disruption, mechanical failure, health hazards, and breakdown caused by natural disaster or terrorist activity.

As long as the sun continues to shine passive solar systems will work. This is a considerably better bet than continued supplies of oil from the Mid-east or even coal distribution within the U.S. As recent events have demonstrated, the rail systems may or may not be able to move coal to the power plants when it is needed. As the rail system continues to deteriorate this problem will become increasingly serious. A strike within the transportation or energy industry can easily disrupt the centralized non-renewable energy systems but it can't stop your passive solar heating system or hot water heater from working.

The smallness and simplicity of passive solar systems remove the element of risk from mechanical failure. If worse comes to worst and your hot water system fails (tanks will corrode, although less quickly than they do in a conventional heater) you can always go next door for a shower. But if a 2000 megawatt reactor goes down at the wrong time it may take the whole system with it, causing regional blackouts—and complete failure of conventional electrical heating systems.

The most favored option of many utilities has been nuclear generating plants. Although their safety record to date has been good they are particularly sensitive to risk of health hazard because they involve such dangerous materials. When even a modest mechanical failure occurs it can cause the regulating agencies to shut all similar systems down until a specific mechanical defect is fixed. The limited number of types of systems has made this particularly troublesome and several cases of generic shutdowns have already occured. In the event of a serious failure or accident public outcry might shut all of the reactors down - leaving everyone desperately short of electricity.

And finally, decentralized energy systems are much less likely to be damaged by natural disasters or terrorists. A large earthquake, such as the New Madrid, Missouri quake of 1812 would quite possibly destroy some of the generating capacity and much of the transmission and distribution system in a broad area. This might take weeks, months, or even years to fully repair. Yet passive solar systems will still be working in most of the houses in the affected area. Houses will stay warm (or cool) and as soon as water lines are repaired people can have hot water. The threat of terrorist acts is also increasing and centralized utility systems are very attractive targets. Passive solar systems are not.

Passive solar systems can provide the security of a comfortable home in an increasingly troubled world.

## 17. Speed of Changeover

It may take a utility 10-15 years to get a power plant designed, approved and built. A passive solar system can be designed and built in three months (for a house), a week (for a hot water heater), or a few hours (thermal drapes or shutters). This is particularly important: when we decide we must change--we can do it in a matter of years, rather than decades.

## 18. More Efficient

Passive solar systems are also a better investment because they are more efficient. As Amory Lovin so clearly stated it, "using electricity to heat water is like using a chainsaw to cut butter." Solar energy falls in usable quantities on almost every house in the U.S. and passive solar systems are ideally suited to the demand for relatively low temperature (less than 200°F) heating demand for space heating, water heating, and some process heat. They can also eliminate the need for air conditioning in many areas. In a recent study of end use of energy in the U.S. almost 30% of the total energy use was found to be for demands which passive solar systems can meet.

## 19. Personal Control

One of the most important reasons for using Passive Design is the fact that it gives you control of your own energy supply. You still may have to wait in line for gas, but at least you'll be able to return to a comfortable and economical home—that is virtually independent from ever incresing energy prices. Do it now.

**RENEWABLE FUELS FOR ENERGY INDEPENDENCE**

**DON'T TREAD ON ME!**

## 20. Survival of America

And finally, passive solar design will prove critical to America's survival. Our energy use has again surpassed what we used before the first oil crisis of 1973 and in fact, our energy situation has gotten worse. Oil imports make up an even larger share of our energy supply—and make us ever more susceptible to political pressure. As several countries demonstrated in 1973 such dependence may cause almost complete political amorality in return for oil. This is as dangerous as heroin addiction—both for the addict and the supplier.

If we wish to remain free and proud, we need to eliminate our dangerous dependence on foreign oil. Passive solar design will play a critical factor in kicking the habit.

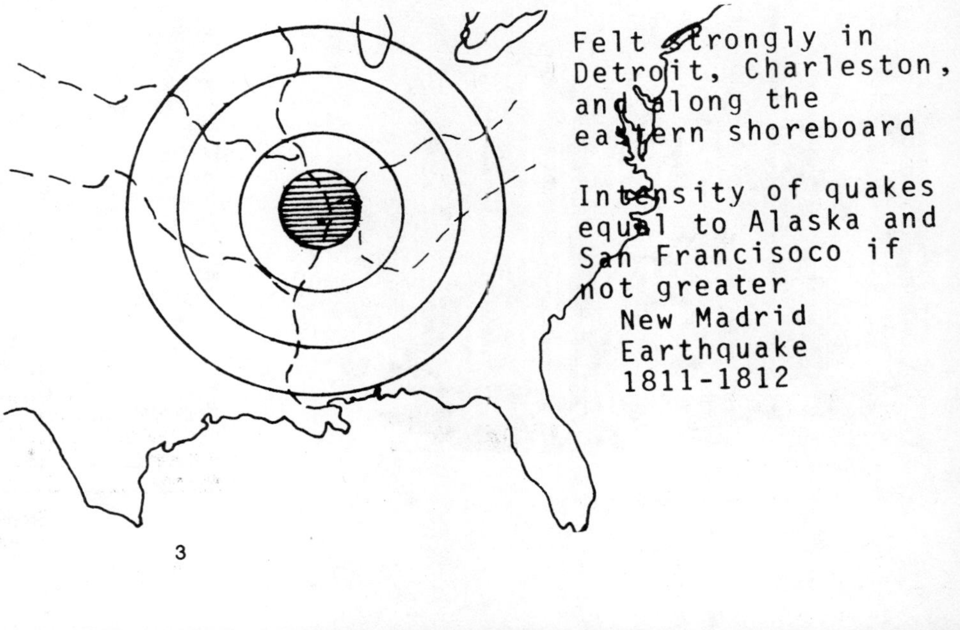

# Chapter 3. WHERE PASSIVE IS NOW

The growth in Passive Solar development in 1979 was spectacular and almost exceeded our own hopes for the year. Passive solar design came of age and is beginning to be included in the market as an accepted and desirable method for space heating and cooling and hot water heating.

While hundreds of passive installations were built in 1978 many thousands were built in 1979 and tens of thousands may be expected in 1980. Major developers are incorporating passive solar design in their standard plans or offering it as an option. In Sacramento, California, five passive subdivisions are underway where there were none in 1978. A development underway in Southern California will have 2,900 passive solar units at completion and many more smaller developments are underway across the country.

*Professional Builder,* the developers magazines, featured passive solar design in many articles in 1979 and now offers several passive solar house plans for sale. And the builders organization have been actively involved in passive solar promotion. Particular credit goes to the California Building Industry Association for their efforts, including an extensive and heavily attended seminar series and a hand book of solar design—including passive.

Professional Builder Plan C

Government involvement also continued to increase in 1979. California completed a Passive Solar Competition which was well conducted and received some excellent submissions. The California Energy Commission supported work on a comprehensive Passive Solar Handbook which will be available in early 1980. Several of the details in Chapter 6 were taken from this excellent manual. And the Tennessee Valley Authority made considerable progress on their excellent passive solar program including: an outreach program to educate area architects, engineers, and builders; development of a Regional Design Manual; and development, demonstration, and construction of residential and commercial projects utilizing passive design. Special credit goes to S. David Freeman for turning T.V.A., the largest electric utility in the country, toward the sun. The table below indicates State incentives available in 1979.

## State Solar Incentives

LEGEND

- ● Active and all passive systems
- ■ Active and some passive systems
- ▼ Active systems only

| | Income Tax Credit | Property Tax Exemption | Business Tax Incentive | Sales Tax Exemption | State Grant Program / State Loan Program | Mandatory Building Code | Mandatory Energy Code |
|---|---|---|---|---|---|---|---|
| Alabama | | | | | | ● | ● |
| Alaska | ■ | | | | | ● | |
| Arizona | ● | ● | | ● | | | |
| Arkansas | ■ | | | | | ● | ● |
| California | ● | | | | ● | ● | ● |
| Colorado | ● | ● | | | | ● | ● |
| Connecticut | | ● | | ▼ | ● | ● | ● |
| Delaware | ▼ | | | | | | |
| Florida | | | | ▼ | | ● | ● |
| Georgia | | ▼ | | ▼ | | ● | ● |
| Hawaii | ■ | ■ | | | | | ● |
| Idaho | ● | | | | | ● | ● |
| Illinois | | ■ | | | ● | | ● |
| Indiana | | ▼ | | | | ● | ● |
| Iowa | | ▼ | | | ● | ● | ● |
| Kansas | ▼ | ▼ | | | | ● | ● |
| Kentucky | | | | | | ● | |
| Louisiana | | ▼ | | | | | |
| Maine | | ■ | | | ● | ● | |
| Maryland | | ■ | | | | | |
| Massachusetts | ■ | ▼ | | ▼ | ● | ● | ● |
| Michigan | ● | ● | ▼ | | | ● | ● |
| Minnesota | ● | ▼ | | | | ● | ● |
| Mississippi | | | | ▼ | | | |
| Missouri | | | | | | | |
| Montana | ● | | | | ● | ● | ● |
| Nebraska | | | | | | | |
| Nevada | | ▼ | | | | | ● |
| New Hampshire | | ▼ | | | | | ● |
| New Jersey | | ● | | ▼ | | ● | ● |
| New Mexico | ● | | | | | ● | ● |
| New York | | | | | ● | ● | ● |
| North Carolina | ▼ | ▼ | | | | ● | ● |
| North Dakota | ▼ | ▼ | | | | | ● |
| Ohio | | | | | | ● | ● |
| Oklahoma | ▼ | | | | | | |
| Oregon | ● | ● | | | ● | ● | ● |
| Pennsylvania | | | | | | ● | |
| Rhode Island | | ▼ | | | | ● | ● |
| South Carolina | | | | | | | |
| South Dakota | | ▼ | | | | ● | |
| Tennessee | | ▼ | | | ● | ● | ● |
| Texas | | ▼ | ▼ | ▼ | | | ● |
| Utah | | | | | | | ● |
| Vermont | | ▼ | | | | ● | |
| Virginia | | ▼ | | | | ● | ● |
| Washington | | ▼ | | | | ● | ● |
| West Virginia | | | | | | ● | |
| Wisconsin | ● | | | | | ● | ● |
| Wyoming | | | | | | ● | ● |

Stephen Dalton (1979) "Dealing with Government", 4th Passive Solar Conference Proceedings, K.C.

The utilities have also been busy. For example, Pacific Gas and Electric in California built several passive solar demonstration houses, initiated a program that will pay up to $1,000 to builders who build passive solar homes, and is now heavily involved in developing an energy audit/solar design program that will provide free home analysis throughout the PG&E service area in 1980. Nearby, the Sacramento Municipal Utility District has also been busy. In 1979, they prepared a very good and very enthusiastic report on passive potential in the SMUD service area. It includes an evaluation of different system designs, costs, and performance. SMUD also sponsored a Passive Solar Institute Seminar on Energy Efficient Neighborhood Design. These are but two of the many utility programs involving passive design now underway.

**California Building Industry Association**

**Energy Education Programs**

Yet passive development was not confined to big business and government and in fact continued strongest at the grass roots, both in new construction and retrofit construction. Activity has been so hectic that no one can keep track of it anymore. My best guess would be that there were several hundred thousand passive projects completed or undertaken nationwide in 1979. As an example, a recent Solar Program in Grass Valley included tours of 12 local solar houses. They represent about 10% of the passive solar projects in this region, population 10,000. The level of interest is indicated by the response to a recent solar program. Planned for 500 people, over 650 were accommodated and perhaps another 200 turned away.

## Passive Solar Homes-- Grass Valley, California

*Courtesy: Foothill Solar Exchange*

Three Story wood frame residence with double-glazed 10' x 25' greenhouse. Active solar hot water system.

Two story wood frame house with single glazed greenhouse. Active solar hot water with wood assist.

Two story wood frame residence with attached greenhouse. Vented mass wall for thermal storage.

Two story wood frame house with solar greenhouse. Shading, ventilation, and swamp cooler for cooling.

Research and development in passive solar design was also much more extensive than it was as recently as 1978. Much of this work was represented in the 3rd and 4th Passive Solar Conferences, including almost 1,700 pages between them. Of particular value was the upswing in interest in passive cooling, an area where much still remains to be done.

Passive solar water heating also began to blossom. Several subdivision scale installations were made or begun including a 80-unit development by Sun Energy Builders in Sacramento. Several plan sets became available, including PSI's, and probably several thousand systems were installed nationwide. At year's end, two major efforts were underway to manufacture passive solar water heaters, on the scale of thousands to tens of thousands of units/year.

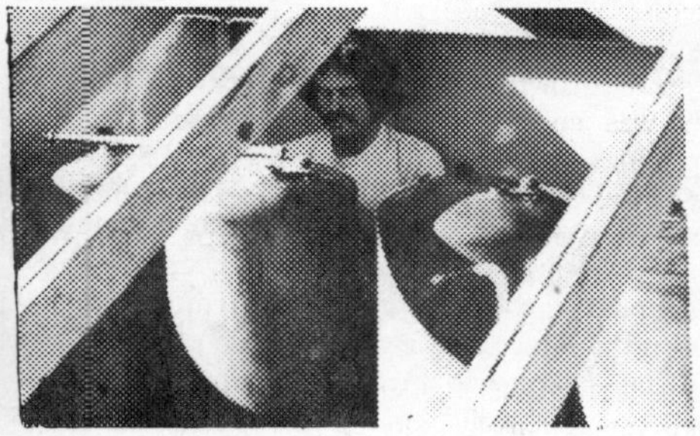

And finally, manufacturing was up several times from the level of activity in 1978. For example, major manufacturers of insulated drapes and shutters grew from 3 in 1978 to above 10 in 1979. And the products that were already out were improved. The increase in activity can be seen by comparing the listings in the 1st Passive Solar Catalog with the listings this year. The economics have become so compelling that the market is roaring.

All in all, it was a good year. The year to come will see even more rapid progress and quite possibly will reach the milestone of the 1 millionth passive solar installation. Let us hope so because it may be vital for America's survival.

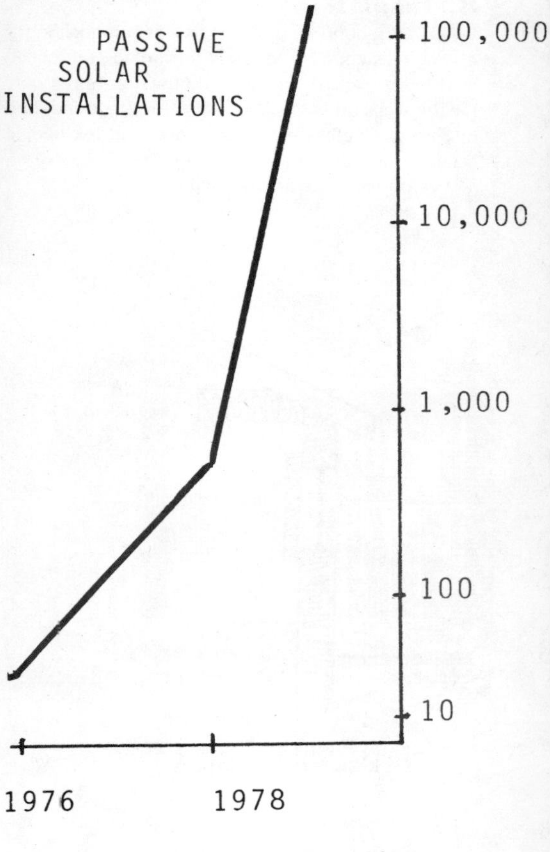

# Chapter 4: The Basis of Passive Solar Design

This Chapter provides a quick review of the basics of passive solar heating and cooling and hot water heating. For a more detailed look at these topics, as well as daylighting, solar access, the history of passive solar design, and much more please refer to the First Passive Solar Catalog.

This chapter is broken into three sections, covering respectively: energy conservation, passive solar heating and cooling, and passive solar water heating.

## Energy Conservation

The first step in passive design is improving the thermal performance of a building. Improving the energy efficiency of a building beyond the standard level is remarkably easy. In most areas it is reasonable to expect at least a doubling in the energy efficiency with only a modest increase in cost. The following steps should be included: A) orientation, B) insulation, C) weatherization, D) shading, E) modifying the environment.

### A. Orientation

Good orientation is essential for an energy efficient structure. The effects of bad orientation can be remedied to some extent — but doing so can be expensive. The best orientation is for a building to be longer east-west that it is north-south, with a major yard to the south, and most of the windows facing south and a few facing north, east, or west. This orientation will maximize solar heating in winter and minimize summer overheating by making best use of the difference in sun path. Even a structure with less than ideal shape will work well if the glazing is predominantly on the south side.

### B. Insulation

The second step in building or remodeling for energy efficiency is reducting unwanted conductive heat loss (or gain). Insulation is the key — not only for the walls and ceiling, but also for the foundation or slab perimeter, windows, doors, and the people inside. Insulated drapes and shutters are very effective on windows and sky lights.

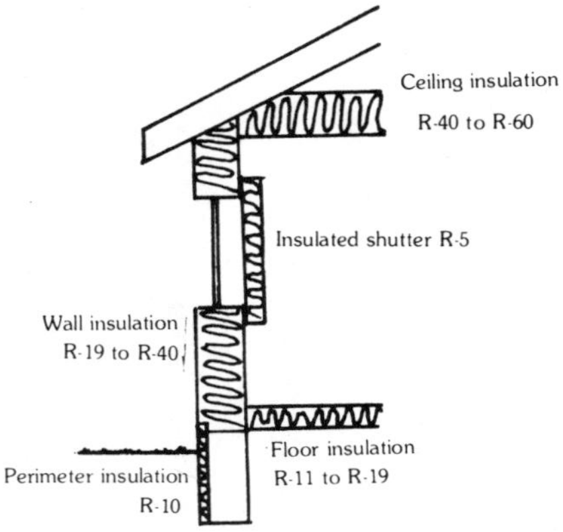


Desired Insulation

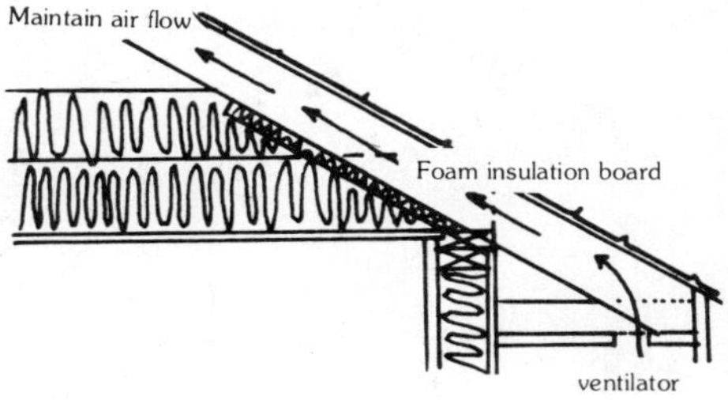


### C. Weatherization

Infiltration losses are as important as conductive losses and careful weatherizing is necessary. This includes both the obvious problems of weatherstripping doors and windows and also the more general problems of caulking and sealing of building joints, access holes, and other areas where unwanted infiltration occurs. Infiltration may easily account for 1/2 of the heat loss in a well insulated but poorly weatherized house.

The infiltration rate on a typical house is around 1.5 air changes/hour. It can go us significantly if a high wind is blowing. However, with careful attention to detail this can be reduced to about 0.2 air changes/hour. In colder areas an air-to-air heat exchanger is desirable for ventilation. It warms incoming fresh air with warm stale outgoing air.

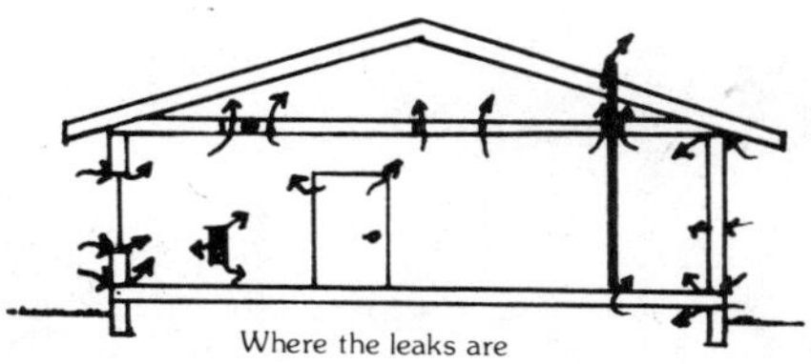

Where the leaks are

### D. Shading

The importance of shading is best illustrated with an example. In the Central Valley near Sacramento, 55 square feet of west facing glass will add 55,000 BTU to a house during the afternoon. A ton of air conditioning is required to offset this gain. And to run a typical air conditioner at this peak period of electrical demand requires almost 2 kw of added generated capacity. Total cost $3.000+.
Or a shade costing $20 could be put up.

Shading or solar control is relatively easy if the house is oriented properly. Overhangs and wing walls can shade the south windows. The other windows can be shaded with any of the exterior shades or shade screens.

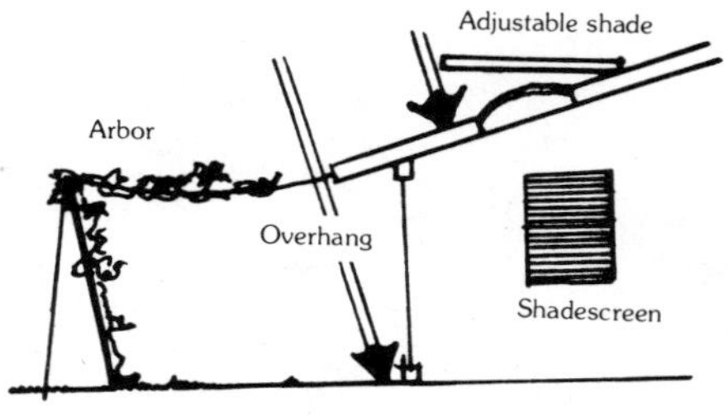

### E. Modifying the Environment

Very substantial savings in energy use for heating or cooling may be realized by altering the environment the house sits in. Two possibilities are particularly attractive: landscaping and berming or undergrounding. Landscaping can provide shading in the summer and wind protection in the winter. Berming or undergrounding can greatly reduce heating and cooling loads.

## Passive Space Conditioning

After improving the energy efficiency of a building, we can apply the following principles of passive solar design to make it a naturally heated and cooled house.

Passive solar systems are very effective for space heating. Passive heating systems differ widely in configuration, performance and cost, but do have several features in common. These include: glazing to admit the sun's energy, glazing control (insulated drapes or shutters) to reduce unwanted heat loss or heat gain through glazing, and thermal mass to store the sun's energy. The following rules apply to all passive solar heating systems.

### Six Rules for Low Cost Passive Heating:

1. **Make sure the building is weather tight and well insulated.**
2. **Make sure the structure is oriented properly, with most windows on the south.**
3. **Use the minimum amount of glazing necessary for heating.**
4. **Make glazing control simple, reliable, effective, and convenient to use.**
5. **Use the simplest, smallest, and most economical method of providing needed thermal mass.**
6. **Consider cooling demand as well and integrate design to achieve lowest cost heating and cooling.**

PASSIVE HEATING

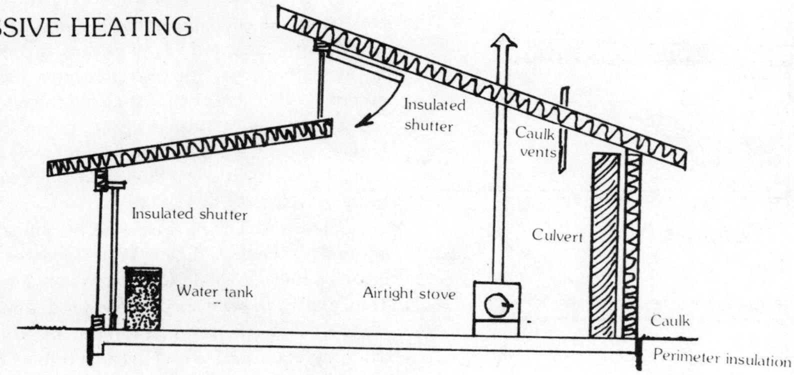


For passive solar cooling, the structure must emphasize different, although complimentary things. These include reducing unwanted heat gain in summer with shading, light colors, and similar factors. Cooling is provided by night ventilation (with cool night air), night sky cooling (to the sky vault), and/or evaporation.

The following rules apply to all passive cooling systems.

### Six Rules for Passive Cooling

1. **Make sure the building is oriented properly with most windows on south and north, few on east and west.**
2. **Provide shading for all windows in summer (awning, overhang, etc.)**
3. **Use light roof color.**
4. **Provide windows, vents, and fans (if necessary, for good ventilation and convective cooling.**
5. **Use thermal mass (water tanks, rock or concrete) to store nighttime cooling for use during the day.**
6. **Consider heating demand as well and integrate design to achieve lowest possible cost.**

PASSIVE COOLING

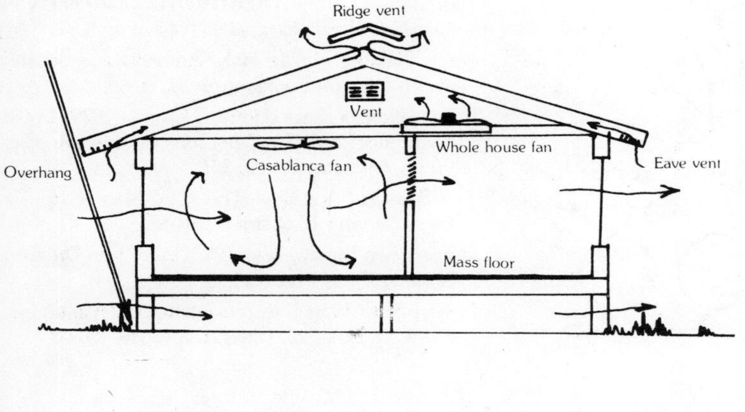


Five of the most promising passive solar heating and cooling systems are the simple passive house, solar greenhouse, water wall house, trombe wall house, and the roof pond house.

## Simple Passive House

Every standard frame house can be a passive solar house if it is properly designed and exposed to the sun. To work well it should be oriented so major wall and most windows face south. Added insulation and double pane windows may be desirable. Slab edge insulation is desirable. Exposing the slab will increase storage capacity. Use tiles integral color, linoleum, or parquet, (not carpet) to increase heat transfer. East and west windows must be shaded in hot areas. South window can be shaded with overhang or arbor. Interior insulated drapes or shutters are helpful.

**Rules**

Most windows face south (2/3 of total)
Windows shaded in summer
Insulate thoroughly
Weatherstrip thoroughly
Use light color roof (in areas with signicant cooling demand)
Provide good cross ventilation (whole house fan may be desireable)
Overheating in winter possible, may require window to be opened to cool off.
Thermal drapes desirable

## Solar Greenhouse

The solar greenhouse or solarium is particularly well suited for colder areas where heating is more important than cooling. Several versions are now being built the type shown here is the easiest to retrofit. A solar greenhouse may also provide cooling if it is shaded, well ventilated and has sufficient thermal mass.

**Rules**

Face South
Insulate thoroughly
Weatherstrip thoroughly
Thermal drapes desirable
Exterior shade desirable in summer (arbor?)
Open plan structure for good air circulation
Greenhouse area 1/5 floor area of house
Vent high and low for good circulation into house
Mass desirable in greenhouse for storage, use water tanks or drums

SOLAR GREENHOUSE

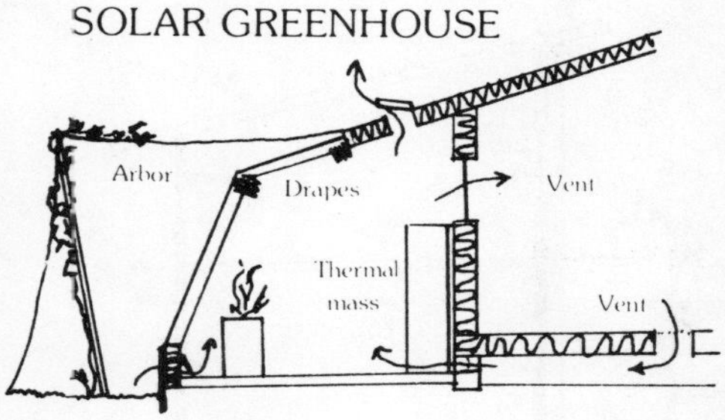


## Water Wall House

Adding termal mass to a simple passive house can greatly improve its performance for both heating and cooling. A custom steel water tank is a good way to provide this thermal mass. Other water containers have also been used, including 55 gallon drums, culverts, bottles, and cans. The water tank stores the sun's heat for nighttime heating in the winter and the night's coolness for use on hot summer days. They can be very comfortable and economical. Good exposure of the tanks is desirable. Thickness less than 2' preferred. The tank can be coupled to a wood stove for backup heating.

**Rules**

House has major wall to south
Most windows (2/3) face south
Overhang for south windows
4-6 gallons of water per square foot of south glass
Use insulated drapes and shutters
Provide good cross ventilation
East and west windows shaded with exterior shade.

WATER WALL HOUSE

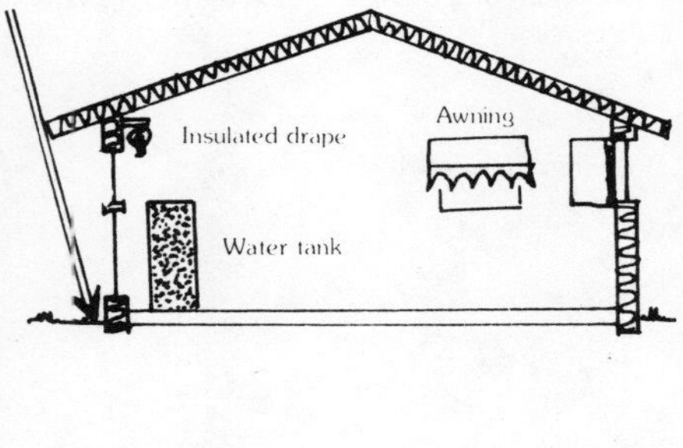

## Trombe Wall House

Concrete or masonry can be used to store the sun's energy instead of water. This can be more expensive but is better for very cold areas. Because concrete and rock is less than one-third as effective as water, much more mass is needed. Because masonry is fixed and has resistance to heat flow the surface to volume ration must be kept at 1 ft 2/3 ft $^3$ or more. A typical installation will cover the entire south wall with concrete. This type of solar system has been used in very cold areas with considerable success.

**Rules**

Major wall faces south — mostly glass
60-100 + lbs. of concrete/sq. ft. of south glass
Vent area in wall appropriately sized for good convection
Wall shaded and vented outside in summer
Insulate thoroughly
Weatherstrip thoroughly
Windows for good natural ventilation in the summer

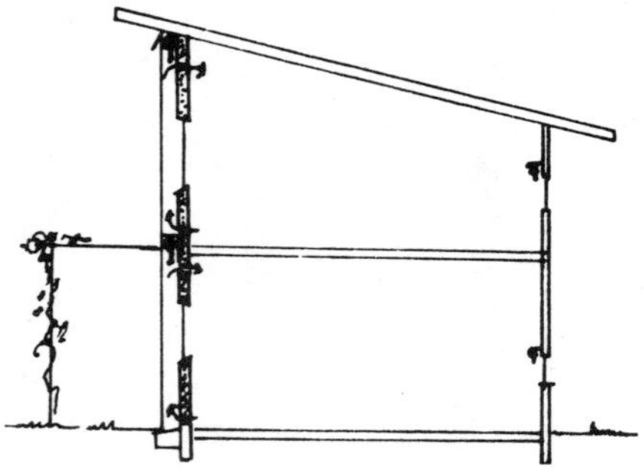

TROMBE WALL HOUSE

## Roof Pond

Thermal storage can also be placed overhead in a roof pond. This type of house allows a more flexible interior plan. It is particularly effective where the need for cooling is dominant. Insulated lids are vital for the Sky therm system. They are open at night in summer and during the day in winter. Lids can lift (as shown here) or slide and stack. In cooling only areas lids may be replaced with full shading in a cool pool house. Alternatively the water can be moved to an insulated space and pumped into the pond as needed. Concrete has also been used overhead for storage but is less effective than water.

**Rules**

Lid opens to give a good view of winter sun for heating
Open plan for good heat transfer if not full roof
South glass and exposed slab still desirable
Pond from 1/2 to full roof area
Lids designed for maximum wind load
Insulate thoroughly
Weatherstrip thoroughly
Good heat conducting material between roof ponds and living space

ROOF POND HOUSE

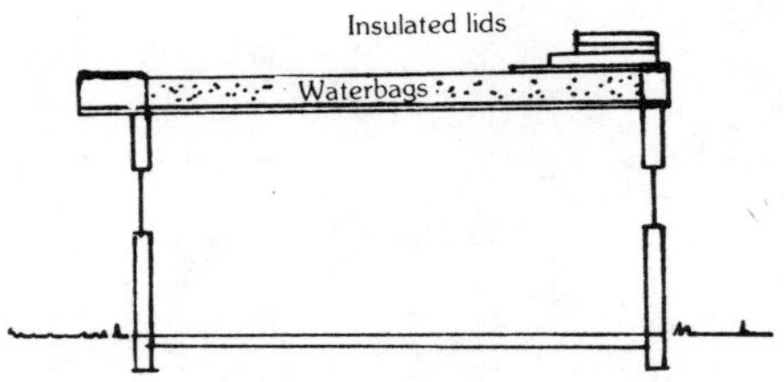


Summer Night

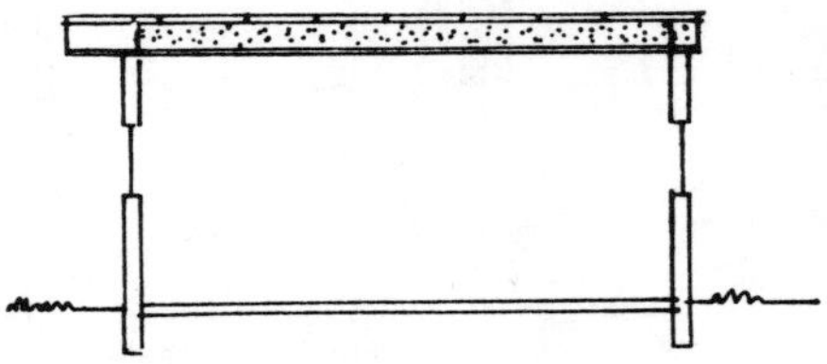

Summer Day

With these systems, passive solar design can provide full heating even in the cold north. Several houses in Regina Saskatchewan have almost full passive heating. They begin with super insulation, weatherizing, and an air-to-air heat exchanger ventilator. The heat from a 4 slice toaster is enough to balance heat loss on a cold winter day so the passive solar contribution doesn't need to be very high.

On the other hand, even in the hot, hot desert passive systems can provide full cooling. A skytherm system roof pond house in Phoenix stayed cool and comfy all summer long. Backup cooling will be needed only in the very humid southeast U.S.

## Passive Solar Water Heating

Passive solar water heaters are basically water containers painted black and set in the sun. They have been used longer than any other solar water heaters and are receiving increasing attention because of their simplicity, low cost, and reliability. Improved performance can be obtained from this type of heater if it is set in an insulated box with south glazing rather than just set on the roof. No controls or pumps are required as the tanks are on the hot water feed line. As hot water is used, more water is drawn into the tank where it is heated for subsequent use.

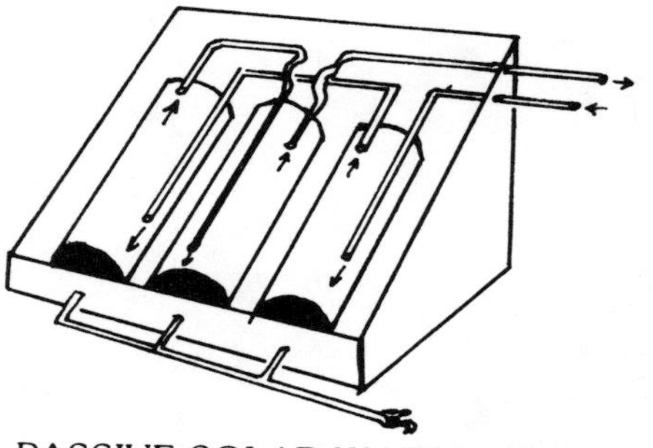

PASSIVE SOLAR WATER HEATER

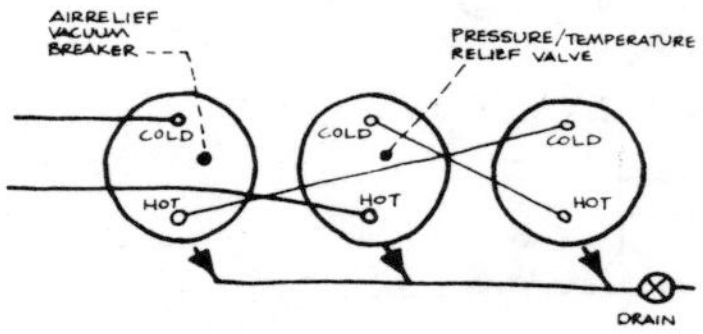


The performance of these very simple heaters can be quite good in temperate and hot climates without any insulated lids or covers. In colder areas an insulated cover for night protection is virtually essential. It is also desirable to put the passive solar water heater in a heated space, either in a greenhouse, or in the house itself. The insulated shutters can be reflective on the inside to provide reflective gain in the daytime.

Standard Oil of California is introducing an interesting Passive Solar Water Heater in 1980. It uses a thermic diode to prevent heat loss at night. They expect to have six plants in production in 1981. Several other groups are gearing up for production of different types of PSWH in the near future. Howe , it is still cheaper to build your own. Plans for a three tank PSWH are available from PSI for $10.

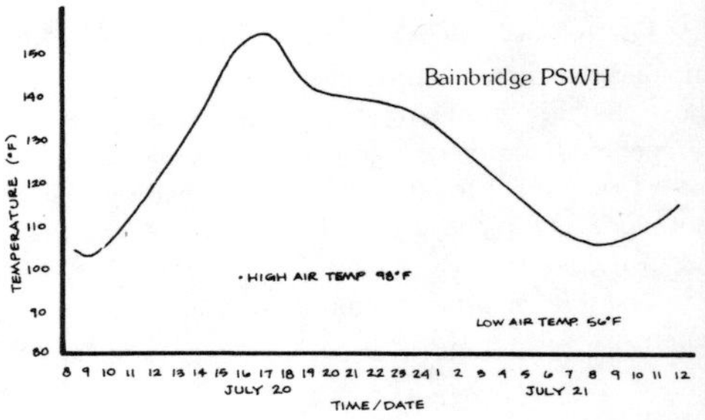


PERFORMANCE RESULTS

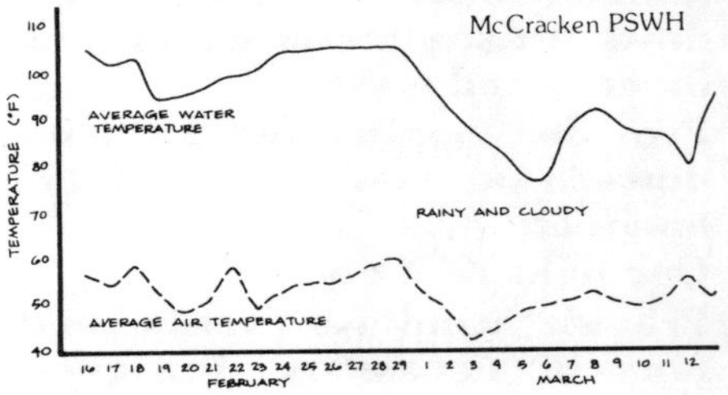


## Do It Now

The various state and federal tax credits in Chapter 3 are an excellent incentive build your passive solar system today, rather than waiting for energy prices to rise even higher. A passive solar house will save you money, be more comfortable, provide security in an even more troubled world, and help America toward energy independence.

## For more information:

The First Passive Solar Catalog (1978) D.A. Bainbridge, PSI, P.O. Box 722, Davis, CA 95616. $7.50

Natural Solar Architecture (1979) Dave Wright, Van Nostrand Reinhold.d $7.95

Solar Heating (1978) Holy Antolini, ed. Sunset, Menlo Park (at bookstore). $2.95

Village Homes' Solar House Designs (1979) David Bainbridge, Judy Corbett, John Hofacre, Rodale Press, Emmaus, PA $6.95

The Passive Solar Book (1979) Ed Mazria, Rodale Press (at your bookstore) $10.95

Homegrown Sundwelling (1977) Peter Van Dresser, Lighting Tress Press, $5.95

Passive Solar Water Heaters (1980) D.A. Bainbridge, PSI, P.O. Box 722, Davis, CA 95616. $5.00

Winter heat load calculation:Worksheet

CONDUCTIVE HEAT LOSS

| | Wall area(-window) | R-value | Area/R-value Hourly Heat loss | Window area | R-value | Area/R-value Hourly Heat loss |
|---|---|---|---|---|---|---|
| SOUTH | ______ | ______ | ______ | ______ | ______ | ______ |
| WEST | ______ | ______ | ______ | ______ | ______ | ______ |
| EAST | ______ | ______ | ______ | ______ | ______ | ______ |
| NORTH | ______ | ______ | ______ | ______ | ______ | ______ |

Total hourly heat loss per degree F ________ BTU (Walls) ________ BTU (Windows)

Total hourly heat loss per degree F, walls plus windows ________ BTU

Roof or attic area ________ R-value ________ Area/R-value Hourly heat loss ________ BTU

Slab floor, perimeter in feet ________ times 75(unins.) times 37(ins. R-5) ____ ________ BTU

Suspended floor area ________ R-value ________ Area/R-value Hourly heat loss ________ /BTU

Correction factor, .5 uninsul. perimeter, .2 insul. perimeter wall

Suspended floor hourly heat loss per degree F ________ BTU

TOTAL CONDUCTIVE LOSS, wall, window, floor, and roof ________ BTU

CONVECTIVE HEAT LOSS

Volume of house x Air changes hour x .018 = Hourly heat loss degree F

________ x ________ x .018 = ________ BTU

Correct factor varies with tightness, use 2ach for leaky house, 1.5 for standard house, 1 for good house, .5 for tight house, .2 for very tight house.

TOTAL HOURLY HEAT LOSS°F=Conductive + Convective Heat Loss ________ BTU

TOTAL SEASONAL HEAT LOSS=Total hourly heat loss x heating degree days

________ x ________ = ________ BTU

WINTER DAY HEAT LOSS=Total hourly heat loss x av. temp.(in-out) x 24 hours

________ x (______ - ______) x 24 = ________ BTU

Local climate information can be obtained from the local utility, University, weather station, or from : The National Climatic Center, Federal Building, Asheville, Tn 28801. State climatic information is 50¢. The Climatic Atlas of the U.S is $6.

You may wish to keep records at your site with a max/min thermometer to check against the nearest station.

# Chapter 5: Retrofitting for Passive Solar

Photo Courtesy Kalwall

Almost every residence in the U.S. can be retrofitted for passive solar heating, cooling, or water heating. The type of retrofit, the expected performance, and savings will depend on: the construction of the house, apartment, or mobile home, the climate, the orientation and solar exposure of the structure, the ability of the occupant to do some or all of the work involved, the availability and cost of financing, the cost of competing energy, and the applicability of State and Federal tax credits. Given the many factors involved, it is difficult to generalize; however, there is little doubt that a passive solar retrofit is the best investment you can make in most areas in the U.S.

This Chapter begins with a section on the opportunities for energy conservation, the first step in any solar design. It almost always makes more sense to reduce loads to the minimum before trying to meet them with solar design. Many of these techniques can also be used for new construction. The next section concerns adding more south windows to improve winter heating. This is followed by a section on adding more mass to store this heat for use on cloudy days or cold nights. These are followed by a look at the possibilities of building a passive solar addition to improve your total house performance. The next section looks at backup systems and the Chapter closes with a section on life-cycle cost and a methodology to evaluate the economics of retrofitting passive solar systems.

Retro-fitting for energy conservation and passive solar design isn't limited to those who own a residence, although they will usually be able to do more. It is also possible and cost effective for a renter to make improvements if the utilities are not included in the rent. And even the worst case, where utilities are included in the rent, it will often be possible to get the landlord started on a retro-fit program by demonstrating the savings involved.

There are many business opportunities in retrofitting passive solar systems. These include the obvious tasks of building and remodeling, insulation, and weatherstripping; as well as such things as manufacturing insulated drapes and shutters, thermal mass water tanks, and passive solar water heaters; and the more immediate need of distributors to handle energy conservation and passive solar products. The opportunities are almost unlimited, and can be the basis for a new business or a new sideline for for an established one.

The chart below suggests the energy conservation and passive solar retro-fit actions that will generally make the most sense. The order they are listed in is a ranking of importance based on cost and difficulty. Those near the top are essentially given, everyone should do them, those near the bottom should be evaluated more carefully.

## Retrofit Chart

- Lower thermostat: Winter (65°F night), Summer (78°F day/night)
- Weatherstripping
- Add insulation to water heater tank, turn down water heater thermostat to 120°F.
- Caulk and seal
- Clean furnace filters and air conditioner, refrigerator coils
- Install setback thermostats
- Buy comforters (down or fiber fill)
- Storm windows - cold areas, shade screen - hot areas
- Awnings - hot areas, South, East, West windows
- Add attic insulation (R-38 temperate areas, R-50 cold areas) and under suspended floor (R-19)
- Place airtight stove in fireplace with outside air intake if possible
- Add vents and small fans for circulation
- Insulate pipes, ducts, etc, and seal carefully
- Add ceiling fan if possible
- Landscape for shade, wind control
- Add Passive solar water heater, where feasible
- Add whole house fan in hot area with cool nights
- Insulate walls where feasible
- Paint roof light color, hot areas
- Double walls, cold areas for added insulation
- Roof insulation, reroof light color, roof deck add overhang if necessary
- Double roof, cold areas
- Add overhang as required
- Add south windows (remove north windows)
- Add air lock entry (cold areas)
- Add thermal mass floor, tank, etc.
- Add solar greenhouse
- Add passive solar addition with south windows and mass
- Add wood fired boiler and hydronic system
- Add active solar hot water heater
- Add backup active solar air system
- Add backup active solar liquid system
- Add windmill heating system, cold windy areas

The rest of this Chapter describes how to carry out these suggested improvements. To determine which is best for your particular situation, an evaluation of your building is in order. Use the Energy Evaluation Sheet and the methods described in Chapter 4 to estimate Winter Heating Load

These are approximate for a more accurate technique, I would suggest that you use the methodology in ASHRAE'S **Handbook of Fundamentals.**

# Energy Conservation:

Energy conservation comes first. Begin with the simple things: caulking/sealing and weatherstripping, retrofit insulation for the water heater tank, etc. These are all simple, low cost, effective, and within the capabilities of even the most unhandy homeowner.

CAULKING IS IMPORTANT

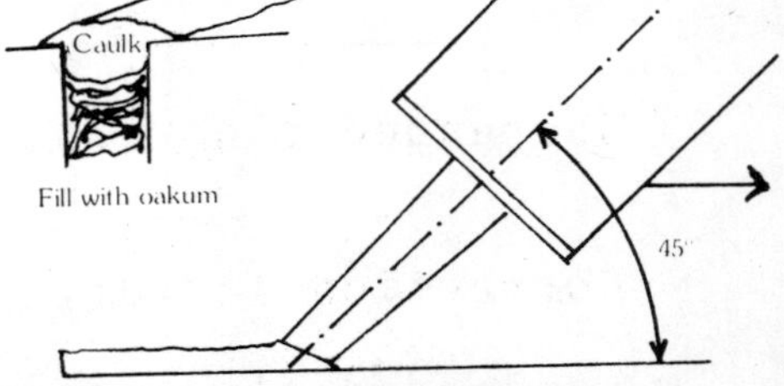


## Stopping Infiltration

Caulking and sealing cracks is the most economical and effective place to start. The accompanying diagrams show where the leaks occur. These tips should make it easier for you to caulk and seal your house.

First, clean old caulk dirt and sealent from cracks and joints. A putty knife, razor knife, wirebrush, and sand paper will help. Prime any exposed wood, metal, or masonry with appropriate paint. Fill deep cracks to within 1/2" with oakum or fiberglass scraps.

Second, insert the caulk cartridge into the gun, trim nozzle at an angle. Size of opening depends on crack size, bead should overlap. Then on shallow or narrow cracks run a bead of caulk by holding gun at 45° angle toward direction of movement. (Don't caulk on cold or wet days if you can help it.) Bead should adhere to both sides. On deeper cracks you may find holding the gun at an angle opposite to movement will help force the caulk in deeply.

After you have finished, plug the nozzle with a nail or wood plug, smooth joints with a finger (use a rubber glove if tube suggests you may react), wooden knife cut to fit joint, or putty knife.

And finally, clean up with water or solvent as required. Pay particular attention if you've used silicone caulk because paint doesn't stick to it.

You can also seal cracks using clear tape made for the purpose. It is often a good idea to seal many of the windows with this tape and then very carefully weatherstrip those that will be opened in the winter.

# Weatherstripping

There is an almost infinite variety of weather-stripping available. Special types are available for every type of door and window, even for jalousie windows. The best type for you will depend on your available money, (a renter might choose the cheapest), the application (an often used door v/s rarely used door) etc., and your skills (some are easy to fit, others hard).

The best weatherstripping system for doors is the interlocking metal variety backed up with a vinyl covered foam strip. The interlocking channels are expensive and require use of a router. Cost for professional installation can be $100+ per door. For a door that is not used frequently the adhesive back foam strips work fine—simply measure the gap and buy foam strip to fill it up. The squishier the foam the better. On doors that are used more often, a springmetal strip or the vinyl covered foam works well. In very cold areas, keep the foam on the inside of the door so it will stay flexible and seal well.

The bottom of the door should also be sealed. The interlocking metal system usually includes the threshold. You can also buy a rubber strip for the door bottom that fits into an aluminum threshold. The easiest and cheapest seal is a sweep type that simply presses against the threshold. Lower it as the sweep wears out and replace it if it cracks or no longer seals. A similar seal is available that flips down ony when the door shuts and folds up as the door opens. You can also make a foam and sand filled snake to lay at the bottom of the door to seal out drafts. Drape it over the door knob when you use the door. Don't forget to weatherstrip attic and basement doors.

The double hung windows on older houses are often very leaky. Special weatherstripping is available for them. One type includes a spring metal strip for the sash channel, a strip for between the upper and lower window, and a foam or metal strip for the bottom. Work carefully and make sure the window will slide. Bending the ends of the spring strip down will help prevent the sash from sticking as you slide it up. The old standbys of adhesive backed foam, vinyl covered foam, and felt can also be used on double-hung windows.

For metal windows and sliding wood windows the foam insulations can be used as required. A special aluminum weatherstrip is available for metal easement windows but can be hard to find. For jalousie windows a vinyl weatherstrip is available. It fits on the edge of each glass pane to make a better seal.

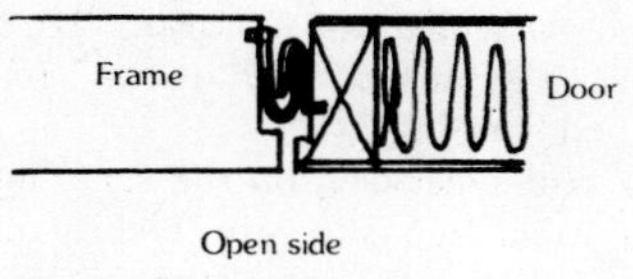


INTERLOCKING WEATHER STRIP

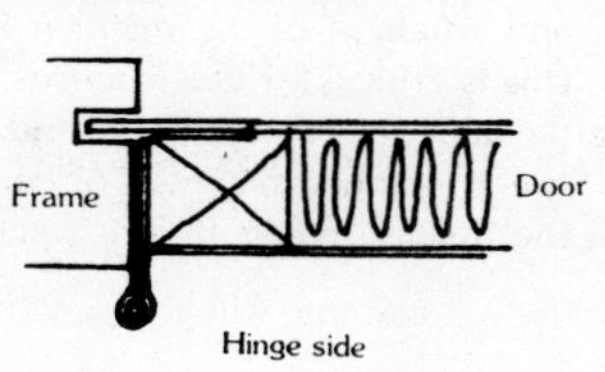


STORM WINDOWS HELP

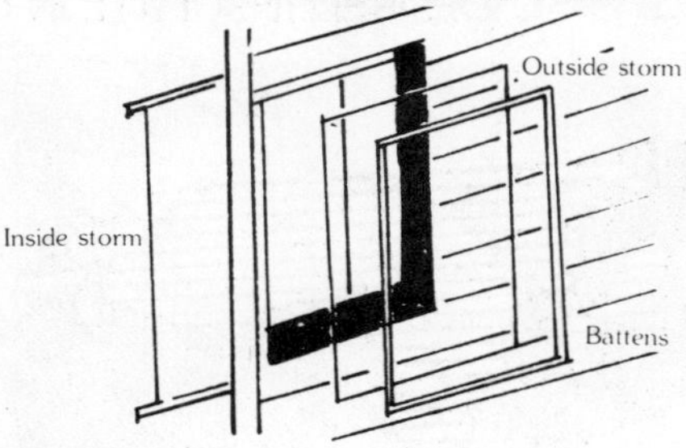


# Storm Windows and Doors

Often storm windows and doors are desirable to further reduce infiltration. You can buy these and install them yourself, have them installed, or better yet make them yourself. Keep in mind the desirability of maintaining fire egress as you add storm windows.

The simplest storm window is made simply by stapling plastic over the window. These may look ugly, flap in the breeze, and last only one season—but they do save energy and money and are very inexpensive. Use a wood or vinyl batten to prevent the plastic from ripping at the staples. These can also be made on separate frames, like a standard storm window.

The next step up, and one of the most attractive in terms of cost, simplicity, and performance is the inside storm window. These can be cut from acyrlic, weatherstripped, and then fastened inside the windows. Plaskolite's Insider (see product listing) is one of the commercially available models.

Acrylic or polycarbonate exterior storm windows are also an attractive alternative. They are lighter and stronger than glass and that can be a real asset when you take them off and replace them every year. The shorter life (15 years in direct sun) of acrylic is not as important in storm windows because they are inside, out of the sun during the summer. You can make the frames from wood or aluminum.

And finally, wood or metal frame glass storm windows may be used. These should be well built and fasten securely, with weatherstripping to make a good seal to the house.

In some cases it may be easier to add another permanent window, either single or double pane. This will work as well as a storm window, without the need to remove it every year. This is easiest when aluminum frame sliding windows are used on a fairly wide sill.

Storm doors can also be purchased or built. Most of the available storm doors are uninsulated so if you are at all skilled in building you may wish to make an insulated storm door. Even better than a single door is an air lock storm entrance. This can often be made by weatherstripping an existing door on the inside of a closed-in summer porch or entry room. If this isn't possible another door sometimes can be added to the entry hallway to make a mudroom. Or an exterior entrance way can be built. If your house is situated properly, the entry way can also provide some passive solar heating.

Another source of unwanted infiltration is the cracks and gaps around bathroom vent stacks, bathroom and kitchen exhaust fans, and furnace and fireplace exhausts. Vent stacks should be sealed using a neoprene roof flashing cut to size, caulking, and/or insulation. The bathroom and kitchen vents should be fitted with an insulated cover and rerouted through a side wall if possible.

A fireplace should be converted to an air tight stove with either a fireplace insert or stove set in, or in front of the hearth. A furnace vent may warrant an automatic flue damper or heat recovery unit. These are covered in more detail under the section in this Chapter entitled backup heating and cooling.

If any remodeling or additions are contemplated then consider using the best air infiltration stopping techniques. In addition to those already mentioned a vapor barrier is desirable. After the walls are framed, but before the sheetrock is installed, wrap the house in plastic film (4+ mil polyethylene) sealing all the joints with sealent. Where electric boxes are placed a plastic sheet or pan is used to seal off the outlet area.

VAPOR BARRIER DETAILS

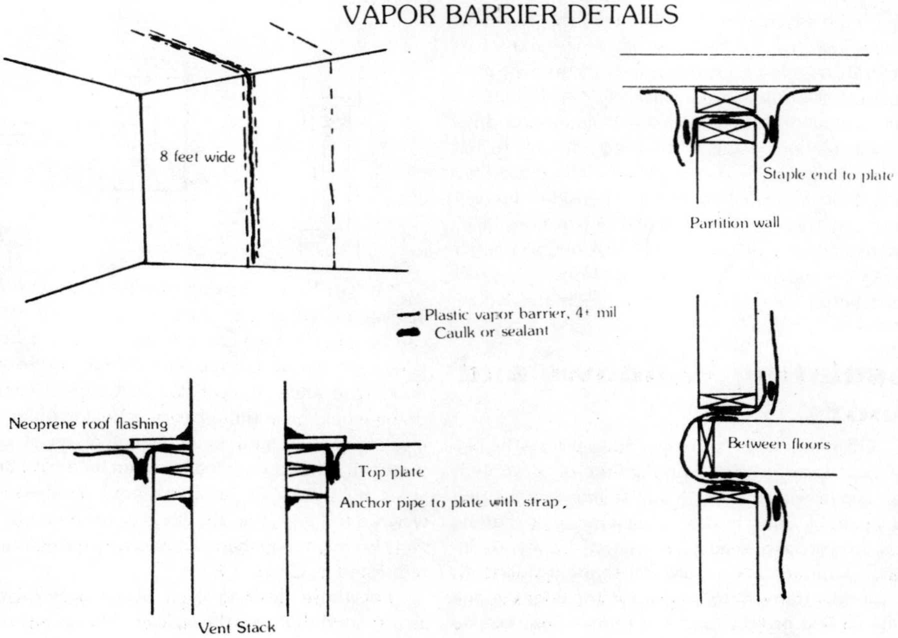

# Insulation:

Much of the energy loss in winter from a typical house is lost by conduction. Insulation has low conductivity or a high resistance to heat flow. Most older houses have been built with little or no insulation and even now the code requirements in most areas should be considered an absolute minimum. A recent study showed that R-19 insulation in the walls and R-38 ceiling insulation was economical virtually everywhere in the U.S. The accompanying map is PSI's recommendation for insulation level in the U.S. You can retrofit your home or build an addition on new homes to these standards using the information in this section and in Chapter 6. The following materials are commonly used for insulation and building. We have not included urea formaldyhyde foam because it requires expert application to prevent serious health problems. Its use is now banned in Massachusetts for that reason.

DESIRED INSULATION (Minimum)

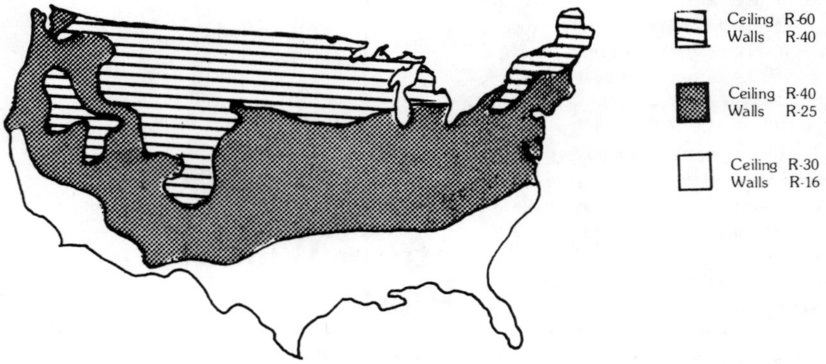


## R-Value of Common Insulation & Building Materials Per Inch

| | | Advantage/Disadvantage |
|---|---|---|
| Cellulose | 4 | Low energy cost, easy to handle, settles easy to retrofit |
| Mineral wool | 3 | Moderate energy cost/use mask when cutting |
| Expanded polystyrene | 4-5 | High energy cost, deadly fumes if burned. Never expose expose inside, use mask when cutting |
| Expanded polyurethane | 6 | High energy cost, deadly fumes if burned. Never expose inside, use mask when cutting |
| Perlite | 3 | Good R-value, high cost |
| Vermiculite | 2.1 | Poor R-value |
| Fiberglass | 3 | Moderate energy cost, use mask |
| Wood | 1 | Low R-value |
| Brick | 0.2 | Virtually transparent to heat flow |
| Stone | 0.1 | Virtually transparent to heat flow |
| Double pane window, winter | 1.8 | Should always have inside |
| Triple pane window, summer | 1.5 | Drape or shutter |
| Plasterboard | 0.9 | Low R-value |
| Concrete | 0.1 | Virtually transparent to heat flow |

In addition to the material involved in a wall, each air space and surface also provides some insulating value. The value depends on the orientation of the air space, its thickness, the direction of heat flow, the temperature difference and mean temperature. However, the area of the framing cuts the value of insulation considerably. At the studs, plate, and sill the R-value is less because wood replaces insulation. Studs at 24" are preferably to 16" for that reason. For the simplified method used for heat calculation in Chapter 4, we consider the wall and ceiling to have the R-value of the insulation alone. If no insulation is included use R-3. After calculating the conductive heat loss for your house, you will have a better idea where you need more insulation.

# Insulating Windows and Doors

For many houses, the windows will be the biggest heat loser in winter and source of unwanted heat gain in the summer. You can improve the performance of your house considerably by cutting these losses with shades, insulated drapes, or insulated shutters. A simple roll shade installed in the window frame with guides for the edges is one of the easiest projects and can reduce heat loss by 50% if it is made with Foylon or aluminized mylar. The accompanying detail will help you install shades properly.

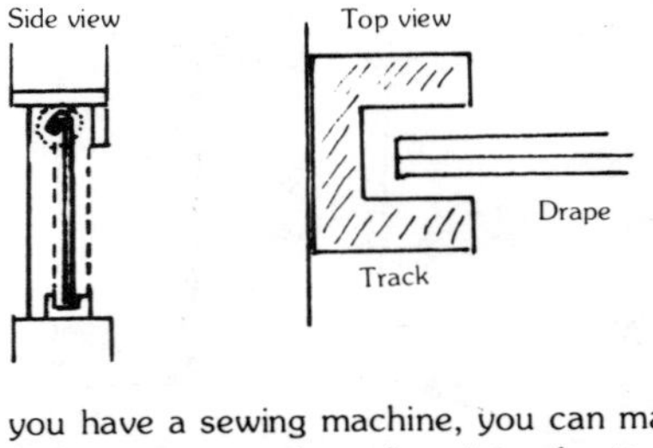


If you have a sewing machine, you can make insulated drapes or curtains. A variety of patterns can be used as long as: the top, sides, and bottom edges are sealed and the material has a reasonable good resistance to heat flow. A typical insulated drape would have three layers, a Foylon outer layer, insulating fiber, and inner facing of an attractive material. The top is usually sealed by a valance, the sides by tracks or fasteners, and the bottom by weights to keep it on the floor or window sill. You can also buy ready made drapes from the manufacturers listed in Chapter 8.

For those desiring even better performance, an insulated shutter is the answer. These can be purchased or you can make them yourself. Because we feel these are the best solution in most cases, we will describe how to build them in considerable detail. Insulated shutters can be added to either the outside or inside of a window. We prefer to put them on the inside for the following reasons: they do not need to be weather proofed, they are easier to operate, and they can be installed even in inclement weather. The following directions explain how you can build interior insulated shutters. In some cases, particularly in milder climates or in verv severe climates where a shutter might be closed all winter, exterior shutters might be desirable.

Insulated shutters are relatively easy to build. They require only basic, commonly available materials, and simple tools and make a very good project for a cold winter weekend. We recommend building wood frame wood skin shutters with fiberglass insulation. Simpler shutters can be made, but although several groups have recommended panels of plastic foam for interior shutters, we feel that the added fire safety and durability of wood shutters are worth the added expense. Wood also uses renewable resources and is much more pleasant to work with.

The first, and often most difficult step in building insulated shutters is deciding what style to build. Will they open up, down, to the side? Will they be single, bifold, trifold, double bifold or? Will they be in a frame or just attached to the existing frame? We describe a shutter that is attached to the existing window frame because that will be the most common design for retrofits. For new windows, a built in framed shutter will be easier. For double wall houses, the shutter can be installed between the walls. How will they be finished? This can be particularly hard on your first shutters - because you won't know quite what they will look like. Try something relatively easy at first, such as the bifold shutters described here.

After deciding on the style, the second most important step is to very carefully measure the window frame so the completed shutter will fit. Several measurements are necessary to determine the size of the opening, the squareness of the opening on the inside, and trueness of the opening from front to back. This is critical for older houses which gradually settle and deform openings and new houses where "green" lumber may twist or bend out of true even if the builder makes it very square when it is built.

Begin by measuring the width and heighth in 3 or more places. Then measure the diagonals. They should be the same. Next with a carpenters square, check the sides and top for trueness, and then check the trueness of the opening from front to back with the trisquare. And finally, check the sill and side that the hinges will be on for level and plumb.

With all the measurements, carefully noted draw up a sketch (to scale is desirable) of the window opening. Quickly check your drawing against the window frame using your tape. Then check it again.

Design the shutters to fit the opening. Allow 3/16" for clearance on each side and on the top and bottom and 1/8" for the centerfold. When things are out of square, split the difference between pieces so they will be equal size when folded. Check all your additions and subtractions again! Remember, you can't add wood but you can always trim—so add maybe a 16th or 1/8th to your finished size.

The internal framing should also be designed at this time. Use butted joints on the frame—they are the strongest type. Add reinforcing blocks for the hinges on larger shutters. A rib or reinforcing block in the center of a larger shutter is also desirable. Latches may also warrent reinforcement. If you want to add a small window to the frame, do it at this time.

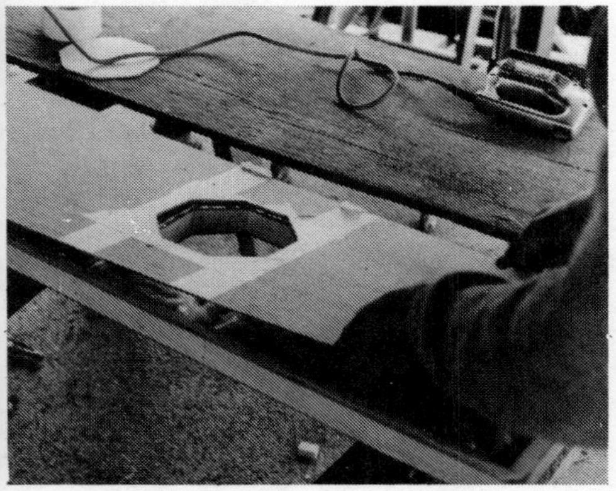

Make a list of all the pieces you'll need. Length, size, etc. Check it against the drawing once, then again. Lumber will come in lengths by 2', so figure which lengths will minimize waste. Prepare a cut list.

After checking everything carefully, order your materials. Doorskin, a 3mm or (1/8") mahogany, is available at most lumber stores (here in California anyway). Remember, that you'll want to match grain direction if the wood will be exposed.

The frame can be pine, spruce, fir or redwood. 1" x 4" (actually 3/4" x 3-5/8") wood is a good dimension to work with. You may have to ask the store to rip it for you because you'll want square edges - not the rounded edges you'll get with S4S (surfaced four side) lumber.

Insulation can be picked up at the lumber yard or hardware store. If you'll need only a little bit you can probably find some in the scrap pile of a nearby construction site. Ask before taking anything!

You can also buy your hinges, latches, screws, nails (4d finish and 5/8" brads) and white glue. Make sure that the hinges are strong enough (figure 1" of hinge per 12" of length of shutter) and that the mounting side is long enough to pass the sheetrock. Shutter hinges are available with one side longer than the other.

You may also want to check your tools to make sure you hvae everything you will need:

**Minimum**
trisquare
hammer
nail set (1/16)
saw (8-10 pt. inch)
tape measure
sandpaper 120 grit
chisel 3/4"
screwdriver - standard/phillips to fit hinge screws

**Desirable**
carpenter square
trisquare
level
tape measure
circular saw with plywood blade
chalk line
finish sander
belt sander
screw set punch
nail set
chisel 3/4"
screw driver
razor knife - retractable blade

With materials and plans in hand you're ready to cut the pieces. Take your time and measure everything at least twice. Think about which sides will show - and make sure that you get the nicest side out. Mark cuts with a trisquare and sharp, firm pencil. Remember to cut to the line, not on the line or you'll be consistently off by the width of the sawblade. Adn after you've cut the pieces, mark them so you can put them together correctly.

When cutting the door skin, a chalk line will mark the long cuts. A board clamped on as a saw guide will help keep things straight if you're a little wobbly. A plywood blade can be purchased for a smooth clean cut.

When cutting door skin with a circular saw keep the finish side down. Masking tape may also be desirable to reducing splintering. Keep the saw set only a bit deeper than the wood so as many teeth are in contact with the cut as possible which reduces splintering and provides a smooth surface — and use a sharp saw. Before putting the cutting tools away, check pieces against the cut list. You may also find putting all the pieces together helpful as a final check.

Building the shutters is easy and rewarding after all the careful measurement and cutting. The frame is assembled first and then the skins are added to clamp it all together and make it rigid.

First, set the 4d finish nails in the edge pieces. (For very dry wood or old wood, blunt them first by epending them on a hard surface and tapping them with the hammer - this will reduce splitting problems). Make sure you have the nails in the right place.

Then apply plenty of glue on the inside piece (the end grain will suck it up) and nail them together. Leave the nail head a little above the wood because it will be set with the nail set later on. Wipe up any glue you migth have spilled or that drips out of the joint when it is nailed. Try not to smear it out as it will affect the staining properties of the wood. Then add the second inside piece and finally add the other side. Fancy joints are not necessary, the simple butt joint will do with the glue and the door skin holding it together.

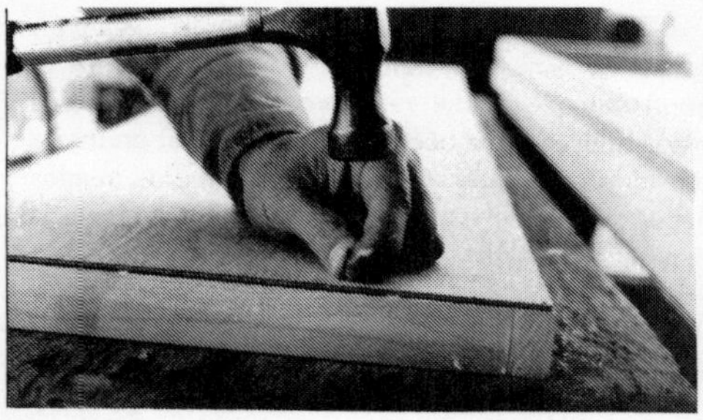

After you have put all the frame together, it is time to add the skins. lay a bead of glue all around the frame and then set the skin on. (Make sure it is the right one with the finish side out!) Then start at one corner and nail off one side using the 5/8" brads about 6" apart. Match and nail the other corners next and then nail off the sides.

After one side is on turn it over, fill the space with fiberglass, mineral wool, or properly fireproofed cellulose insulation and nail on the other cover using the process decribed above - set aside for the glue to dry.

After the glue is dry, overnight in most cases, sink the brads and frame nails and nails with the nail set. Then fill the holes with putty matched to the wood. Try to keep putty just in the hole because it will stain the wood a bit. A screwdriver blade can be used (like a putty knife) to help put it in. Set aside to let dry.

When dry, sand the faces and edges with 120 grit sandpaper. Always rub with the grain.

Figure hinge placements (you should have done this roughly in Step 1) at 1/4 of length in or 1/3. Mark the actual placement using the hinge. You can make them all the same if you line up the center edge of the hinge (by the pin) with the edge of the wood every time). Then mortise the hinge mounting spot with a chise). Cut the edges first with the chisel held vertical to the wood, beveled side toward where the hinge will go. Tap it firmly. Then using the chisel with the bevel down, very carefully remove the wood until the hinge will set flush with the wood. Sliding the hinge in from the side rather that laying it in will help.

Next, using a set punch, mark and begin the screwholes. Use a good solid blow for fat screws, a lighter tap for thinner screws. And then drive in the screws. Make sure you set all the hinges the same way, i.e. all two tang sides to the middle - or what ever. Put the two halves of the shutter together.

The completed shutters can be stained or painted to suit. We prefer an oil stain or Watco oil to acent the beauty of the wood. They could also be painted, wall papered, or used for mounting posters or prints.

If stain is used, mix it thoroughly so the color is uniform. Then apply it with a rag and then rub off the excess with a second dry, lint free, rag.

After the easy rapid progress in assembly and finishing the shuttering, mounting may seem tedious. But the end is in sight so take your time and do it right.

Begin by setting the shutter in the frame. Use shakes as wedges underneath the shutter to achieve equal spacing all around it. Make sure it will operate without binding and clear inside features like window cranks and locks. You may have to reset the cranks on casement windows so that the shutters will close.

Then mark the hinge placements. With an assistant holding the shutter so the hinges are in the proper place (the same wedge underneath will help) punch the holes and drive the hinge screws in. Use a shim if necessary to get the hinge side plumb. Make sure it swings freely. It should if everything was measured right but if it doesn't, you can sand the edges until it does. Then set the stop strip so the shutter is flush. Add the latch and you're nearly done.

The final step is relately easy but requires careful attention to detail. The shutter type will determine the type of weatherstrip that is required. The example described here used very soft foam strips on the edges and center joint and a sweep type strip for the bottom and top.

Virtually the same techniques can be used to make insulating doors, for entrance ways, to the attic, to the cellar, and for covering vents. You can also purchase insulated doors. Many of these are steel and very strong and may be desirable for your main entrance. You can also insulate an existing door by gluing on 1" of foam and then covering it with doorskin material or masonite. Trim with wood to match door pattern. On a solid care door, you can increase the R-value from 1 to 6.

Insulated Door

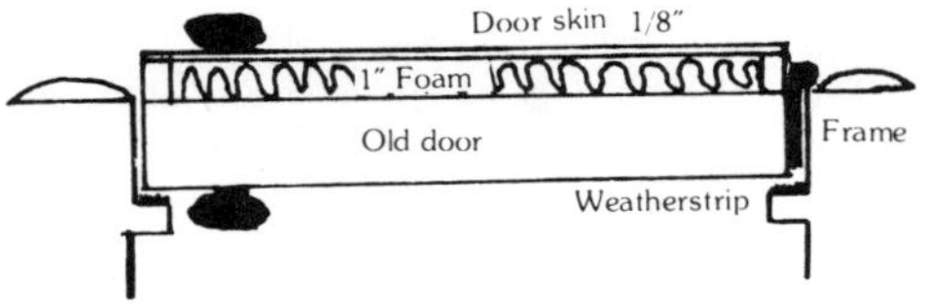


## Insulating Attics and Roofs

Adding more insulation to an attic is usually straightforward and very effective. Use blown in cellulose properly fireproofed to Federal Spec. HH-1-515C , or add mineral wool or fiberglass batts to the desired level of insulation. Remember to keep insulation away from recessed lights and heat sources and leave room for air circulation from the eaves. Simple shields will keep the insulation away from lights if you use loose fill. If the space at the edges is restricted you can use foam board to provide the desired insulation—yet still preserve ventilation.

The vaport barrier should always be on the warm side. If multiple layers ae used, the vapor barrier should be broken on all but the innermost layer. If you install it yourself use a good mask, wear gloves and long sleeves, and watch out for nails sticking through the roof. Don't try to move wires around and step carefully. It can be painful as well as embarrasing to step through the ceiling to the room below. Building elevated crawl ways is desirable when the insulation will cover the joists.

If you have a deck type roof with exposed beam ceiling or a house without an attic, you can either put insulation between the exposed beams and sheet rock over it, put up finished foam insulation specially made for the purpose, include 2" or 4" inches of foam on the outside the next time you reroof, or put on a new roof with a rafter space for insulation above the old one. Incolder areas, you may wish to add a double roof and add insulation inside. A double roof is often desirable because it can prevent possible trouble from ice daming.

Double Roof

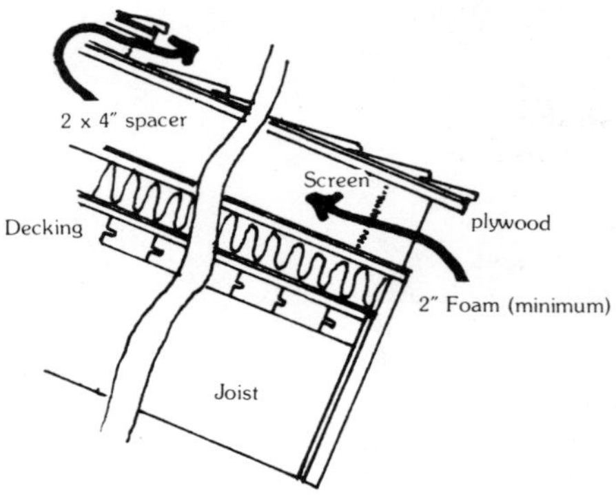


If you live in an area with heavy snow loads you may need to reinforce the roof after you insulate it. With the uninsulated roof, the snow may have melted rather than built up. It is cheaper to reinforce the roof and add insulation then it is to buy energy to melt the snow.

Insulating under the floor is also important. It is easiest with a suspended floor but be forewarned, it is not very pleasant crawling around in the dank, dark, dust, and spiders while working overhead with insulation. On a suspended floor you can usually use mineral wool or fiberglass batts. Remember, the vapor barrier goes up toward the warm side. These can be held in place by wood spring strips, wire mesh, wire strung between nails, or in some cases, glue.

Floor Insulation

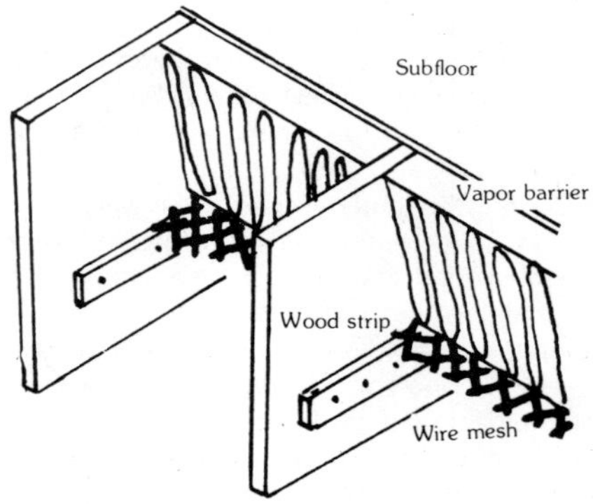


In addition to the floor itself you should insulate the walls of the foundation. You can either nail wood strips to joints and staple batt insulation to them or use foam insulation. Stuff batts between the joists out to the header or sill. The vent holes in the foundation should be fitted with covers for use during the winter in colder areas.

If you have a slab floor or basement or don't relish crawling around under the house to insulate the stem walls you can add waterproof foam insulation to the outside. You will have to dig down all around the outside of the foundation to do this properly. If you have any seepage problems, you can attend to them at the same time. Flash everything properly if you run the insulation up to the siding.

A less satisfactory but much easier method of insulating the foundation is to run the insulation down the wall and out horizontally from the house. Slope it away from the house and put in a trench drain if drainage is a possible problem. Needless to say, a water proof foam should be used. If it will be subject to assault by garden spades and shovels, put a layer of pressure treated 3/8" plywood on top of it. A pressure treated wood box full of insulation can be used to insulate between the wall and the foam.

Foundation Insulation

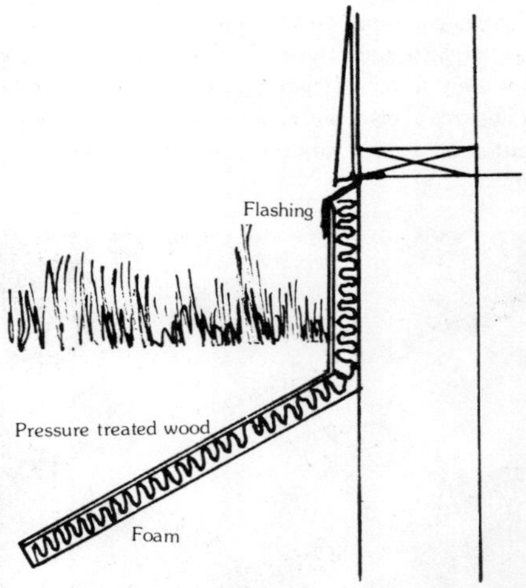


For rural homes or rented homes, hay can be used to insulate the foundation. Stack bales all around the foundation after running a plastic sheet along the wall. Straw or weedy hay is fine. The grain of the bales should run horizontally not vertically so they'll stay drier. Use loose hay to fill cracks and crannies. In the spring, all the hay can go into the garden or onto the compost pile.

Economy Insulation

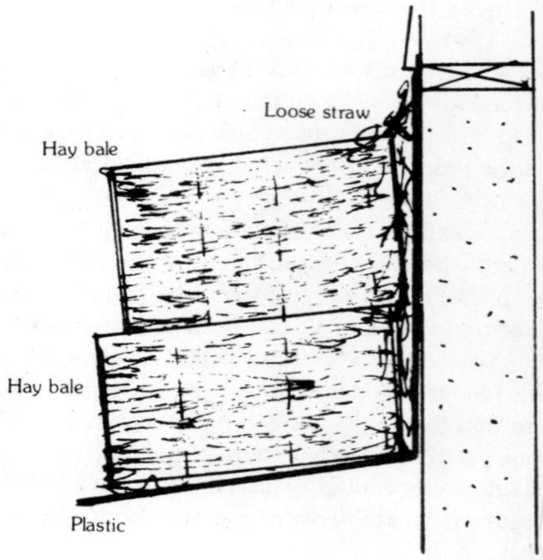

# Landscaping

The choice and maintenance of your landscaping can also help reduce your energy use. A hedge row or group of evergreens can minimize the impact of the cold winds of winter. A shelter belt works best if it is reasonably wide, somewhat porous, and has both tall and short trees. Savings on the exposed cold plains have been up to 20% in experiments with shelter belts. In the fall, you'll also want to prune grapes on arbors and other obstructions to the south to let as much sun shine on the windows as possible.

For summer cooling, you will want to pamper those big shade trees to the East and West as they provide maximum summer shade. A hedge, esplaniered trees, arbor or shrubs to the East and West will also provide relief from the hot summer sun. These might also be designed to channel cooling breezes through the house. An arbor on the south is also worth developing if overhangs are not wide enough. A wire frame arbor is desirable to minimize winter shading.

# Solar Control

In a warmer climate, solar control is often as important as insulation. By controlling the impact of the sun comfort can be increased while energy bills are minimized. The most critical factor in most cases is the windows. The East and West windows are most important although if the summer is long then even the south windows can be important.

The cheapest solar control is aluminum foil taped to the inside of the windows. If you like to be able to see out the windows, you can provide exterior shade using an awning, shadescreen, arbor, or extended overhang.

A roll down awning of canvas or bamboo is not very expensive and can be very effective. The savings can be very substantial, both to the homeowner and the utility because the summer peak is the biggest power demand for many utilities. For example, in Sacramento, every 55 square feet of west facing glass requires the equivalent of 1 ton of air conditioning. This in turn requires the addition of almost 2kw of generating capacity at the power plant. The difference in cost betwen a canvas awning ($100) and an air conditioner and added generating plant capacity ($3,000) is substantial.

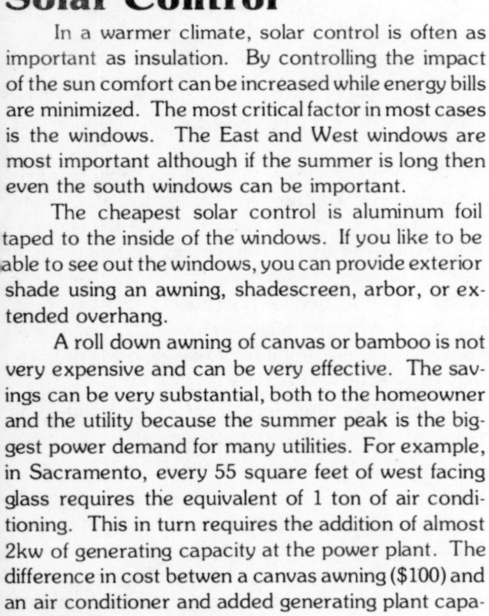

Shadescreen is also effective and relatively easy to install. Two basic types are available, metal screens with louvres and fiberglass webbing. The metal screens are more attractive, more expensive, and more sensitive to mechanical damage, but are very durable otherwise. The fiberglass is cheaper and more resistant to damage but has a limited lifetime. Both are excellent investments in hotter climates. Wooden louvres can also be used to make shades. The final option for wndow shading is the addition of an overhang. This can be a fixed aluminum or canvas awning, a wooden awning, or part of the roof. Extending the roof overhang is permanent and will also help protect the south wall from rain.

The choice of roof color and material can also reduce overheating. A light roof can be 60°F cooler than a dark roof. Shingles, shakes, and tile are cooler than tar and gravel or metal roofs.

# Ventilation and Circulation

The ventilation of a house is usually much more important in summer than in winter; however, an extremely tight house may get stuffy unless some ventilation is provided. An air-air heat exchanger is preferred. This machine warms incoming air with exhaust air to reduce heat loss. These can be purchased or built (more details in the section on backup systems).

In summer, cross ventilation can be important for natural cooling. Keep summer needs in mind as you plan solar improvements for winter. Both natural, stack, and induced ventilation help cooling in summer. Natural ventilation uses the cool night breezes. Stack ventilation uses the tendency of warm air to rise for ventilation by providing low inlets and high outlets. Induced ventilation adds special solar chimneys to increase ventilation. The First Passive Solar Catalog includes more details on ventilation.

Circulation within the house is important in both winter and summer. An open plan helps both in cooling and heating. Different levels also help. An upstairs loft may be too warm in summer yet just right for reading and socializing in the winter while a basement may be too cool in winter but a great hideaway in summer.

A ceiling fan (casablanca fan) can help keep you cool in summer by increasing evaporation. In the winter, it can circulate hot air from the roof peak back down to the floor. Small fans may also be used to move air from the hot spots through ducts to the colder areas of the house.

Vents, grills, and ducts can also be very useful in both summer and winter. In the summer, they can improve cooling by increasing ventilation. In the winter, they can be used to reinforce or set up thermosiphon loops to equalize temperatures throughout the house.

### How to Install A Whole House Fan:

The installation steps listed below should enable you to install a whole house fan in a typical house, with attic and sheetrock interior. The steps are essentially the same for houses with attics and other interior finishes but may require different tools and techniques.

A whole house fan can also be installed in a house without an attic. It can be mounted in one of the sidewalls or can be mounted in an enclosure added to the roof. The added expense of doing this may be warranted for larger houses with many interior walls.

The details will also vary with the whole house fan used, but most fans are installed in much the same way. Ask for the manufacturers installation detail sheet for your model.

**Tools Required:**
Hammer
Screwdriver
Drill (hand or electric)
Sabre saw, circular saw, or hand saw
Keyhole saw
Pencil or Pen
Extension cord
Ladder (step)
Wire stripper/wire cutter
Utility knife
Tape measure
Putty knife
Flashlight
Spackle (small box)
Stud locator
Test lamp
Fish Tape (for wire, optional)
Goggles (optional)
Gloves (optional)
Mask (optional)

**Materials Required:** (Purchase only after Step 1)
Whole house fan
12 gauge wire (or wire required by local code)
Junction box (may not be required)
Switch box (single or double as required by controls)
Electrical connections
Screws (#8 x 1-1/2" wood)
Gable vents; grills
2" x 6" or as required for blocking around fan and grill
2" x 4" as required for blocking around
Nails 16d
Switch or fan control (if not supplied with fan)
1" x 2" moulding for your choice for trim
Hardware cloth or chicken wire (if access to attic is easy for animals or children)
Thumbtacks or push pins

The following steps should be undertaken in the order listed. Read through them carefully before you begin. Take your time as you work, measure everything twice or three times, take a break when you're tired, and think safety!

Use the material in the text to help choose the proper location for the fan. A clearance of 2' + is desireable between fan and roof. The hallway width (and house size) will determine the fan size. Measure the rafters spacing to determine the amount of blocking material you'll need. Refer to manufacturers sheet to determine need for added trim, and measure size needed for chosen location.

Look around the chosen location for an electric circuit you can tap into. Indentify it by tracing where it goes and check for the service panel (circuit) breaker box) to make sure it can carry the load of the fan, typically 100-500 watts. Most bedroom circuits will be sufficient for more efficient fans if they are for 20 amps. If a larger fan with high start demand (to 30-40 amps) is used, a separate run from the service panel may be required.

The switch box should be located in a convenient location in the living space. It may be possible to add it near the existing light switch. The use of a single or double switch box will depend on the fan control. Make sure you get the right size and type. Switch boxes for retrofit include either tabs which will lock on to the wall or a flange which can be nailed to the front of a stud after chipping out a hole. Put the control at about 4'.

Now take a look around the attic and figure out the existing vent area. Compare it to the manufacturers recommendation for the fan you are using. Many houses will need more vent area. Gable vents are relatively easy to add. Place them near the top of the attic. Added blocking may be needed to install them, measure the distance between studs to determine how much block material you'll need.

You are ready to begin installation after after you have assembled all the tools and materials. The first step in cutting the hole for the fan is accurately locating the fan location (making sure you are where you thought you were!) Drive a nail into the center of your chosen fan location from below. In the attic, with your gloves and mask on, feel the insulation until you find the nail. Remove the insulation in the area (use the knife to cut it). Relocate any wires that run through the area.

Using the nail as an approximate center measure the length needed for the fan installation. It should fall on existing joists. (If it doesn't you'll have extra work boxing it in). From above drill a hole flush to the inside of each end joist and at each joist in between. Now from below mark out the opening using the two end guide holes. (Most fans will come with a cardboard template or you can make your own). Tack it in place. Measure to hallway sides to make sure it's parallel. Then mark it out.

Now drill bigger holes at the corners of the marked opening. Then saw out the opening (mask and goggles are desirable for this). Leave the interior joists in place for now. Use the utility knife to score the sheetrock over the joists and break out the piece.

Now back in the attic, cut off the interior joists, keeping your weight off the joist you are cutting. In most installations, you will cut it off 1-3/8" from the edge of the opening to make room for the header. The circular saw or sabre saw will make much of the cut. Then reverse the handle on the hole saw and complete the cut.

Now cut headers to box the edge of the opening. These should be the same dimension as the joist and cut very carefully. Nail in place. Lift the fan up through the opening now—careful, it's heavy.

Box Rafters

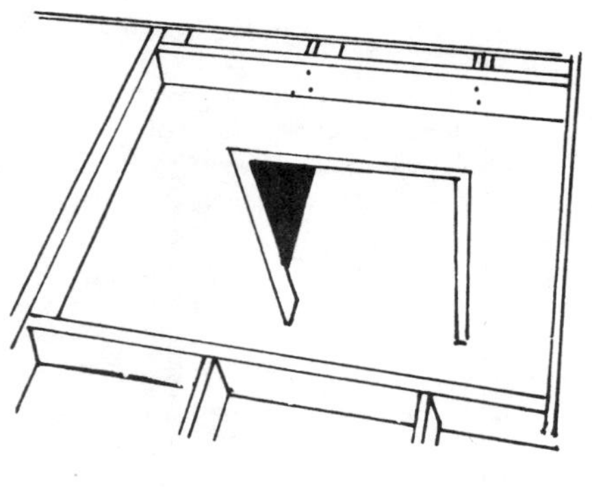

Some fans will now fit into the opening. Others will require a special frame, consult your directions. If a mounting is required, build as suggested and then center and mount fan. Most are nailed loosely with a rubber gasket to avoid vibrating the house. Check the fan for clearance and check bolts and nuts and belt.

While you're in the attic you can also cut the holes for the attic vents and frame them. Locate the desired vent placement and carefully mark the hole size. Drill starting holes in the corners and use the sabre saw or hole saw to cut it out. This may be easier from outside. Now cut blocking for the edges and nail it in. Now, from the outside mount the vent.

Gable Vent

The fan will also have a shutter assembly which is mounted from below. Usually it will be mounted with screws into the joists. Lift it up (with a helper) and mark the mounting holes. Drill starter holes if necessary and then mount. Some shutters will require adjustment of closing tension before mounting—check your instruction sheet. Most will look finished as installed but some may require a wood frame. If so, measure and then cut trim to fit.

Before starting work on the wiring, turn off the circuit by either switching the circuit breaker or removing the fuse. Make sure you've got the right circuit with the tester or voltmeter. Hold both probes in one hand. Make sure switch/controls are off.

If you're in luck, there will be a junction box that you can wire to. If not, you'll have to add one. Use the wire cutter and stripper to prepare the wire for joining, and the wire nuts for the connections. The diagram below is typical of the circuit you'll be installing, but use the instruction sheet that came with your fan.

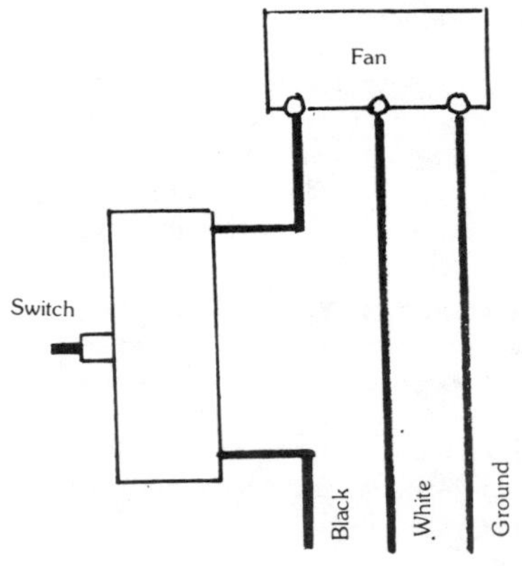


Wiring Diagram

The switch will require additional work below. Drill a 1/2" hole in the plate of the wall above where it will be mounted. Then go below and make a template for the switch box you will be installing and use it to draw the location on the wall, next to a stud.

Drill a starter hole near a corner and saw it out with your keyhole saw. Feed the wire down from above (use the fish tape if needed) and into the box. Now mount the box. Wire the switch, per instructions and screw to switch box. Put on cover plate and make sure controls are off and turn circuit back on. Now the moment of truth, turn on the fan, being ready to turn it off it it doesn't sound right. If you've done your work carefully, you should be rewarded with a fully functioning whole house fan. Make sure shutters open fully when fan is on. You're almost done.

If the attic is easily accessible to critters or people make a chicken wire or screen (big mesh) box to go over fan.

Screen cover

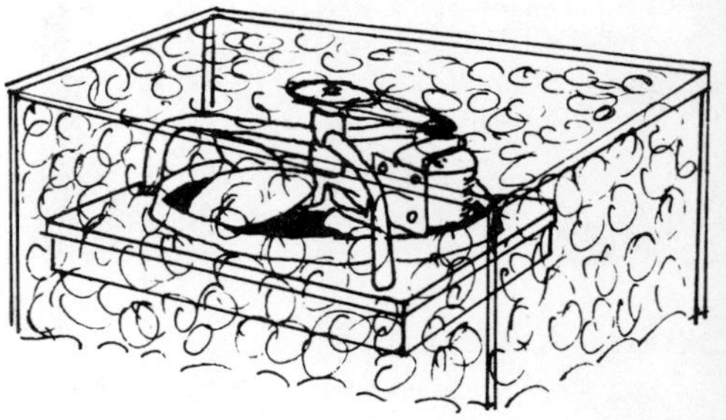

A final trip to the attic will enable you to replace insulation around the box and do a final cleanup and check.

In colder climates, you will probably want to build insulated shutters for the added attic vents and for the ventilation. You can make the vent covers from bead board (styrofoam or polyurethan foam). The shutter for the ventilator could be made with door skins over bead board with 2 screws to hold it in place—or if you built a screened box on top of the fan, you could drape an old sleeping bag or fiberglass batts over it.

Lubricate motor shutter pivots and bearings as indicated. Check belt tension (about 1/2"). Place and remove insulated shutters and fan cover seasonally.

These are some of the most important steps you can take to reduce the demand for energy for space heating and cooling in your home or business. We'll talk about some others in the section on backup heating, cooling, and hot water. The energy conservation details have been described in depth because even if your house is in the shade of trees or other buildings, you can still do a great deal to increase your comfort while you reduce your energy bills. If you are more fortunate and your house is exposed to the winter sun you can do much more. First we describe how you can add more windows to increase your solar heating, then we describe how to add thermal mass to store this energy for later use.

# Passive Solar Retrofits

## Adding Windows

Windows, particularly south facing windows, act as the collectors for a passive solar building. Once you've completed your energy conservation improvements, you can consider adding more south windows. The area of window you need can be determined roughly by calculating your average January day heat loss in BTU and dividing by the average day January solar insulation in BTU/SQ FT. This will give you a reasonably good estimate of how much window area you need.

If you have done a good job of reducing heat loss you may find you need little or no additional window area. If you need more window area, remember the lessons from earlier in this Chapter and make it a tight window, with an insulated shutter or drape, summer shade, and potential for summer ventilation if additional ventilation is needed. Also, read the last section of this chapter to determine how much thermal mass is desirable to prevent overheating during the day and storage for heating at night.

The final requirement, of course, is that the window be exposed to the sun during the winter. You can determine this by making or by making or buying a solar site evaluator.

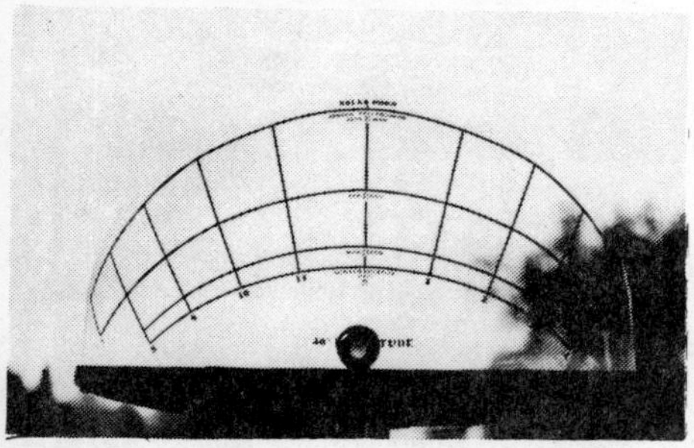

If your windows will be shaded by the bare branches of deciduous trees, make an estimate of total shade by holding up a sheet of ruled paper and counting squares with sun and those in shade. A dense branching pattern may block 50% of the sun or more. Factor this into your square feet of glass required. After you have done this, you are ready to proceed.

## Make Existing Windows Work Harder

Before you start carving up the house to add new windows and skylights, make sure you are using your existing windows to their fullest. Keep the glass clean. Take screens down in the fall. Selectively prune trees to let more sun through. Consider adding a horizontal reflector to south windows and vertical reflector to East or West windows. A reflective surface, including fresh snow, may add 50% to the heat gain through a window.

The reflector may be incorporated in an external insulated shutter or more simply built as a fixed reflector. For a south window the reflector should be titled down away from the house, both for drainage and to increase collection area. On the East and West, the reflector should extend above the window and sit at about a 60° angle to the South. The precise angle will depend on house orientation and latitude. After you've improved your use of existing windows, you can consider adding more window area.

## Add A Simple Window

The difficulty, ease, and desirability of adding a window depends in large part on the type of construction, the condition of the building, and the size of window to be added. If the house is masonry or stone, skip directly to the next section—retrofitting a trombe wall. If you are a fair to middling carpenter you can retrofit windows in wood frame construction. If the project is too much for you, it is worth paying someone to do it for you.

The simplest case is the relocation of a window from the cold north to the south side of the house. Using the tools described in Chapter 12, you should be able to remove the window without damage. Fill up the opening by putting in studs, insulate, and cover to match the rest of the north wall. Use the dimensions of the frame to figure out the finished opening needed on the south side. Then mark out the location on the wall and remove the interior surface, including enough extra to put in the header, sill, and cripple studs to support the header and sill.

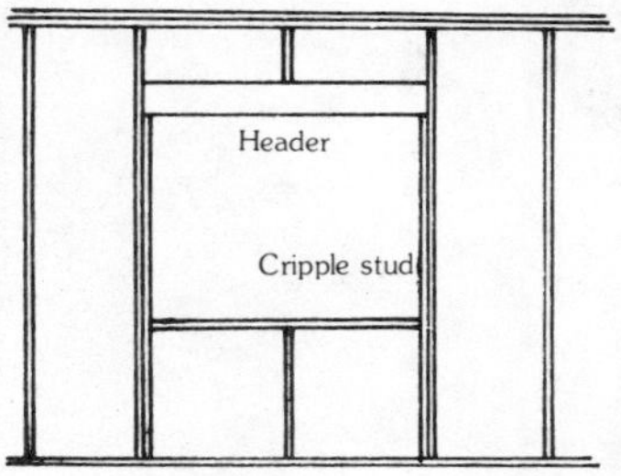


After you've measured everything twice, make sure your marks are square and in the right spot, you can frame up the window opening on the inside, check the fit of window, and finally saw out the outside wall to make a snug fit for the window. Then flash, caulk and set the window. Caulk and seal all cracks around frame and finish with trim to match.

If you are adding a number of windows to the south side you may find it easiest to remove all the exterior siding. This can be done with little damage to most siding materials if you are careful. This will also make it easy for you to insulate the wall as you install the new windows.

The use of skylights has often been highly recommended but I am less than enthusiastic about them, unless the sun shines only on the roof. Although installation is relatively straight forward and skylights should rarely leak—they often do. They are also more difficult to shutter properly and pose more difficult problems for overheating in the summer. Use them if you must, but use them wisely, with an insulated shutter and summer shade. A movable shade/reflector can be very helpful.

## Make A Trombe Wall

If you have a brick, stone, concrete, or concrete block house, adding windows may not be worth the trouble because you can more easily make the whole south wall into a solar collector and storage system. This is done by adding a support frame to the outside of the wall and double glazing it. Existing windows are left in place and serve to let the heated air into the house. Additional vents may be needed as well, both at the base of the wall for cold air return and at the top to let the heated air into the house. An exterior vent for summer ventilation is usually desirable. Typically 1%+ of the wall area should be vents. Where possible or as necessary for fire egress, an operable window should be included in the glazing in front of existing windows. An insulating drape between wall and glazing will improve performance still further. For more complete details, refer to Chapter 6. If you don't have a high mass wall which can be turned into a trombe wall and you can't retrofit windows easily, you may find a solar greenhouse is in order.

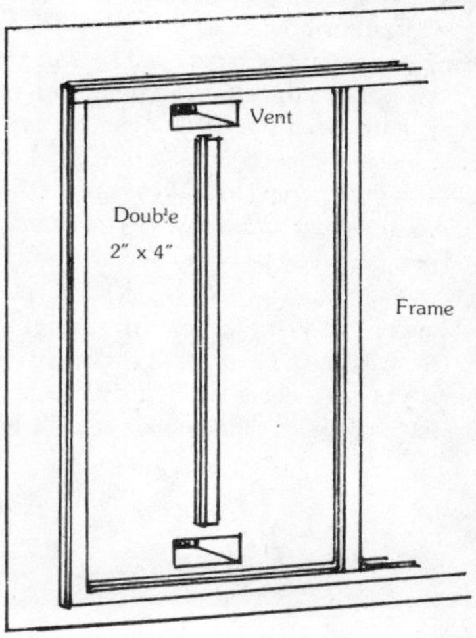


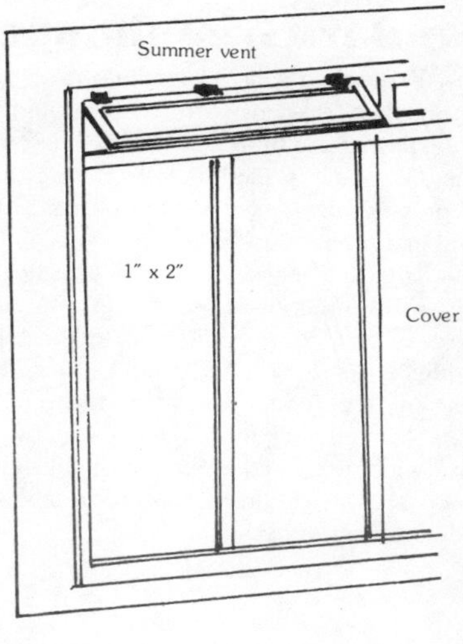

*Courtesy Vegetable Factory Green Houses*

# A Solar Greenhouse

An attached solar greenhouse is often the easiest and least expensive method of adding south glazing to a house or mobile home. The greenhouse glazing is usually sloped to collect more winter sun and provide better light for the plants. The added space in a greenhouse is also an excellent place for adding thermal mass, both to moderate temperatures in the greenhouse so that plants don't cook or freeze and to provide heating for the house. In addition to saving energy used for heating a greenhouse can also reduce energy use by providing fresh vegetables in the winter. Like any window, a greenhouse window should be weathertight, have an insulated shutter or drape, and have some form of shade in the summer.

The construction of the greenhouse can vary tremendously depending on the money or materials available, the builder's skill, and performance desired. A simple greenhouse built of plastic on a simple post foundation could be built for less than $1/sq. ft. and would in most cases pay for itself within a year. It might be the choice for a renter if he couldn't get the landlord to pay for it. At the other extreme, a commercially built triple pane greenhouse set on an insulated concrete foundation costing $30 sq. ft. might be the choice for a business person without the time to build a greenhouse. It will pay for itself many times over during its lifetime.

The key points to remember when building a greenhouse include: 1) placing it where it will be in the sun, 2) building a strong foundation, 3) tying it into the house securely, 4) providing a means to get heat into the house (vents, windows, or doors), 5) providing thermal mass for heat storage, 6) making provision for insulating the glass with shutters in drapes on winter nights, 7) providing shade in summer (roll down awnings arbor, or . . .) and 8) placing vents in roof and end walls for good ventilation in summer to prevent unwanted heating.

I particularly recommend: **Solar Greenhouse for Canada** (1974) $4.50 from Solplan, P.O. Box 86627, North Vancouver, B.C. V7L 4L2 and **The Solar Greenhouse Book** from Rodale Press, Emmaus, Pa. or at your bookseller. You might also wish to consult the additional sources listed in **The First Passive Solar Catalog.**

# Adding Windows - Not So Simply

In some cases you may find that your solar exposure or situation will not enable you to build a solar greenhouse, add a simple window, or make a trombe wall. Don't despair, you may still be able to add window area that will provide solar heating and a good return on your investment. Be forewarned however, that most of these retrofits are no longer simple do-it-yourself-on-a-weekend projects. They may require more sophisticated design skills, more advanced carpentry skills, and more time and money.

The most attractive alternative if the house is oriented properly is the addition of clerestory windows to the roof. Even if you have been able to add windows to the south, you may wish to consider this because it can help heat the north side of the house.

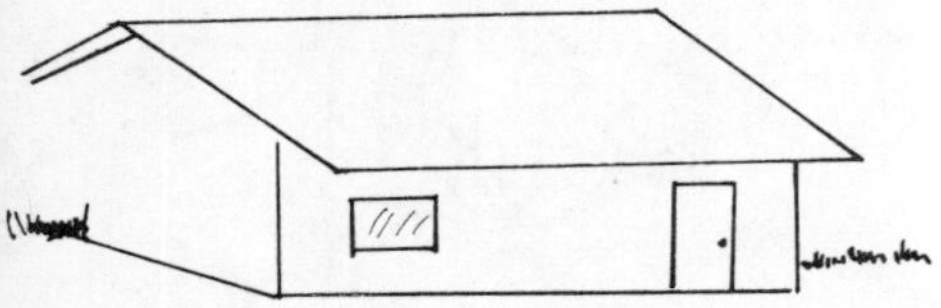

Clerestory Windows

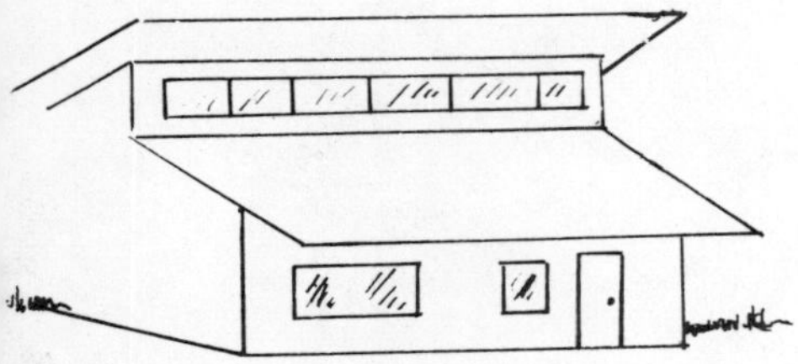

Clerestory windows are better than skylights because they are easier to weatherproof, easier to shade, and easier to shutter. Like operable skylights, they are also good for stack ventilation in the summer. Unfortunately, retrofitting them is more expensive than building them in. The added cost can be justified more easily if the remodel also provides additional living space in what was the attic or in a specially constructed loft.

The basic steps involved in a clerestory retrofit are: 1) careful design and planning, 2) assembly of materials and tools, 3) removal of ceiling sheetrock and insulation, 4 with a prediction of several days of dry weather removal of south roof, 5) extension of north roof rafters and framing of new south roof, roof trusses, etc., 6) sheathing south roof, 7) installation of windows, flashing, trim, 8) reroofing south roof, 9) finishing inside to suit. You might have a professional crew do steps 5, 6, 7, and 8 while you do the rest, or you might wish to do it all yourself.

If you aren't up to quite such a large project, you can add some high windows by building a dormer or two. They will also enable you to make more living space upstairs. You might also add major skylights in the roof if you can figure out a good system to shutter and shade them.

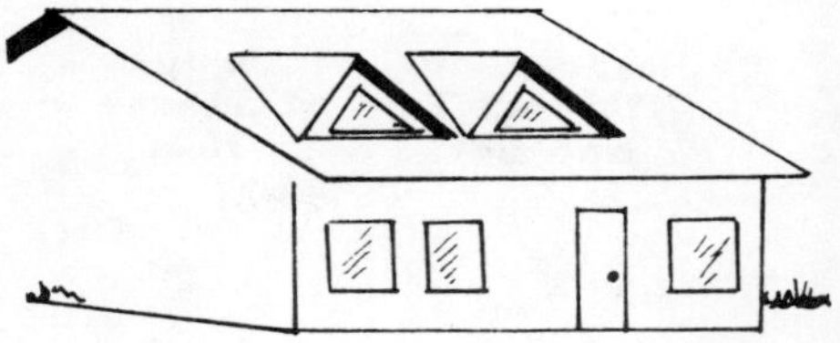

Add A Dormer

If the roof runs the wrong direction (N-S) you face a more challenging problem. The best solution, if you can afford it, may be to remove part or all of the old roof and put in a second story with clerestory in a large cupola or build a solar room at the south end of the roof as shown below. These are really more in the form of a solar addition. Some of the possibilities for a solar addition are described later.

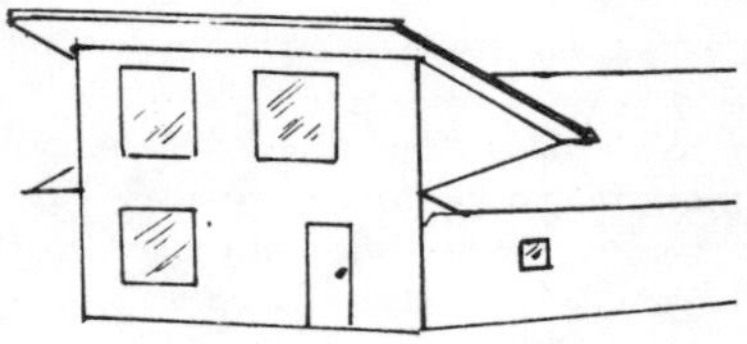

Clerestory Addition

More glazing area can be added to an exposed mobile home by replacing the aluminum skirting with fiberglass on the south side. Adding insulation to the north, east, and west, and adding thermal mass in water containers under the trailer can greatly improve performance.

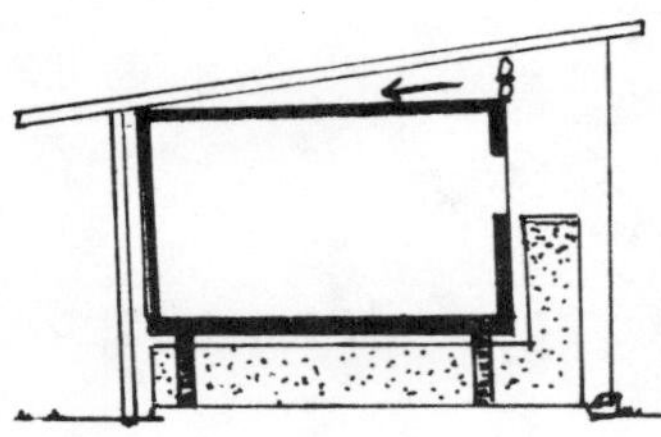

Mobile Home Retrofit

# Adding Thermal Mass

The value of thermal mass is described in several chapters of this book and the First Passive Solar Catalog. This section reviews some of the important points on how and when thermal mass is important and describes how you can add thermal mass to your residence.

As you will recall from Chapter 4, thermal mass can help store the energy from a sunny day for later use at night or on a cloudy day. For this purpose, thermal mass works best when it is all in the sun. As a general rule of thumb, mass in the sun is worth three times as much as mass in the same room but not in the sun and ten times as much as mass elsewhere in the structure. The thermal storage of the mass should be sufficient to prevent over heating even on a bright winter day. For cooling, on the other hand, the mass can be placed anywhere in the house as long as the cooling breeze will reach it at night. It will be most effective if it is widely distributed or designed to exchange heat quickly.

Although thermal mass is usually desirable, for both heating and cooling and comfort, there are situations where it is not desirable. The most common case will be a recreational cabin or house which is only used ocassionally and shut up the rest of the time. For these houses thermal mass can be detrimental. When the house is shut up it will get cooler and cooler (or hotter and hotter in summer). When you arrive even a hot fire may take hours to heat all of the mass up again and it will feel cold in the house while the radiant surface temperatures are low. By the end of the weekend it will be very comfortable again—but after you leave it will slowly cool off again. Houses like this are better built with little thermal mass but super insulation. Then they will warm up very rapidly when you arrive.

Thermal mass can also work against you if your house is thermally inefficient. If temperatures get too high in summer then thermal mass may keep it hot too long into the night after it would otherwise have cooled off. And conversely, if the house is too cold in winter, added thermal mass may keep it cold longer. Therefore, before adding thermal mass, you should improve your house's thermal efficiency. This will include both increasing insulation, sealing up cracks and leaks, and improving shading in summer and solar exposure in winter.

The use pattern of a building will also affect the type and placement of thermal mass. A commercial building that is used only from 8-5 will have considerably different requirements for thermal mass than a home that is used all day. For commercial buildings, less mass is often used to allow heating to occur rapidly in the mornings. The cool off at night is faster too, but no one is there. For a home, however, more mass is used to store heat for evening use and on through the night.

And finally, the seismic stability of the area you live in may determine the type of mass you use. The map on page 20 is a recent map of earthquake risk in the U.S. and will indicate where special attention is needed for structural design for earthquake safety. The earthquake codes are generally inadequate and should serve only as a bare minimum. Earthquake acceleration and loads are much higher than was previously thought. In the darker areas of the map, you should probably investigate local faults carefully (by reading and discussion with the USGS and your State Geological Survey) and consider a lateral acceleration of up to 1 g (or equal to the force of gravity) possible. You will also probably want to improve the earthquake safety of your home in general by reading **Peace of Mind in Earthquake Country.**

After you've upgraded you building's thermal performance, you can consider adding thermal mass to make it work even better. The thermal mass acts like a storage battery for heat or coolth. The size and type of mass you add will depend on the function of the mass (heating, cooling, or both) the design of the house, and the use pattern of the building.

The heat storage of different materials varies considerably. It depends primarily on the heat capacity of the materials, the product of the specific heat times the density, but must also consider its ability to exchange heat. For this reason water storage is considerably more than twice as effective as concrete per cubic foot. Common materials used as thermal mass are listed in Table 1.

## Table 1: Properties of Heat Storage Materials

| Material | Specific Heat BTU/Lb °F | Density lb. Ft. 3 | Heat capacity BTU/Ft. 3/°F No Voids | Ability to Exchange Heat |
|---|---|---|---|---|
| Water | | 62 | 62 | Very Good |
| Scrap Iron | 0.12 | 490 | 59 | Good |
| Concrete | 0.16 | 140 | 22.4 | Fair |
| Stone | 0.21 | 90 to 170 | 18 - 36 | Fair |
| Brick | 0.21 | 123 | 26 | Fair |
| Eutectic Salts | | | 240 | Fair |
| Glass | 0.18 | 154 | 27 | Fair |
| Wood | 0.65 | 30 | 19.5 | Poor |
| Earth, Damp. | 0.40 | 100 | 40 | Good |

Seismic Risk - East

Seismic Risk - West

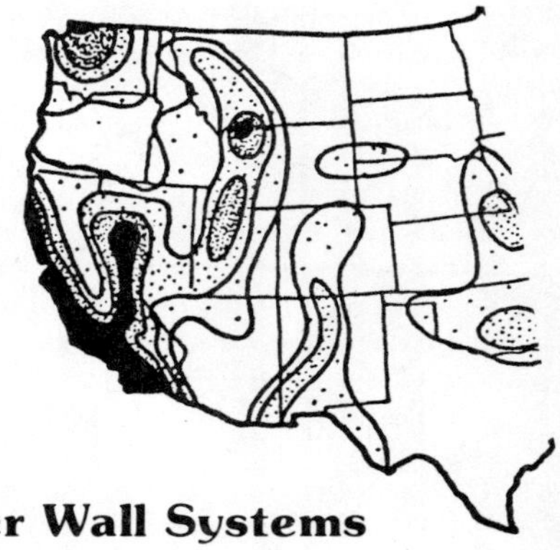

## Adding Thermal Mass

The easiest house to add mass to is a brick, stone, or concrete block house that you can retrofit with insulation on the outside. Usually, this is done by adding urethane or styrofoam plastic insulation covered with stucco. Several commercial systems have been developed for doing this. This insulation is usually added to the west, east, and north side of the building—with double glazing added to the south. The building will go from being a very inefficient building to an excellent passive solar building with these additions. Although the mass in this type of building is better distributed for cooling than heating the overwhelming amount of mass will enable it to perform well for heating as well.

Retrofit Masonry

If you are in a standard frame house, you may also have considerable mass in the house you can use—if you have a concrete slab floor. To expose this mass to the sun or cooling night time breezes you should replace the carpet with tile, parquet, linoleum, or vinyl flooring. You needn't do this throughout the house but only when the sun reaches the floor in winter. Entrances, kitchens, living rooms, and living rooms are also easier to care for with the tile so you'll gain two benefits.

Installation is straight forward following manufacturer's directions. This will usually involve carefully cleaning the floor, measuring the flooring to fit, and gluing it down although some of the new vinyl floors are just laid out. To lay tile, a layer of mastic or grout is spread on the floor and the tile is set in that. Then after it has set, grout is laid in the joints.

Even if you can use the thermal mass of your slab floor you will probably need additional thermal mass for better heating and cooling. There are three areas for adding thermal mass: as a freestanding thermal mass, either water tanks or masonry; adding mass to the existing walls; or adding mass to the floor.

Adding water mass is often the easiest technique to use. Many different containers have been used for holding water for thermal mass. The most common ones are listed below in Table 2. along with cost, advantages, and disadvantages.

## Table 2: Containers For Water Wall Systems

| Type | Source | Size | Volume | Cost/Gal. | Installed Cost | Notes |
|---|---|---|---|---|---|---|
| Tanks | local welder | any 1-½ x 3 x 6 ft. recom. | any | 30¢ to $1.50 | 50¢ to $1.50 | aesthetically appealing, easy to install, effective |
| Drums | drum mfgr., chemical supply, etc. | | 30 or 50 gal. | 10 to 45¢ | 20¢ to $1.00 | cheap, readily available, hard to clean, must be stacked carefully. |
| Culverts | pipe supply, scrap yards | 12 in. + diam. | depends on length | 30 to 50¢ | 50¢ to $1.50 | tough, attractive to some, help where floor space is tight. Make sure installation is seismic-safe. |
| Kalwall cylinders | Kalwall | 12 in. & 18 in. | depends on length | 40¢ to $1.45 | 50¢ to $1.70 | translucent, easy to install and move, easy to damage. Best where traffic is light. |
| PVC Pipe | Ag Supply & Pipe Co. | 6 in. & 12 in. | varies | 10 to 20¢ | 30¢ to $1.20 | light, durable, heat tansfer not as good, must rack or brace to mount. |
| Glass bottles | various | varies | up to 10 gal. | | | cheap, readily available, must seal carefully, moving difficult. |
| Metal cans | container co., restaurants (olive oil) | varies | to 5 gal. | to 60¢ gal. for 5-gal. can | 0 to $1.00 gal. | good for furniture modules, slow, would need bracing for tall stacks. |
| Steel or aluminum pipe | scrap yard or supply | 6 to 12 in. | varies | 10 to 40¢ | 30¢ to $1.50 | readily available, time-consuming to build racks, hard to clean. |
| Plastic bottles & drums | various: grocery stores, etc. | | commonly 1 to 5 gal. | 10 to 50¢ | 15¢ to $1.00 | cheap, may provide pleasant light, no rust. Slow to fill and installation weak. |
| One Design | | module | $1.00 + gal. est. | $1.00 gal. est. | $1.50 gal. | Easy to install and ship. Good potential for cooling. |
| Tabline Tanks | | 46 in. x 4 ft. x 16 in. | 90 gal. | $1.00 + gal. est. | $1.50 gal. | Easy retrofit, good interface with stud construction. |

The best option in most cases will be a custom made steel tank. These are simple to make, easy to disguise, and easy to work with on hybrid systems because the mass is all in one or two big containers, not in many small ones. Plan where it will go, usually in front of south windows. The detail sheet in Chapter 6 descibes the construction of a steel water tank. A tank about 2' x 3' x 6' is relatively easy to handle but bigger tanks can also be moved if you're careful. Water tanks can be effective for mobile home heat storage. They can be placed under the trailer frame with one end exposed to the sun. Or they can be extended as an "L" with a solar green house over the end of the "L".

Figure out how you will move your tank into the house before you order it. Make sure the floor will hold its weight. A slab floor usually presents no problem, but a suspended floor may, see the final section of this section on structural considerations. If you have a twisting flight of stairs you may wish to use an alternative method of heat storage.

It is usually possible to get your local welding shop to make up the tank and deliver it. Have them set it on short pieces of 2" x 4" when they drop it off. This will make it easier for you to finish the tank following the methods described on the detail sheet. This leaves only the move into the house up to you. If you kept the tank(s) to about 2' x 3' x 6' you will find it weighs about 400 lbs. empty. This is relatively easy to move (a piano weighs more) but is still heavy enough to hurt you if it falls over or if you strain too hard while moving it. Be careful. Always think movements through, lift with your legs (not your back), and whenever possible use levers and rollers rather than muscles to move things.

Assuming the tank has been dropped in the driveway and you have a reasonably level concrete walk to the house you'll find it's easy to get the tank in. First, check the entrance door—will rollers or wheels make it across the threshold? If there is a lip or drop of more than 1/2" or so they probably won't and you'll need to find some 3/8" plywood to make a ramp over the threshold. You can pick it up on the way to the rental store to pick up a piano dolly.

Get a couple of friends over to help lift up one of the ends of the tanks so you can place the piano dolly under the center of the tank. Block the downhill wheels of the dolly if you are on a grade. If you are alone you can raise it up using a 2" x 4" or long crow bar as a lever. Block it up with bricks, 4" x 4" blocks, or whatever as you go. After you have it centered, you can simply push it across the drive way, up the sidewalk, over your plywood ramp and into the house. If you have a fine hardwood floor, you may wish to lay down paper or tarps to protect it from dirt on the dolly wheels. If you have to go across a deep carpet you may want to put down 1/4" or 3/8" plywood or masonite runways.

When you get the tank in place take a break. Gather the friends and family and make sure its exactly where its supposed to be. Then lift up one end and pull out the dolly and set it back down. You're done except for any touch up on the tank finish.

If you have to pick up the tank from the shop, don't despair. Its harder than simply just moving it in but its still relatively easy. First, if you don't have one you'll have to find a pickup truck or trailer. If you're able to drive up to the front door, and the weather is good, you might want to arrange to keep the truck for a day or two so you can finish the tank in the truck.

Roll the tank out of the shop on a dolly or pipe lengths until its sitting behind the truck. You'll need help from the welder and a couple of friends for the next step. Have the two strongest characters pick up the front end (legs not back). The third person pushes on the back of the tank when they get it up to tailgate level. Rest. Then all three can push up and in on the back as it slides up and into the truck. Put it up on blocks by levering up one end and then the other. Finish it when you get home (be careful of spills if it is a borrowed truck). Then reverse the process, setting the tank down on the dolly as it comes off the truck.

You can also get the tank into the truck by yourself if need be. Set three 8' 2" x 4"'s (or two 2" x 6"'s) at an angle from the bed to the ground. Lift or lever the end of the tank over the boards and then wrap a loop of strong polyproplene rope around it and run it over a cushion on top of the cab to a come-along or winch. Then winch the tank into the truck onto blocks. Home you go for a break while you finish it.

Lower it by reversing the process. If possible drive up to the house and slide the tank down the rails, across the threshold, and into the house. Then lever it onto the dolly and roll it into place. And finally fill it up.

In place where a tank is not feasible or desired many other water containers can be used. Used 55 gallon drums are the cheapest storage containers and are good to work with. Reconditioned drums are easiest to work with and at about $15, each are a good bargain. If you can find drums locally, you can use them but make sure you know what was in them and don't inadvertently poison yourself. The detail sheet in Chapter 6 sheet describes many of the ways drum can be used. Two people can handle a drum easily or you can move them yourself with a drum dolly or appliance dolly or by rolling them along. You can roll them off the truck without damage if you put down some old tires (without wheels) as a landing platform. Set them in place and then fill them. If you are stacking them, make sure you have access to the bungs and that the bung is up before you strap, block, or clamp it into place.

Square five gallon cans (metal or plastic) are even easier to handle and can be used where access is difficult or where structures can't handle the weight of drums or a big tank. You can pick them up new or collect them from your local fast food restaurant. Clean them and then paint them. On metal cans, use red lead primer inside and outside and then finish the outside with acrylic enamel. Test for water tightness (a few leak) and discard or repair those that leak. Then fill them up. Now you can move them in and stack them. They make good bookcase building blocks, or can be laid up like bricks. Plastic drums are much easier to use. Simply add food coloring to water and put in place. Milk bottle plastic is too thin and doesn't hold up well in direct sun.

In some cases, water pipe or culvert may be the best way to add mass. They can be incorporated in furniture, set vertically in front of south windows, used structurally, hung off roof trusses, or set in the attic lengthwise if the joists will hold them. The most fool proof installations use welded plates to seal the bottom end of vertical sets and both ends of a horizontal set. A sheet metal cone cap will work on top of the vertical set. The detail sheet in Chapter 6 describes culvert installations.

The culverts should be the helical welded variety, not the riveted type. Ask for water tight ones as they vary some in quality of the weld. You can have the supplier reform the ends to make welding easier if you wish. Plain pipe can often be found used at prices considerably less than new.

The pipes or culverts can be mounted directly on a concrete slab with expansion bolts through a mounting plate. On a suspended floor they can be bolted through the floor to a steel plate below the joists. The high loading of culverts or pipes will almost always require additional bracing on a suspended floor. You might find it easier to make a mini-slab below the floor and mount the culvert on it. If the subfloor area is uninsulated, you'll want to wrap the base of the culvert with insulation.

Moving culverts and pipes around is easy for the smaller diameter sizes but can be a bit tricky with 18" or 24" diameter pieces 14' long. A dolly helps as does as assistant or two. A 14' 18" culvert may weigh only 190 lbs., but can be very awkward to move around. Make sure you can set up the culverts at the location you have chosen by leaving enough clearance at the top. In earthquake country, mounting should include both mounts at the top and bottom with considerble cross bracing to resist shear.

You may also use fiberglass tubes to hold water. These are light, easy to handle, quite attractive, and strong enough to hold the water. However, they are not very resistant to impact damage and are perhaps best used where little (or not so little) kids can't get to them. They are also more difficult to brace properly because they are flexible and don't have bases that can be bolted down. A timber frame or nylon strap system would seem most promising for these tubes. They are available from Kalwall, see Chapter 8.

A number of other containers have been used and certainly new ones will be developed over the next few years. The local availability of containers, the house, desired location of mass, function, and your own feeling of aesthetics and construction skills will determine which is best for you. If water storage doesn't fit into your house retrofit plans well, you might like to carefully consider using a high mass floor.

Adding mass to the floor has already been mentioned for a slab foundation. It is also practical to add a high mass finish to a suspended floor. The amount you can add will determined by the existing structure and your willingness (or lack thereof) to reinforce it. Most buildings, except in earthquake areas, can add considerable mass without reinforcement.

The highest mass, lowest cost floor we've seen is built using masonry wall toppers. These 1-1/2" thick concrete pieces were mentioned in the Haggard House retrofit and are more attractive than you might suspect. The detail sheet in Chapter 6 explains how they are installed and used. They add about 15 pounds per square foot to the floor.

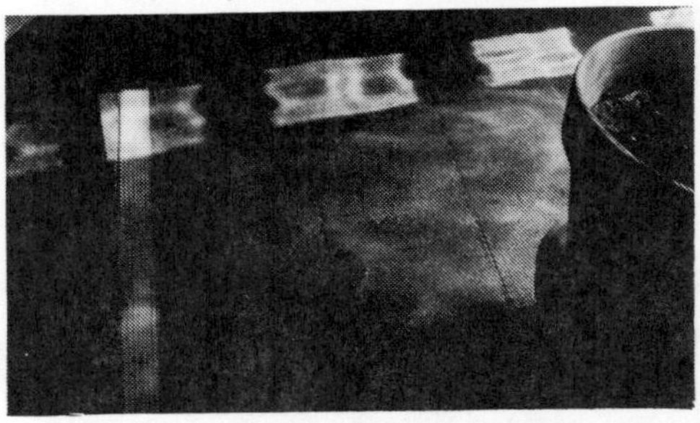

A high mass floor can also be added by pouring a thin slab of concrete. The thickness of the slab will determine the weight added and the advisability of adding thermal mass. By laying polybutylene pipe in the floor, the slab can be heated using a solar heater or stove water coil. Construction paper or tar-paper is usually laid down over the floor to prevent cement leaks dripping thru during installation. In a major retrofit you may find it desirable to make cut outs in the finish floor and subfloor to make a concrete beam support reinforced with steel.

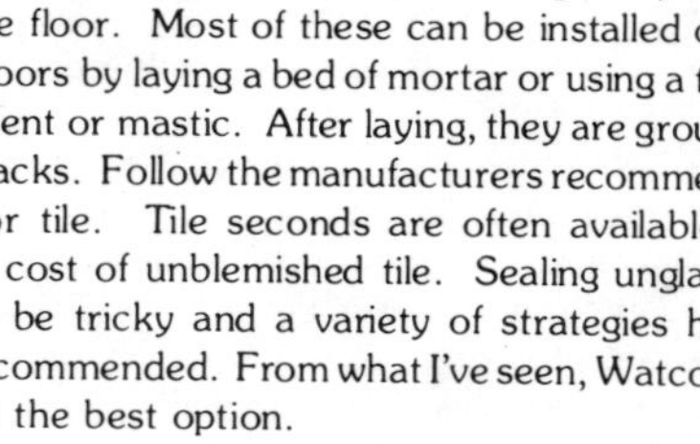


High Mass Floor

Mass can also be added by installing a tile, brick, or stone floor. Most of these can be installed over wood floors by laying a bed of mortar or using a thin-set cement or mastic. After laying, they are grouted to fill cracks. Follow the manufacturers recommendation for tile. Tile seconds are often available at 1/2 the cost of unblemished tile. Sealing unglazed tile can be tricky and a variety of strategies have been recommended. From what I've seen, Watco Oil is about the best option.

If you have hardwood floors or don't want tile or concrete floors, you can include the needed mass by adding stone, concrete, clock, adobe or bricks to existing walls or a new wall or room divider. The high loading from these walls will often require additional bracing particularly if it is not over a bearing wall or main beam.

The easiest high mass wall to build is a surface bond concrete block wall. Simply stack up the blocks and then trowel on the surface bonding in accordance with manufacturers instructions. For added mass, the cells can be filled with sand. In earthquake country, you'll probably need to add some steel reinforcing and tie the wall to both the foundation and frame. Added stability can be gained by building an "L" shaped wall.

A natural stonewall is more difficult to build but can be much more attractive. A layer or two is added using mortar. Let this set, then add another layer or two. Ties nailed into the wall will help secure the rock wall.

Artificial stone is easier to work with because it is uniform in thickness and lighter weight. It is usually glued to the wall and then grouted. Most artificial stone is concrete based with a surface finish that looks like stone.

The final option for adding thermal mass is the use of eutectic salts or other phase change materials. These are now available from several manufacturers and have excellent storage capability (up to 4 times water). They are expensive however, and must generally be used with forced air to achieve good heat exchange. They also typically change phase at 81°F — a bit warm for most uses except in an isolated greenhouse. Keep track of the developments in this field because improved prices, operating temperatures, and heat exchange will undoubtedly come along soon. Some of the exising suppliers are listed in Chapter 8.

These then are the more common materials and methods use to add mass for improved passive solar heating and cooling and better comfort. Further information on the fairly complex structural issues of supporting added mass and resisting earthquake movement can be obtained from a structural engineer.

# Supporting Added Mass

A full treatise on structural engineering is beyond the scope of this section but a look at two of the most common structural evaluations involved when added mass is incorporated in a building will be reviewed, along with a more general discussion of earthquake bracing.

The simplest case in a house retrofit is the addition of a high mass floor to an existing house. The formula used for determining the loads imposed on this simple beam are:

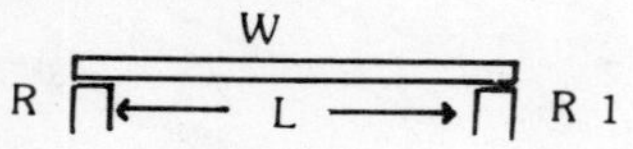


R = reactions, lbs.
W = weight, lbs.
L = length
M max = bending moment, in lbs.

$$R = R1 = \frac{W}{2}$$

$$M\ max = \frac{WL}{8}$$

The basic existing load include, 5 lbs. sq. foot, dead load, and 40 psf, live load (people, etc.). If we add the wall pavers to the floor they will add additional $\frac{1\text{-}1/2'' \times 144'' \times 140 \text{ lbs. cu. ft.}}{1728''} = 17.5$ lbs. sq. ft.

The total load then would be 45 psf + 17.5 psf = 62.5 psf. To check if our existing doug fir 2″ x 8″ joists 16″ oc are sufficient for the 12' span.

$$W = 62.5 \times 1.33' \times 12'$$

$$M\ max = \frac{997 \text{ lbs.} \times 144''}{8} = 17{,}955 \text{ in lbs.}$$

$$S = M\ mass\ /f = \frac{w \times d^2}{6}$$

$$S\ calc. = \frac{17{,}955 \text{ in lbs.}}{1700 \text{ psi}} = 10.56$$

$$S\ act. = \frac{2 \times 6^2}{6} = 12$$

S. A > S. C.

S = section modulus
f = fiber stress in bending, psi
1200 psi pine
1700 psi douglas fir
2000 psi oak
w = width of beam
d = depth of beam

So in this case, the floor is strong enough to carry the load.

The second most common load imposed in a retrofit with thermal mass will probably be the addition of concentrated mass, such as a water tank or rock wall. This can be evaluated as a simple beam with uniform load added to a simple beam concentrated load.

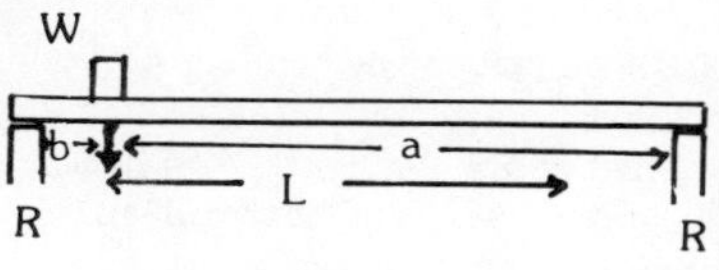


Using our earlier example as a basis, we can then compute the additional load for a concentrated load.

$$R = \frac{Wb}{L} \qquad R1 = \frac{Wa}{L}$$

$$M\ max \text{ (at point of load)} = \frac{Wab}{L}$$

A typical case of a water tank 3 feet high located 6″ from the window wall is illustrated below with 2″ x 6″ douglas fir full dimension 12' joists at 18″ centers.

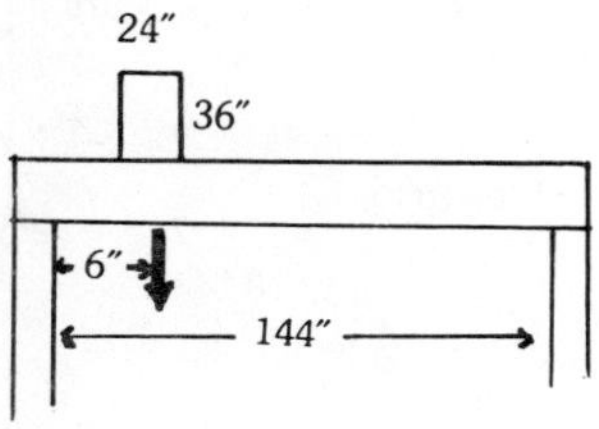


Weight = 3 ft. 3 x 62 lb. sq. ft. = 186 lbs. sq. ft.

$$R1 = \frac{186 \text{ lbs.} \times 18''}{144''} = 23.25 \text{ lbs.}$$

$$R = \frac{186 \text{ lbs.} \times 126''}{144''} = 162.75 \text{ lb.}$$

$$M\ max \text{ (at point of load)} = \frac{186 \text{ lbs.} \times 18'' \times 126''}{144} = 2929 \text{ in lbs.}$$

We add this to our earlier calculation, assuming a concrete tile floor was retrofit as well.

M max = 2929 in lbs. + 17,977 in lbs. = 20,884 in lbs.

$$S\ calc. = \frac{20{,}884 \text{ in lbs.}}{1700 \text{ psi}} = 12.28$$

Sact ≈ Scalc

In this case, the floor joists could handle the weight of the tank as well, although some reinforcement might be desirable. If we used pine joists instead, then:

$$S\ calc = \frac{20{,}884 \text{ in lbs.}}{1200 \text{ psi}} = 17.4$$

Then the load is almost 1.5 times the strength of the joist. Here the joist would have to be bigger. To find out how much bigger, we can set:

$$S = 17.4 = \frac{w \times d^2}{6}$$

assuming a 2″ wide joist

$$S = 17.4 = \frac{2 \times d^2}{6} \qquad 52 = d^2$$

d = 7.2

Because it is hard to buy 2″ x 7.2″ joist, we would go to a 2″ x 8″ joist. In a retrofit case though, the joists are fixed and we can therefore either reduce L by adding a new support under the new load, or we could beef up the joists by adding a 1″ x 6″ to the side of each 2″ x 6″. Adding this extra piece is harder than adding a beam, however, and would rarely be resorted to.

In addition to the load in the joists, you may also need to calculate the load on the bearing wall. The simple methods shown here should enable you to determine whether you can add more mass to the house in a seismically stable area. If earthquakes are likely (or certain) then the problem becomes more difficult because horizontal shear and overtuning becomes a problem.

If you are adding a mass wall or floor you should probably have the resultant shear calculated for the likely acceleration in your area (up to 1 g near major faults). Overturning of a water tank or small wall is a little easier to figure and you can roughly estimate it yourself.

The chance of overturning can be estimated by figuring out the squatiness ratio, or ht/width of the mass. If this is less than 2 it will slide rather than tip over. If it is more than that then steps will have to be taken to tie it into the structure of the house.

For a better evaluation of overturning moment you can draw a force diagram, using the expected lateral acceration and the shape of your tank. If expected acceleration—for example is 0.3 g then the resultant vector is:

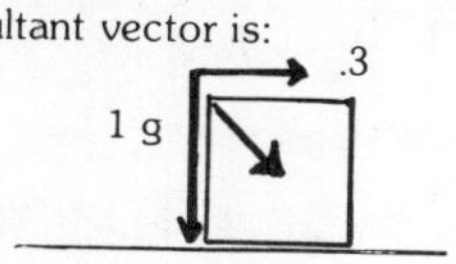


1.045 g at 17°

If the tank is 3' tall and 1' wide its cg (center of gravity) is 1.5' up and 6″ in from edge and the resultant angle for stability is 18.5°. This would be secure.

In the worse case, of expected accelerations of 1 g than the squatiness ratio should be kept at 1.

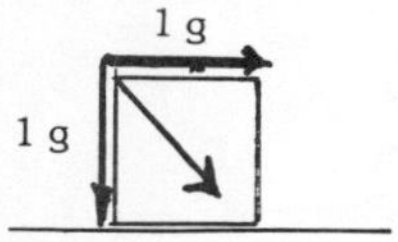


1.4 g at 45°

This is a simplistic way to evaluate earthquake force however, because it neglects complications caused by both vertical uplift and horizontal acceleration at the same time. You may wish to have a competent structural engineer who is experienced with earthquake design take a look at your house. (There are only a few around, particularly if you live in the east or south but try anyway).

## A Passive Solar Addition

In some cases a solar addition may be the best solution. It can add more south glass and incorporate thermal mass for passive solar heating and/or cooling and add living space. It is particularly easy to make a summer porch or screened-in porch into a solar addition. Simply weatherproof, insulate, improve windows, and add thermal mass, perhaps as a water tank or trombe wall.

If no opportunities present themselves then figure out the most needed revision of floor space in your house (compatible with good solar exposure) and build a new room or wing which is super energy efficient and has oversized glass and mass. If it will provide mainly heating it is often helpful to have it below the level of the existing house. Some possible designs are shown below.

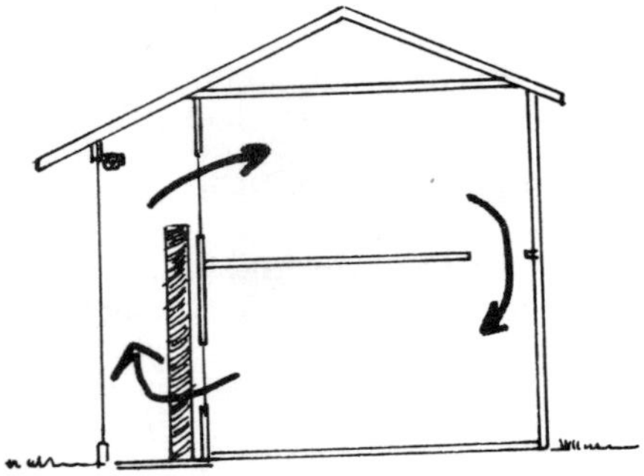

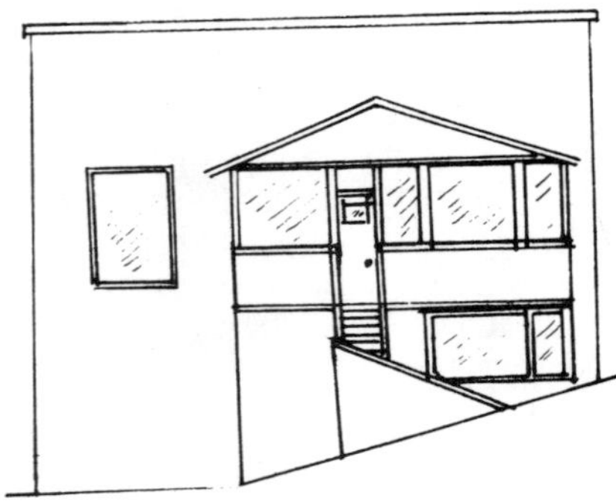

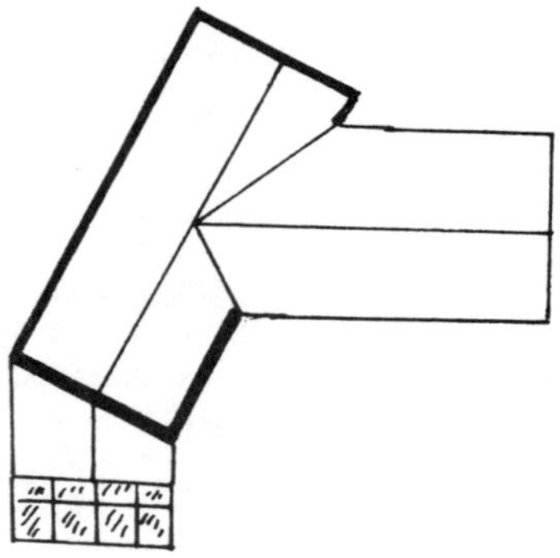

# Backup, Heating and Hot Water, Cooling and Ventilation

In some cases, particularly for retrofits, passive design will prove incapable of providing 100% of the required heating or cooling and some backup equipment will be required. Even where passive design can provide 100% heating to the owners' satisfaction the building code may require backup assistance to maintain 70°F, 3 feet above the floor. And finally, with particularly tight houses, ventilation with an air-to-air heat exchanger may be desirable to reduce indoor air pollution.

## Heating and Hot Water

The most attractive backup system in many areas for both space and water heating is wood. Wood is an economical, reliable, and renewable source of energy. However, wood is not without adverse impacts, particularly air pollution, and should be used only as backup for an efficient solar building. The ideal backup system provides both space and water heating using an airtight stove (50-60% efficiency) with air to water heat exchanger to heat water in the fire box or flue. For smaller houses, a wood kitchen range with water coil may be able to meet all the heating demands.

Backup Heating

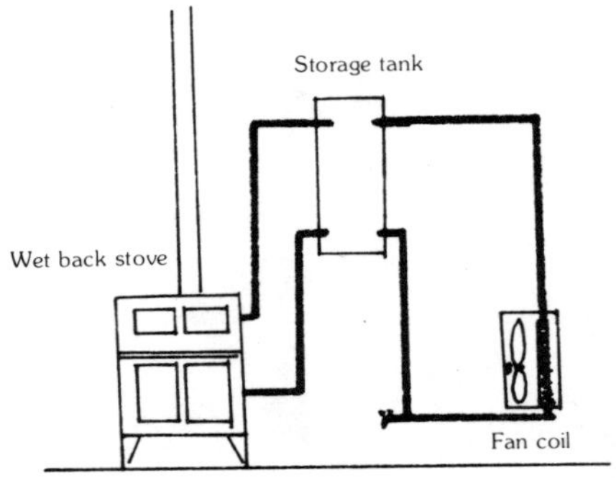


A thermosiphon system is the simplest and most reliable system. A pumped system should be used where a thermosiphon can't be, for example where a second floor stove is to be used to heat water for a first floor hot water heater. A pumped system may also be effective if the hot water from the fireplace is used in a hydronic heating system. For simple cases. a thermosiphon may work but for more involved pipe and radiator runs a pump is usually required. The "Chill Chaser" from Tubonics incorporates pump and controls with the radiator and requires only a simple pipe run.

Wood heating with a fireplace is inefficient at best and loses energy at worst, -10% to +10% efficiency. Even with a heatilator type circulating fireplace, blower grate, and other improvements, a fireplace will rarely reach 30% efficiency. If you have a fireplace, you can improve its efficiency by setting an airtight stove on the hearth and blocking off the fireplace. A wide variety of airtight fireplace inserts are made and are worth a close look. (See Chapter 8)

In areas with good winter sun, an active solar system may also prove useful in a backup role for space and hot water heating. The passive water heater can still serve as preheater to the active solar system which can be optimized for winter performance. an active solar and wood backup heating system is also attractive. If the system is designed well, one pump and controller may be all that is required, offering savings on a hydronic system which would require a pump anyway. A second option, which is less expensive and easier to install is an active air system. The simplest system uses the south wall area as a collector.

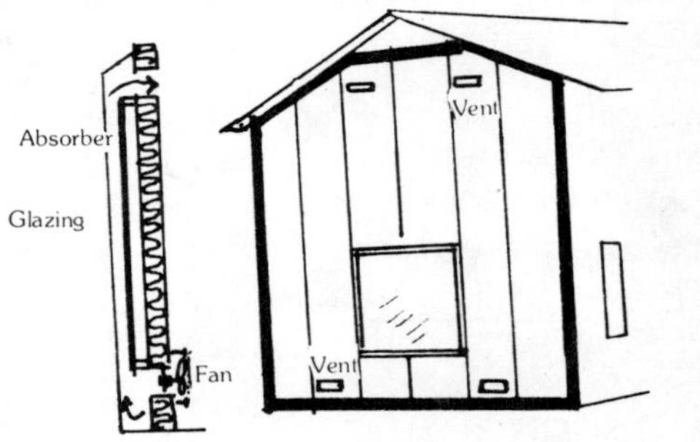


In a windy area, a wind heat backup may be more economical. A fairly simple wind machine/generator will suffice for mechanical or electrical resistance heating. A wind machine for winter heating should be particularly robust and have very reliable overspeed protection.

If none of the above are possible, for whatever reason, it may be necessary to use a more conventional backup system. In some cases the water heating backup can be used for space heating backup as well. Table 3 compares the cost of common energy sources, in early 1980 in California. You can adjust to your area easily by taking the ratio of your raw cost per unit of fuel to the cost shown here and multiplying the result by the cost figure on the right. For example, if your cost is 8¢ KWH v/s the 4¢ KWH shown here then your cost MBTU will be:

$\frac{8}{4} = 2$ 2 x $11.72 = $23.44

The rate of increase in prices for the last six years has been 25% year for natural gas and about 15% year for electricity although fairly considerable variations have occurred. The price of fuel oil has risen 27% year for the last 8 years. During the same period, inflation averaged about 8%. A gallon of oil will cost twice as much in less than five years.

## Table 3: Fuel Cost MBTU

| Fuel | Amount | Cost | Gross BTU | Efficiency | Cost MBTU |
|---|---|---|---|---|---|
| Oak | 1 cord | $100 | 31 MBTU | Air tight stove 50% | $ 6.45 |
| | | | | Fireplace 10% | $32.25 |
| Pine | 1 cord | $90 | 16 MBTU | Air tight stove 50% | $11.25 |
| | | | | Fireplace 10% | $56.25 |
| Coal | 1 ton | $50 | 23 MBTU | Airtight Stove 50% | $ 4.34 |
| #2 Fuel Oil | Gallon | 90¢ | 125,000 MBTU | Furnace 60% | $12.00 |
| Natural Gas | Therm | 30¢ | 100,000 | Furnace 60% | $ 5.00 |
| Electricity | KWH | 4¢ | 3,412 MBTU | Heater 100% | $11.72 |
| Propane | Gallon | 68¢ | 102,000 MBTU | Furnace 60% | $11.11 |

If a convential space heat backup is needed, a through-the-wall gas furnace may be a good choice. With a insulated, tight house the heating load should be low enough for a through the wall furnace to do the job. The through wall heater is more efficient because it doesn't have the flue losses associated with more convential gas heaters.

If natural gas is not available, an electric backup system may be the best choice/although the net system efficiency at 30%-50% is usually lower than gas at 60-70%). The advantages of electric backup include the clean heat, low first cost, and ease of installation. Radiant panels or the electric hydronic baseboard from Intertherm are among the best candidates for an electric heating system. Several electric heaters with mass storage are produced in Europe. These can be rolled from room to room as desired. In some solar/wood hydronic systems, an electric heater may be included in the storage tank. This can work out quite well. A heat pump may also prove workable.

Another attractive possibility worth considering is a multi-fuel boiler for backup hot water and heating. Several models are now available with a variety of features and fuel options, see Chapter 8. An automatic switchover from wood to oil or gas may be desirable as it would meet the UBC heating requirements yet would allow you to use wood as much as you wish. In more sparsely populated areas with little wood a coal option multifuel boiler might be a good choice. Here coal could be used as much as desired with an automatic oil or gas burner for backup.

In cases where the major concern is staying warm at night a down or fiberfil comforter may be a better buy—particularly for renters. Also worth considering are electric bedwarmers and electric blankets. The electric base sheet is more comfortable and efficient than the blankets—but harder to find.

If water heating is the only consideration a simple water coil in the fireplace or stove will often suffice. Alternatively, a wood fire water heater can be purchased. These are now available with kerosene backup and will work well if used properly, i.e. by scheduling water demand so a fire or two a day will suffice.

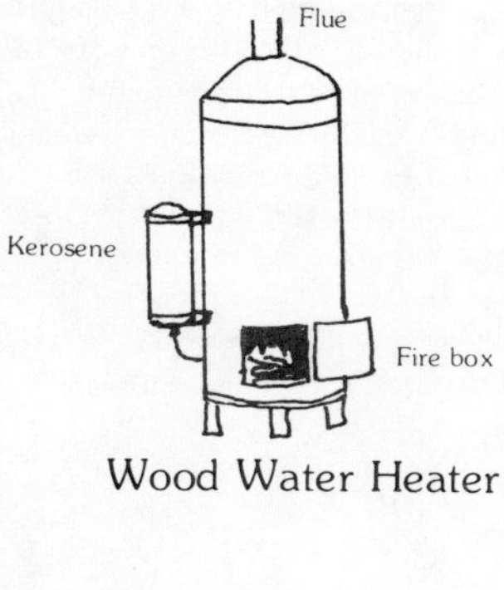

Wood Water Heater

Just gaining publicity is the furnace room with thermal mass to store heat. These systems are a good approach but are bulky and expensive. They offer the possibility of high efficiency with low pollution because they can be run hot for a couple of hours and then left off while heat is drawn from storage. Look for continued improvements in this area.

An important factor in retrofitting any water heating system is reducing demand. This includes using water efficient appliances, use of flow restrictions, and if possible, scheduling use to periods when the solar contribution will be highest. This is particularly helpful with a passive solar water heater where the hottest water is available in the afternoon and evening.

A water heater will also benefit from a thermostat with time setting. These are available for electric heaters and will pay for themselves quickly—despite the initial cost.

A supplemental water heater at the kitchen sink may allow you to get by without using your main backup heater. This might involve a flash water heater or instant hot water tap. The flash heater should be a thermostat controlled model. The fixed temperature rise is adequate only if relatively constant temperature feed water is used. Emerson's Hot Tap keeps a quart of 185°F water in an insulated tank below a third faucet. This can be used for hot drinks and coffee (much more efficient than using the stove) and also used to boost dish washing water temperature if the solar water is a little cool. A seperate switch is useful for the Hot Tap.

Hot Tap

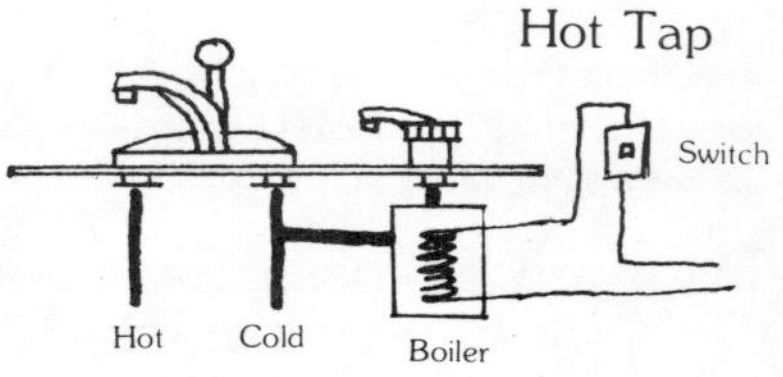


# Cooling

If good passive solar design, with overhangs, light colors, and good stack and natural ventilation proves insufficient then a variety of backup cooling systems can be used including forced ventilation, evaporative cooling, radiant sky cooling, and if absolutely necessary air conditioning.

The most efficient and in most cases most desirable system is forced ventilation at night with a whole house fan. Because this is such an attractive option we have included a fairly complete discussion of whole house fan installation.

If a whole house fan is not feasible, if you are a renter for example, you can use portable fans. These can be used to provide direct cooling (by blowing on you) or house cooling by drawing cool outside air into the house.

A casablanca ceiling fan can also be very helpful. Not only is it an efficient air mover for summer cooling but it will also help carry hot air from the ceiling back to the floor in winter. Ceiling fan installation is simple and fast. A wide variety of models, styles, and prices are available, with and without integral lights.

Indirect Cooler

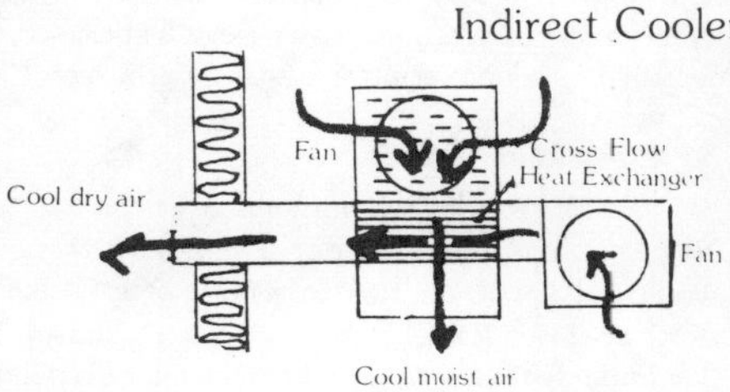


If the whole house fan or smaller fans can't do the whole job then additional cooling can be added with evaporative cooling. This can be either a direct system (the most readily available evaporative cooler, often known as a swamp cooler) or indirect system. The indirect system is preferable because it doesn't add any humidity to the house. You will probably have to build your own however as we haven't been able to find anyone that still markets them. They were widely used and popular in the 40's and early 50's but were driven off the market by cheap electricity and air conditioners. Evaporative coolers can provide cooling equal to air conditioner cooling with about 1/4 the energy use. The evaporative cooler can be inter-connected with the whole house fan so that as the whole house fan goes on the cooler will go off.

It may be possible to use a direct evaporative cooler to cool off the thermal mass in a solarium or living room and then let natural convection cool off the rest of the house. This can reduce the effect of increased humidity considerably.

In a similar fashion a rock bed can be used for cooling by blowing cool humid air out through it, them bringing dry cool air into the house. This is basically an air-to-air heat exchanger with mass. These are called regenerative rock beds and are widely used in Australia.

If an active solar system is contemplated consider making it a summer cooler as well as winter heater. This can be done by using collector glazing that is fairly transparent to long wave radiation or an unglazed collector. Most of the plastic glazings are suitable for this type of system. At night, water is pumped through the collector and is cooled by night sky radiation and conduction. A trickle type collector can be even more effective because it can add evaporation to night sky cooling.

If none of the preceeding options are sufficient then resort may be made to an air conditioner. It is properly sized to fit load (not by rule of thumb) and supplements a whole house fan it may be needed only a few hours a day on the hottest days. Used this way an efficient air conditioner (EER 11+) can be inexpensive to operate. The EER is the energy efficiency ratio, or output in BTU per input watts. A typical air conditioner might use 1,000 watts to produced 9000 BTU of cooling for an EER of 9. Get the best you can afford.

Maintenance and proper location can also help reduce air conditioning costs. Provide shade for it and draw intake air from the coolest spot possible, usually on the north side of the house. Paint the case white for added benefit. Keep coils clean and air filters (if used) clean.

If you will need supplemental cooling in the summer and added heating in the winter you may be able to use a heat pump. These work in temperate climates with just air as a heat sink/source but a heat storage other then cold outside air for their heat source is essential in colder climates. Earth, water, and solar heated water have all be used successfully as a heat source. Again, get a high EER.

## Ventilation

Building codes recommend an exchange rate of 10 cu. ft./min/person. For a family of three that would be 3 x 10 cu. ft./min x 60 min/hr - 1,800 cu. ft./hr or about .2 ACH for a 1,000 sq. ft. house. That would be all that is required in a clean, non-polluting house. Unless exceptionally weathertight design is used no added ventilation will be needed as it is unlikely we will get down below .2 ACH.

However, if the people density is higher, the construction is very tight, and/or the materials or appliances are polluting (gas stove, plastic paneling, plywood, etc.) then more ventilation will be required. Obviously if we crack a window we can get fresh air but only at the cost of adding to our heating bill.

To resolve this conflict, we can use an air-to-air heat exchanger. This uses heat from the hot dirty air exhaust to warm incoming fresh air. These can be bought from BACHO (Swedish) or Mitsubushi (Japanese). The house size model from Mitsubushi costs less than $300. The BACHO is more costly. Efficiency is in the 60-70% range for these compact and fully finished models.

Or for around $60-100 you can build your own from plans prepared by Cooperative Extension at the University of Saskatchewan, Saskatoon, Saskatchewan ($2. ppd), see detail sheet in Chapter 6. This model is bigger and of course has to be built but offers 80-90% efficiency.

Lossnay

# The Economics of Retrofitting

A simple method to compare the value of different energy saving and passive solar retrofits is a calculation of $ million BTU/year. It is less accurate than the life cycle costing method described in the next section but is much easier. Here's how it is done, along with several examples.

Most energy conservation projects involve insulation or infiltration. We can use the information in Chapter 4 and 5 to calculate the energy savings per year.

## Attic Insulation: Example

Say we insulate our 1,000 sq. ft. attic, going from R-3.3 up to R-33. The heat flow reduction is from 300 BTU/HR to 30 BTU/HR. If the heating season has 8,000 Heating Degrees Days then we save:

8,000 HDD x 300 BTU/HR x 24 HRS =
57.6 million BTU

8,000 HDD x 30 BTU/HR x 24 HRS = 5.7 million BTU
SAVING = 51.9 BTU

If the insulation cost $500 then the cost per MBTU is about $10. The comparable cost for fuel oil (Table 3) would be $13 at a fuel oil cost $1 gallon. Because the insulation will last as long as the house, an evaluaton of its life cycle cost is really better and this example is shown later in this Chapter.

## Insulated Shutters: Example

If we add insulated shutters to the windows of a house we can increase the insulation value from R-0.9 to R-10. If we cover 100 square feet of window in our 8,000 HDD climate we can save:

8,000 HDD x 111 BTU/HR x 24 HR = 21 MBTU
8,000 HDD x 10 BTU/HR x 24 HR = 2 MBTU
SAVING = 19 MBTU (if shutter closed all the time)
SAVING = 10 MBTU (if shutter closed only at night)

$200  10 MBTU = $20 MBTU

If we built the shutters, they would cost about $200 for a cost of $20 per MBTU. These would also be a good investment.

## Insulated Drapes: Example
If we bought commercial insu-

If we bought commercial insulated drapes for the same house and had a professional install them they might cost $5/sq. ft. and have an R-value of only 5. Here, the savings would be:

8,000 HDD x 111 BTU/HR x 24 HR = 21 MBTU
8,000 HDD x 20 BTU/HR x 24 HR = 4 MBTU
SAVING = 17 MBTU

$500  17 MBTU = $30 MBTU

These are still a good investment but not as good as the others. This example is also re-examined in the section on life cycle cost.

## Storm Windows: Example

If we add storm windows to a 1200 sq. ft. house with 200 sq. ft. of windows on a house in a 5,000 HDD climate we increase the R-value from 0.9 to 1.9 and drop heat heat loss from 222 BTU/HR to 105 BTU/HR. We may also save by reducing infiltration 0.5 air changes hr.

Conductive Heat Loss:
5,000 HDD x 222 BTU/HR x 24 HR = 26 MBTU
5,000 HDD x 105 BTU/HR x 24 HR = 13 MBTU
SAVING = 13 MBTU

Infiltration Heat Loss:
0.5 ACH x 10,000 CU FT x .018 BTU/CU FT ° x 5000 HDD x 24 = 1 MBTU

TOTAL = 14 MBTU

If we built the storm windows ourselves using plastic film, we could probably do the whole job for about $28. This equals a cost of only $2 MBTU, a very good investment even if you are a renter.

Even if we bought ready made storm windows at a total cost of $600 we would be paying only $42 MBTU. Because they would last many years (while plastic might last only one) this is still a good investment. It is also recalculated using the life cycle cost methodology.

When we consider passive solar systems we have to estimate how much energy they collect during a season. This can then be used to figure the cost per MBTU. Because these are mostly lasting investments the calculation of life cycle cost is usually desirable.

## Adding A South Window

If we add a double pane 30 sq. ft. south window to our house (with insulated shutter) at a cost of $300 we may collect about 5.4 MBTU/SEASON. This is a cost of almost $55 MBTU. This is obviously not as good as adding insulation; however, as life cycle costing will show it is still an excellent investment.

## Add A Solar Greenhouse

A solar greenhouse with mass and insulated shutters can be built for about $10 sq. ft. If we build a 20 x 10 ft. greenhouse, we will pay about $2,000 for 200 sq. ft. of south window. These 200 sq. ft. will collect about 40 MBTU season at a cost of about $50 MBTU. This is a better buy than adding a south window, and a very good investment.

## Add A Roll Down Awning

If we have a house with 110 square feet of west facing glass a roll down awning is worth installing. We can build and install a good one for about $50. Over a summer this awning will stop 86 MBTU from entering the house. A cost of only 50¢ MBTU blocked. Assuming an air conditioner, efficiency 70%, had to run 1/2 the time before the awning was put up this is a cost of only 66¢ MBTU saved which compares to electricity at $11.72 MBTU (at 4¢ KWH). A best buy.

## Life Cycle Cost

For a better appraisal of longer lasting projects we need to consider life cycle cost. This is the cost of the project over its life—say 40 years for a house, 5 years for an awning, 10 years for a drape, etc. Calculating life cycle cost is not easy because you need to know the cost of your money (loan or investment alternatives), the discount (money today is worth more than money tomorrow), maintenance cost, future energy prices, inflation, etc. Even though you probably won't be exactly right, you can still get a fair idea of the value of an investment. You may wish to run up a couple different scenarios—to see what effect different assumptions have on the value of your improvement. Historically, energy prices rose about 1-2 percent year over inflation. However, between 1973 - 1979 in California gas rose at 18% over inflation and electricity rose at 8% over invlation. More recently, between 1976 - 1980, the price of oil (which also determines the price of electricity to a large degree) rose at a rate of more than 20%/year over inflation. We have used only a 10% a year real energy price rise although a higher rate is not unlikely. All comparisons are made against oil, not natural gas or wood, because their prices are rising more rapidly and will probably soon be comparable to oil. Electricity usually is competitive with oil.

Here is how to do a life cycle cost calculation. First, determine the cost of the improvement, including any Federal or State Tax Credits. Second, estimate the life span of your project. Third, estimate the cost of your investment, i.e. what will the money cost as a loan from your bank minus the inflation rate (12% loan rate - 10% inflation = 2% real interest) or alternatively what would you make (or lose) by investing it elsewhere (7% savings account - 10% inflation = -03%, or 12 % bonds - 10% inflation = 2%). *Then add in the discount rate which accounts for the fact money today is worth more than money tomorrow. Use 10% if you are more present oriented or 5% if you expect to be around a long time. Now add investment and discount rate and subtract energy price increase over inflation to determine real interest rate.*

(IC + DR) - (EPI) = RIR

You can then convert the initial investment into an annual equivalent cost using Table 4, add your estimate of yearly maintenance to this cost. Calculate your annual savings in BTU's (as in the last section). Multiply this savings by the cost of energy you are saving or replacing, Table 3. And finally compare the project cost with your energy savings. You can compare the cost of different projects by calculating the benefits of each.

## Table 4. Annual Payment ($) Necessary to Amortize Each $100.

| Amortization Period (yrs) | Real Interest Rate 1 | 2 | 3 | 5 | 7 | 8 | 9 | 10 | 11 | 12 | 15 |
|---|---|---|---|---|---|---|---|---|---|---|---|
| **5** | 20.60 | 21.21 | 21.83 | 23.09 | 24.38 | 25.04 | 25.70 | 26.38 | 27.05 | 27.74 | 29.83 |
| **10** | 10.55 | 11.13 | 11.72 | 12.95 | 14.23 | 14.90 | 15.58 | 16.27 | 16.98 | 17.69 | 19.92 |
| **15** | 7.21 | 7.78 | 8.37 | 96.30 | 10.97 | 11.68 | 12.40 | 13.14 | 13.90 | 14.68 | 17.10 |
| **20** | 5.54 | 6.11 | 6.72 | 8.02 | 9.43 | 10.18 | 10.95 | 11.74 | 12.55 | 13.38 | 15.97 |
| **25** | 4.54 | 5.12 | 5.74 | 7.09 | 8.58 | 9.36 | 10.18 | 11.01 | 11.87 | 12.75 | 15.47 |
| **30** | 3.87 | 4.46 | 5.10 | 6.50 | 8.05 | 8.88 | 9.73 | 10.60 | 11.50 | 12.41 | 15.23 |
| **40** | 3.04 | 3.65 | 4.32 | 5.82 | 7.50 | 8.38 | 9.29 | 10.22 | 11.17 | 12.13 | 15.05 |
| **50** | 2.55 | 3.18 | 3.88 | 5.47 | 7.24 | 8.17 | 9.12 | 10.08 | 11.06 | 12.04 | 15.01 |
| **75** | 1.90 | 2.58 | 3.36 | 5.13 | 7.04 | 8.02 | 9.01 | 10.00 | 11.00 | 12.00 | 15.00 |

Here are some examples

### Attic Insulation: Example

From our earlier example we know our attic retrofit will save 52 MBTU/year. The investment cost is $500 less the Federal tax credit of $100, for a net cost of $400. The lifetime of the insulation is the life time of the house, or 40 years. If we borrow $400 at 12% interest and inflation stays at 10% then the loan cost is 2%. If we use a discount rate of 10% and assume a real energy price increase of only 11% year, the the real interest rate for annualization is 1% year. We go to Table 4 and look up the cost of $400 at 1% over 40 years. It comes to $12/year.

There will be no maintenance for the insulation so the total annual cost for the insulation is only $12. The savings in energy will be 52 MBTU for a cost of 23¢/MBTU. This compares to a cost of $13 MBTU for oil (at $1 gallon and 60% eff.). Yearly savings will therefore be:

52 MBTU x $13/MBTU = $676
52 MBTU x 23¢/MBTU = 12
SAVINGS: $664/year

### Insulated Shutter

As in the earlier example we will assume that our 100 square feet of shutter will save 19 MBTU year. This time however, we will assume we hire a carpenter to make them at $7.50 sq. ft. for a total cost of $750. These can also qualify for a tax credit of $150, for a net cost of $600. The life time of the shutters should be about 20 years. If we take our money out of stocks that were paying a real return of 5% year (15% year - 10% inflation) and we assume a discount rate of 10% and real energy price increase of 12% year then our real interest rate is 3% (15% - 12%).

We look at Table 4 under 20 years, 3%, for $750 and find the annual cost is $50. We add 10% to this for maintenance for a total annual cost of $55. Our energy savings are 10 MBTU/year if the shutters are closed at night. Cost per MBTU is then about $5.50 over the life of the shutters. This compares to a cost of $13 MBTU for oil Yearly savings therefore are:

10 MBTU x $13. = $130 year
10 MBTU x $5.50 = $ 55 year
SAVINGS: $75 year

Now for our passive improvements:

### Add A South Window

If we have the 30 square foot window (with insulated shutter) we used as an example earlier installed by a carpenter it may cost $500. The life time of the window will be 40 years. If we borrow money from our bank at 15%, inflation stays at 10%, the discount rate is 10%, and the real energy price increases 12% year, then we look at Table 4 under $500 over 40 years at 3%. This comes to $25.95 year. We add 10% to this for yearly maintenance of the shutter for atotal cost of $28.60. For $28.60 we get 5.4 MBTU/year for a cost of $5 MBTU. This compares to $13 MBTU for oil, saving:

5.4 MBTU x $13 = $70
5.4 MBTU x $5 = 27
SAVING = $43

And finally, the solar greenhouse.

### Add A Solar Greenhouse

We will use the same 200 sq. ft. greenhouse but assume we have it built at $30 sq. ft. (including shutters, mass, etc.). This calls for an investment of $6,000, less the Federal Tax Credit of 20%, or $4,800. The life time of the greenhouse should be 30 years. Assume the loan costs 2% (12% - 10% inflation) with a 10% discount rate and 11% year real increase in energy cost or 1% real interest rate. Looking in Table 4 we see the annual cost is $146 year. Adding 10% for maintenance pushes the price to $160 year. The greenhouse provides 40 MBTU year heating at a cost of $4 MBTU. Compared to oil this saves:

40 MBTU/YR x $13 MBTU = $520
40 MBTU/HR x 4 MBTU = $360
SAVINGS = $160 year

As a final test you can figure your return based on investent v/s annual savings. For the greenhouse example, this would be: $360/4,800 or 7%. This is a real 7%/year so to equal it with stocks you need to add 10% inflation for a total of 17% year.

# Chapter 6: Details

One of the more persistent obstacles to more widespread passive solar development has been the lack of construction details. This has not prevented the more bold and imaginative designers and clients but it has served as a check on the more conservative elements who might otherwise have used passive design. We are, therefore, very pleased to offer the following passive solar details.

Several of these were developed by David Bainbridge at the Passive Solar Institute. One was developed by Robert Dumont at the University of Saskatchewan, and the rest were prepared at Cal Poly San Luis Obispo under the direction of Ken Haggard and Phil Niles. The rough materials for those were prepared by a group of consultants working for Cal Poly on **The Passive Solar Handbook for California.** This superb book was prepared for the California Energy Commission and should be available from them later this year. The consultants working for Cal Poly included:

David Bainbridge, PSI, Chip Barnaby and Bruce Wilcox, Berkley Solar Group; Polly Cooper, San Luis Obispo Solar Group; Jake Feldman, Cal Poly; Denny Long, Tanks-a-Lot; Bruce Maeda, Davis Alternative Technology Associates; and David Wright, SEAGroup.

All of the details were drawn by Greg Wolters, now working for SEAGroup. Details are used by permission of the authors.

Although we have done our best to ensure that these details are correct and workable, we can assume no liability for them and make no warranty of their performance. We would appreciate any comments, either corrections, compliments or additions to them. Please write: Details, PSI, P.O. Box 722, Davis, CA 95616.

For information on the availability of **The Passive Solar Handbook for California,** please write: The Solar Office, California Energy Commission, 1111 Howe Ave, Sacramento, CA 95825.

More details on super insulation and supertight design are included in: **Low Energy Passive Solar Housing** by the Energy Research Development Group, University of Saskatchewan, Saskatoon, Saskatchewan, available from the Division of Extension and Community Relations for $5; and **How to Build a Superinsulated House** by Ed McGrath, $5 from Project 2020, P.O. Box 81961, College, Alaska 99708.

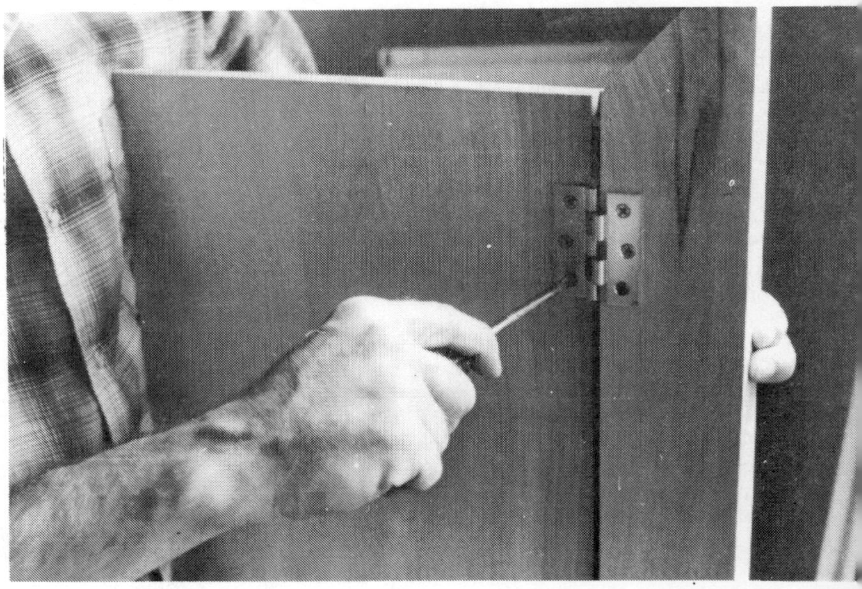

## characteristics

CAN ACHIEVE UP TO R-40+ WALL AND R-50 CEILING INSULATION

## variations

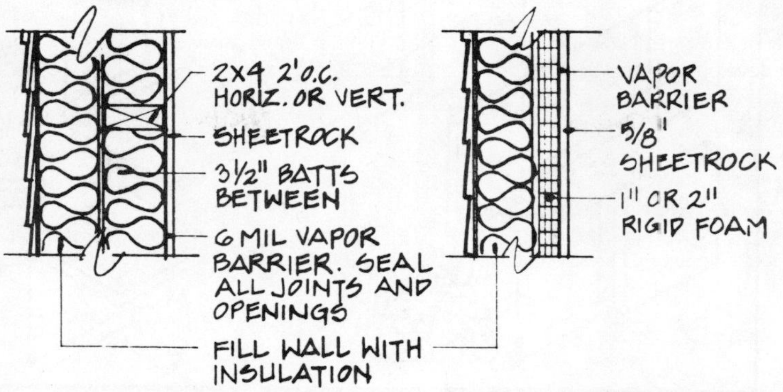


## relation to other components

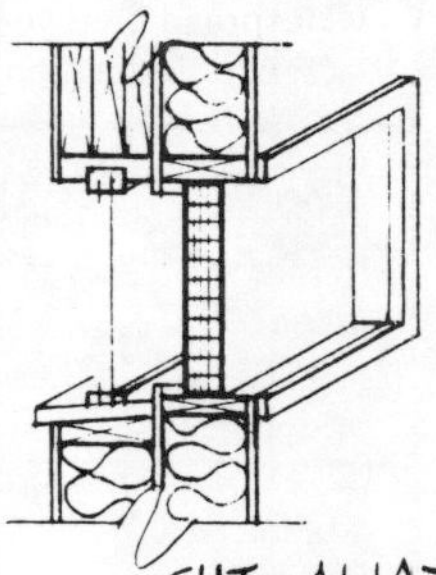

CUT-AWAY VIEW

INSULATED SHUTTER CAN SLIDE INTO POCKET IN DOUBLE WALL. EITHER ON INSIDE OF WINDOW OR BETWEEN INNER AND OUTER WINDOWS.

IF INTERIOR INSULATION IS CHOSEN ALL ELECTRICAL BOXES MUST BE RESET TO NEW WALL.

IF HOUSE IS SUPER INSULATED AND SUPER TIGHT AN AIR-TO-AIR HEAT EXCHANGER MAY BE DESIRABLE.

WITH CATHEDRAL CEILING EXTEND INSULATION TO DECK. INSULATE ON TOP OF DECK BY LAYING RIGID FOAM AND OR SPACERS - 2x6 OR 2x10 - ON EDGE AND INSULATE. THEN DECK OVER WITH PLYWOOD, PUT DOWN 2x2'S VERTICALLY WITH SCREEN VENT AT TOP AND BOTTOM AND RE-ROOF.

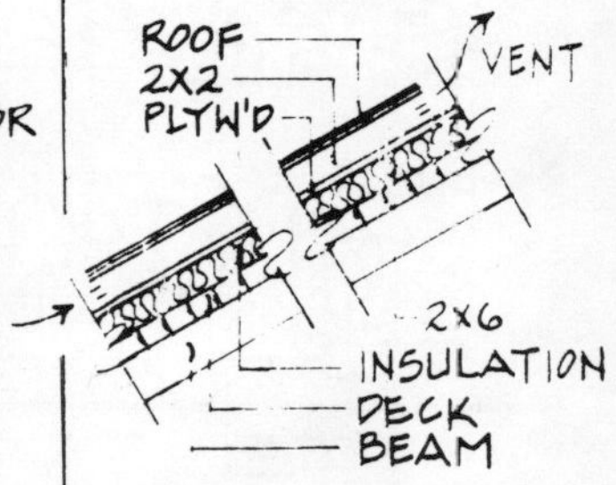


OFTEN BETTER TO INSULATE AND SHEETROCK OVER BEAM

# SUPER INSULATED WALL FOR RETROFIT CONSTRUCTION

1. DRILL HOLES IN EXTERIOR (OR STRIP FOR REUSE)
2. FILL WITH INSULATION
3. ADD VAPOR BARRIER, SEAL ALL JOINTS AND EDGES WITH SEALANT (AND BATTENS AS REQ'D)
4. ADD NEW WALL, BEGIN WITH LEDGER AND BASE AND HORIZONTAL BLOCKING AS REQ'D.
5. BUILD OUT DOORS AND WINDOWS
6. INSULATE EXTERIOR WALL, VAPOR BARRIER FACING IN.
7. INSTALL SIDING

NOTE:
NEW CONSTRUCTION IS SIMILAR BUT USES PLYWOOD PLATE TOP AND BOTTOM. MARK BOTH PLATES AT ONCE.

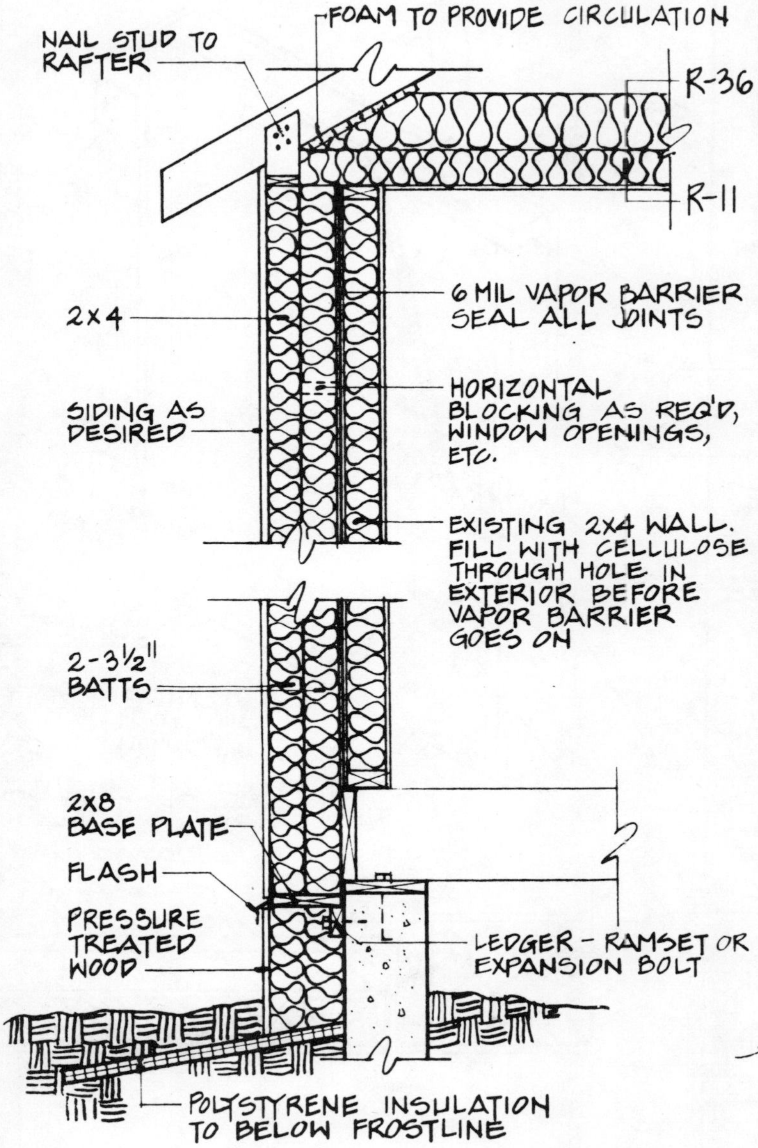


SUPER INSULATED WALL
NO SCALE

1

**applicability:** DESIRABLE IN AREAS WITH 5,000 D.D. OR MORE. COST ESTIMATED TO RUN $2 TO $4 PER S.F. IN NEW CONSTRUCTION, AS MUCH OR MORE IN RETROFIT.

## characteristics

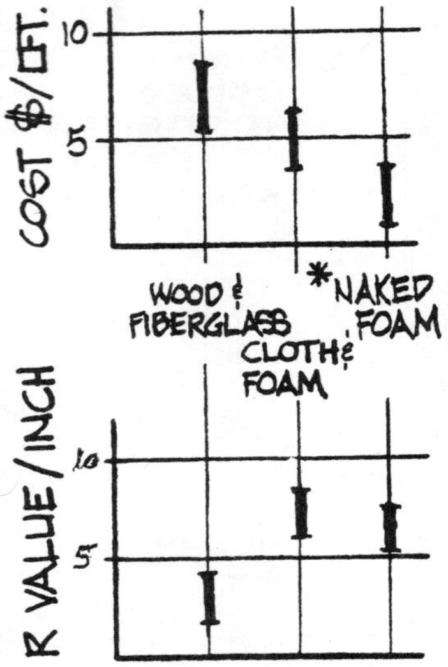


## variations

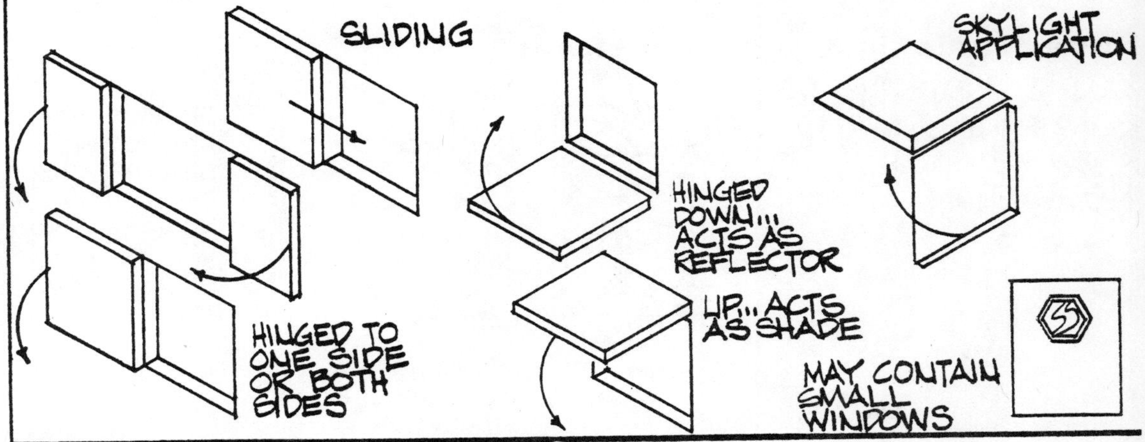


# INSULATING SHUTTER

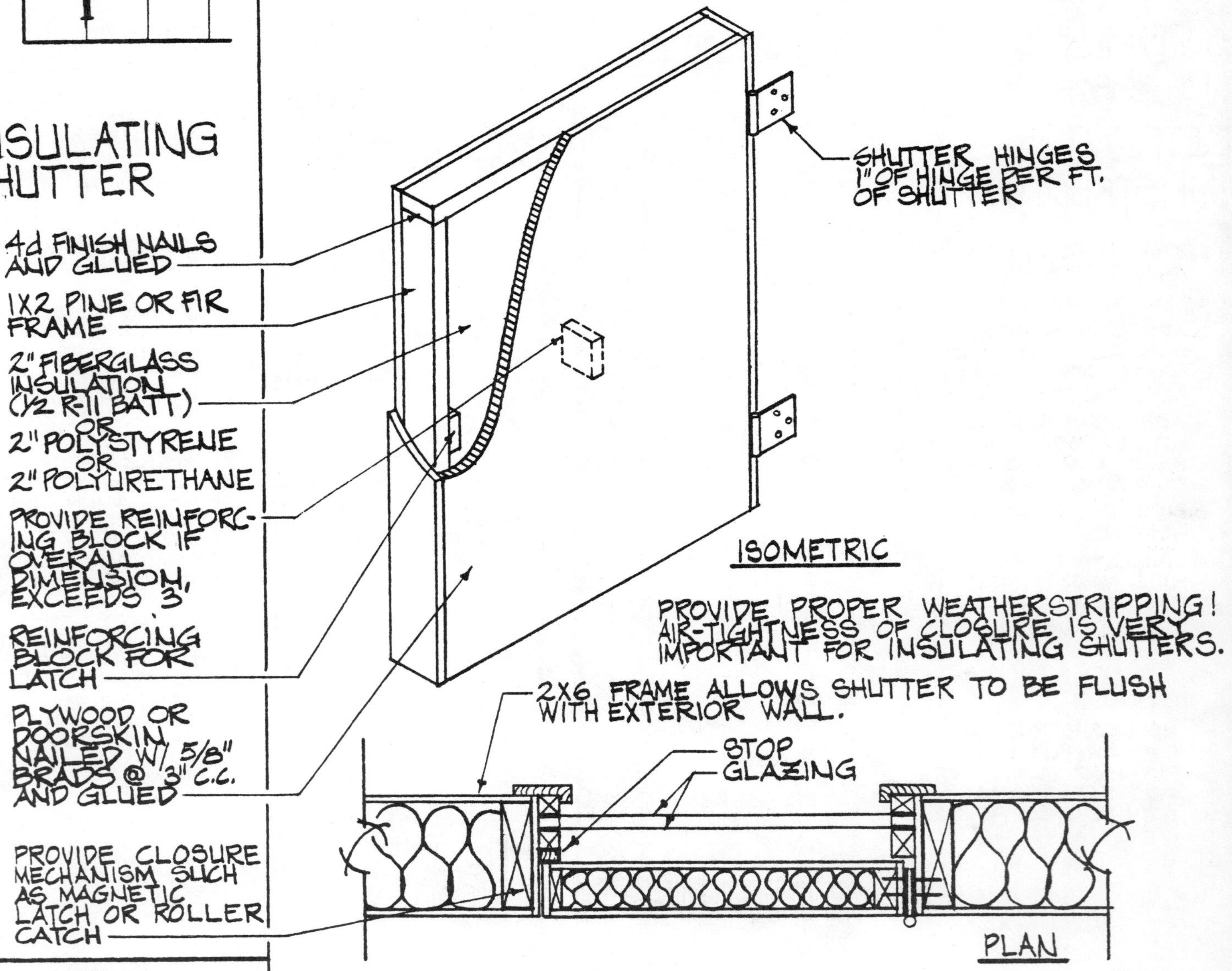


16 INSULATING SHUTTER
SCALE: 1½" = 1'-0"

NOTE
* NAKED FOAM HAS VERY POOR FIRE RESISTANCE.

| 2 | **applicability:** INSULATING SHUTTERS CAN BE 8 TIMES AS EFFECTIVE AS DOUBLE GLAZED WINDOWS, AND 3 TIMES AS EFFECTIVE AS INSULATING DRAPES. THEY ARE MOST USEFUL IN CLIMATES WITH HEATING LOADS OVER 3,000 DEGREE-DAYS. |
|---|---|

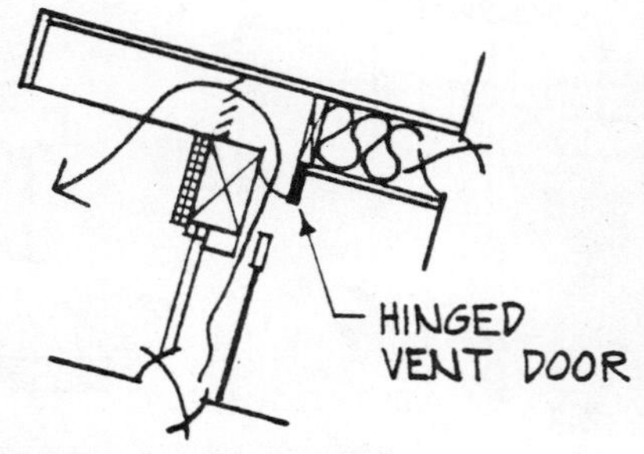


VENT ABOVE WINDOW

EVEN BETTER THERMAL CONTROL OF SOUTH WINDOW SURFACE CAN BE PROVIDED BY USE OF VENTS BOTH ABOVE AND BELOW WINDOW TO ALLOW EXTERIOR SURFACE OF INSULATING CURTAIN TO BE VENTED TO THE EXTERIOR.

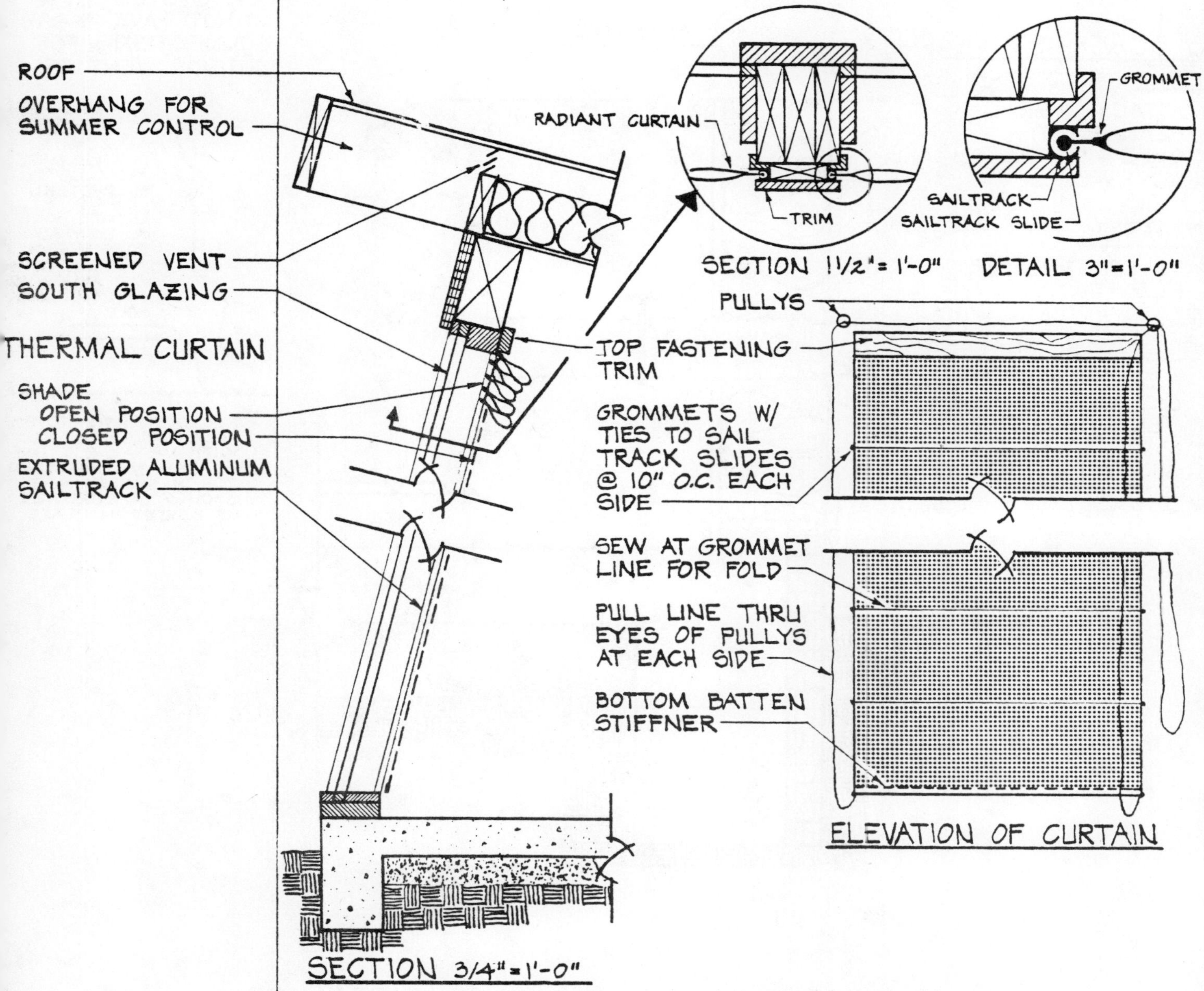


## RADIANT THERMAL CURTAINS: SLOPED WINDOWS

SCALE: AS INDICATED

**applicability:** SLOPED WINDOWS ARE AN OPTIMUM WINTER COLLECTOR FROM A GEOMETRICAL STANDPOINT, BUT OFTEN REQUIRE SHADING DURING OTHER PARTS OF THE YEAR. THIS DETAIL SHOWS AN INEXPENSIVE WAY OF PROVIDING THERMAL CURTAINS ON SLOPING WINDOWS.

**3**

## variations

IF STOVE IS NOT NEAR EXTERNAL WALL, VENT MAY BE TAKEN THRU SLAB OR SUBFLOOR

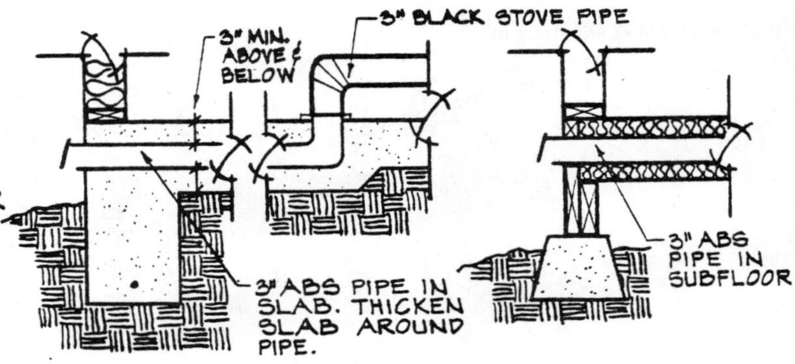


## relation to other components

ALLOWANCES SHOULD BE MADE IN FRAMING AND FORMING DETAILS FOR PASSAGE OF 3" VENT PIPE.

**3" BLACK STOVE PIPE**

ATTACH OVER FLANGE WITH SHEET METAL SCREWS

**DAMPER**

TO CONTROL AIR FLOW

**3" DRYER VENT**

A SLIGHT MODIFICATION OF DRYER VENT SHOULD BE MADE TO ALLOW AIR TO ENTER BUT NOT EXIT THE PIPE

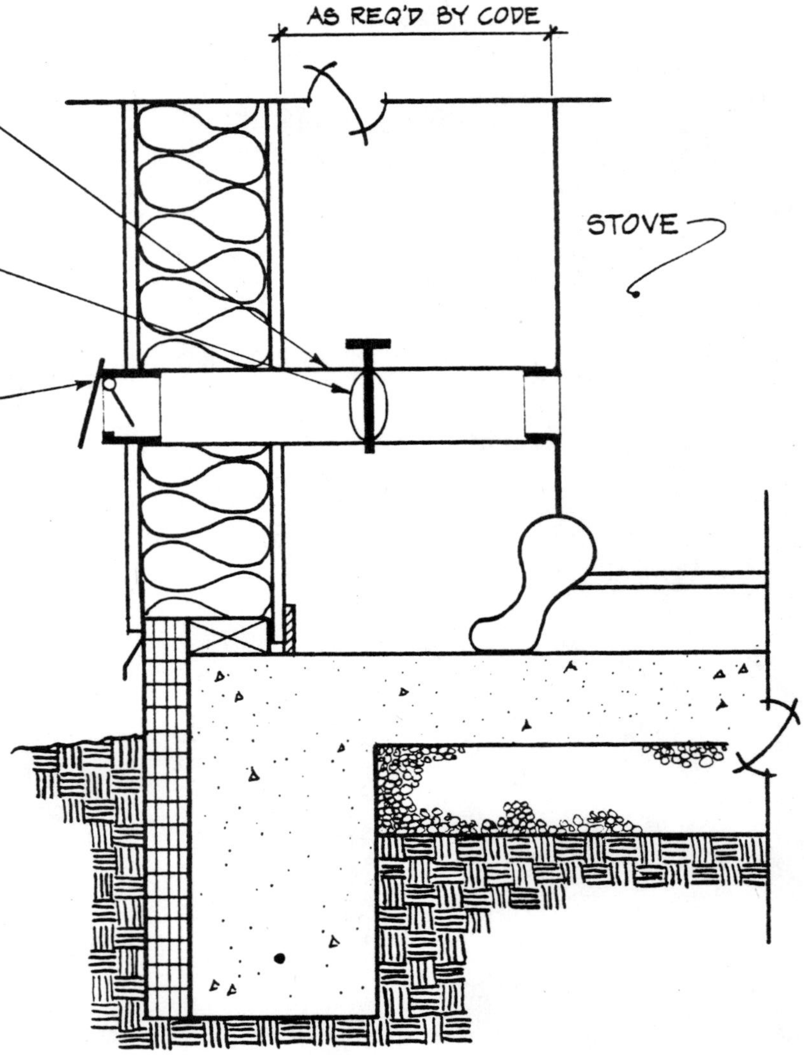


FOR STOVES THAT DO NOT HAVE CONNECTIONS FOR OUTSIDE VENTING

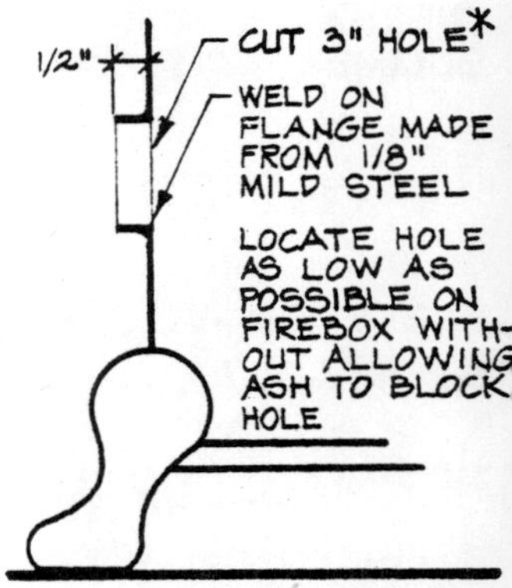


* FOR CAST IRON STOVES HOLE SHOULD BE CUT WITH METAL CUTTING BLADES RATHER THAN TORCH AND FLANGE SHOULD BE SCREWED OR BOLTED IN PLACE

OUTSIDE VENT FOR STOVE OR FIREPLACE

SCALE: 1 1/2" = 1'-0"

| | |
|---|---|
| **4** | **applicability:** WOOD STOVES AND FIREPLACES THAT USE ROOM AIR FOR COMBUSTION INCREASE AIR INFILTRATION, SINCE REPLACEMENT AIR MUST BE DRAWN INTO THE ROOM THROUGH CRACKS AND OTHER LEAKS. THE ACCOMPANYING HEAT LOSS IS OFTEN GREAT ENOUGH TO NEGATE THE HEAT PROVIDED BY A FIREPLACE. DRAWING COMBUSTION AIR DIRECTLY TO THE FIREPLACE OR STOVE VIA A VENT FROM THE OUTSIDE REDUCES THIS INFILTRATION OF COLD OUTSIDE AIR. |

## characteristics

TIGHTNESS OF INSULATION IN CLOSED POSITION IS LIMITING FACTOR IN EFFICIENCY OF SKYLIGHT. MOST MOVABLE INSULATIONS ARE ONLY ABLE TO ACHIEVE ABOUT R-5 DUE TO DIFFICULTY IN ACHIEVING A VERY TIGHT SEAL.

## variations

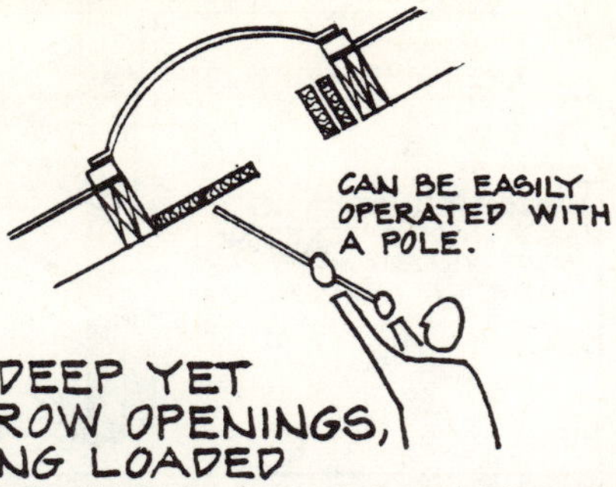


FOR DEEP YET NARROW OPENINGS, SPRING LOADED BIFOLD INSULATING PANELS MAY BE USED.

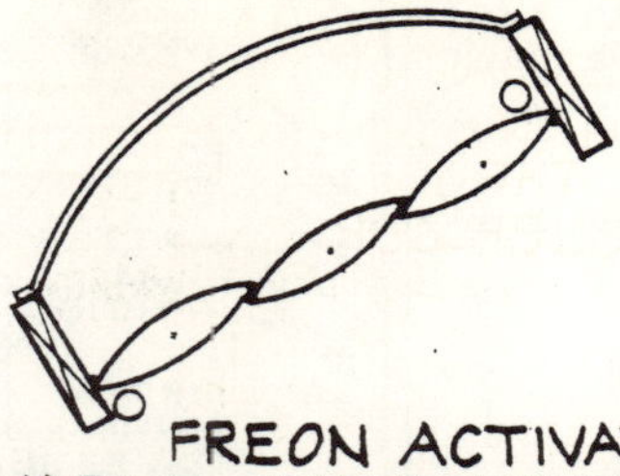

FREON ACTIVATED MOVABLE INSULATION LOUVERS ARE COMMERCIALLY AVAILABLE FOR SKYLIGHTS.

* SEE PAGE 35.

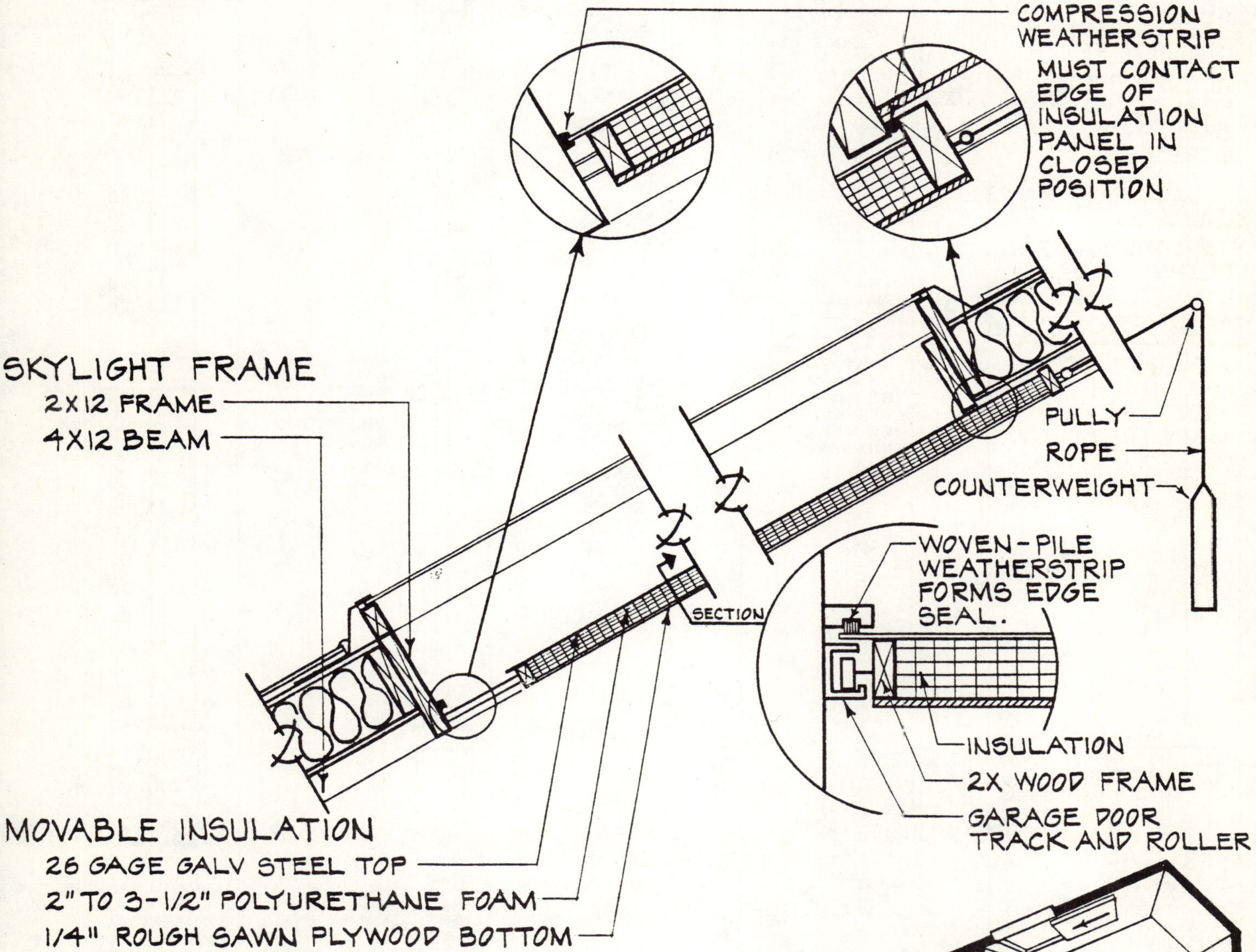


MOVABLE INSULATION FOR SKYLIGHT

SCALE: 3/4" = 1'-0"

**5**

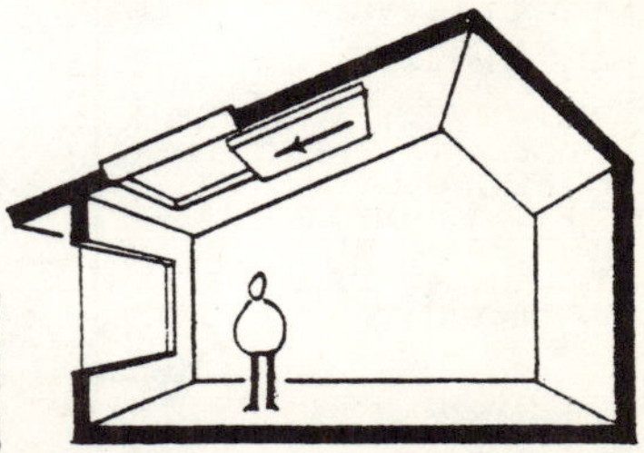

**applicability:** SOUTH-FACING SKYLIGHTS GIVE GOOD WINTER SUN, BUT ARE HIGHLY VULNERABLE TO OVERHEATING IN THE SUMMER. MOVEABLE INSULATION FOR THESE SKYLIGHTS ALLOWS NEEDED SUN CONTROL AND ALSO INSULATES THE SKYLIGHT AT NIGHT TO REDUCE HEAT LOSS.

## variations

SHADES CAN BE SPLIT BAMBOO, WOODEN STRIPS OR HEAVY CANVAS.

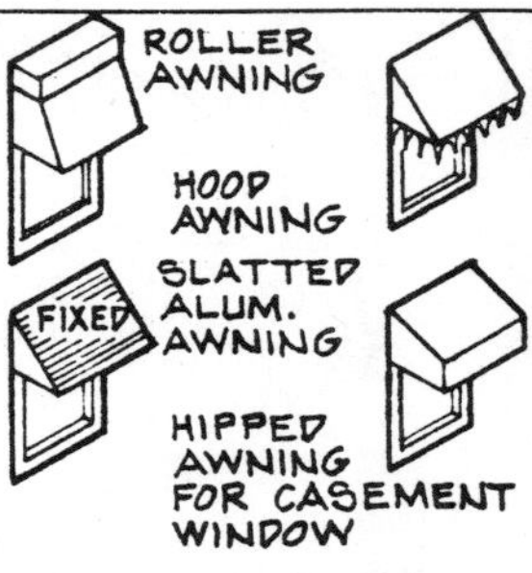


INEXPENSIVE VERSION IS MADE WITH FLAT WEBBING.

LONGER LASTING AND MORE EXPENSIVE VERSION IS MADE OF ALUMINUM.

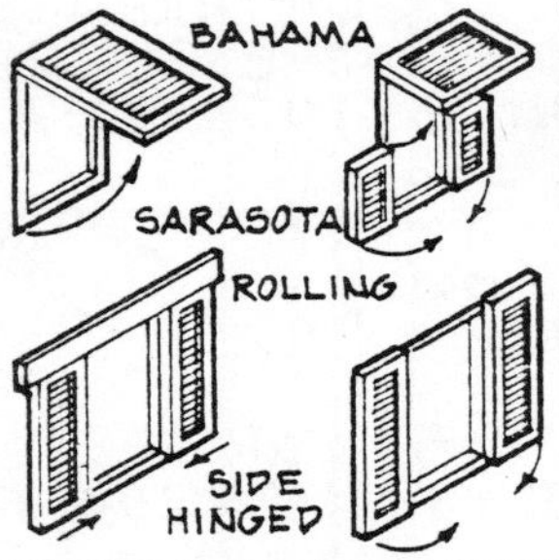


AVAILABLE IN WOOD, VINYL, OR ALUMINUM WITH DIFFERENT SIZED SLATS.

| RELATIVE COST $/□' | | | | |
|---|---|---|---|---|
| 2.00–4.50 | 3.50–4.50 | 3.50–5.50 | 5.50–6.50 | 6.50–10.50 |

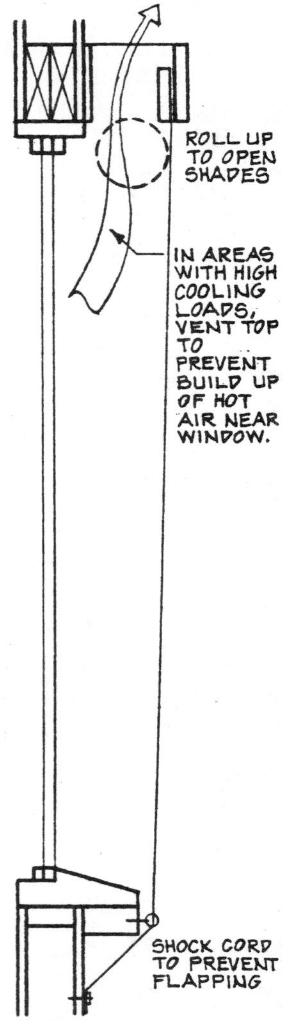


VERTICAL SHADES

RELATIVELY INEXPENSIVE AND EFFECTIVE ESPECIALLY IN HOT AREAS WHERE SHADING FROM DIFFUSE SUNLIGHT CAN BE IMPORTANT.

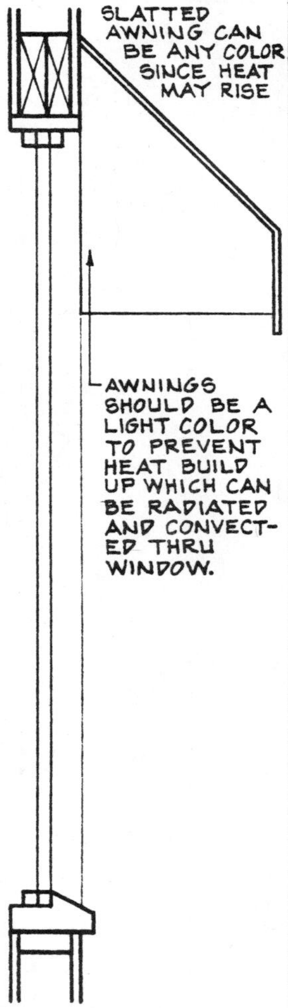


AWNINGS *

TRANSMITTANCE

| | DIRECT | DIFFUSE |
|---|---|---|
| CANVAS | 0 | 0 |
| PLASTIC | .25 | .15 |
| AL. SLATS | 0 | .20 |

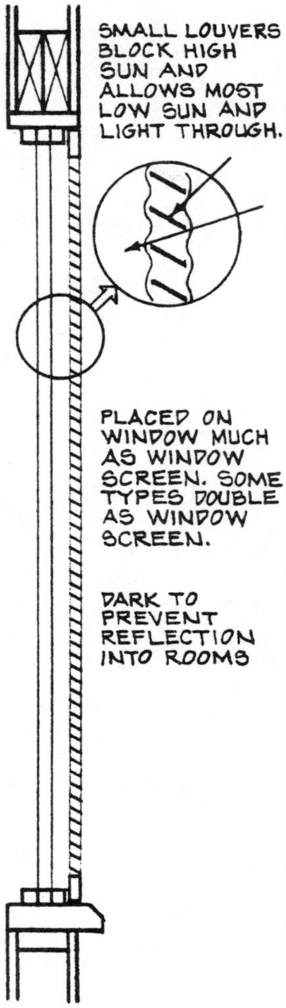


LOUVERED SHADES *

CAN SHADE FROM DIRECT SUNLIGHT YET PROVIDE VIEWS OUT.

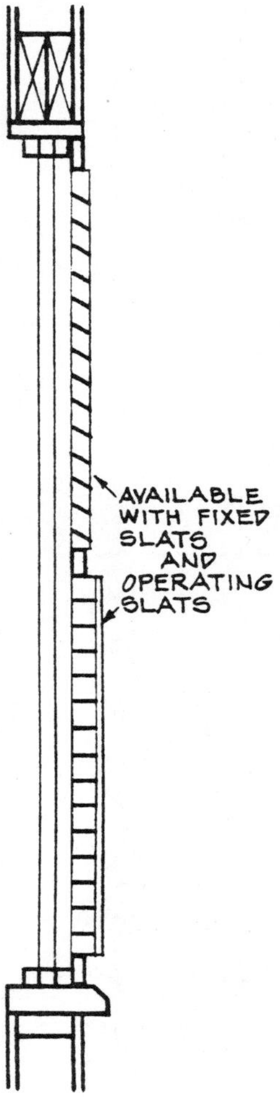


SHUTTERS *

CAN BLOCK BOTH DIRECT AND DIFFUSE LIGHT. CAN REDUCE HEAT LOSS BY PROTECTING GLASS FROM WINDS AND PROVIDING INSULATING AIR FILM ON GLASS.

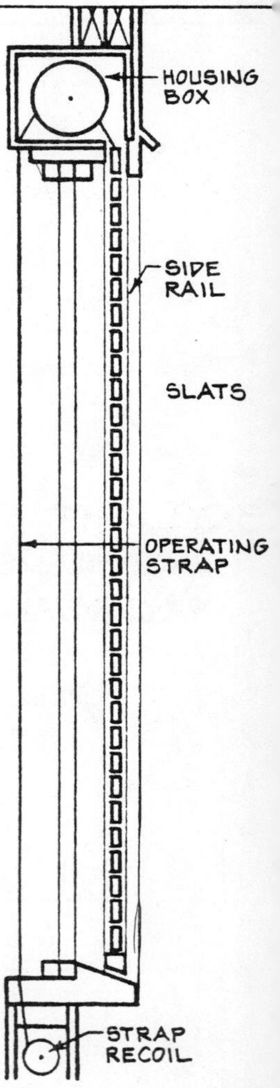


ROLL BLINDS *

USED IN EUROPE FOR MANY YEARS... NOW COMMERCIALLY AVAILABLE IN U.S. IN SIZES UP TO 12 FT. WIDE AND 10 FT. HIGH.

EXTERIOR SHADING DEVICES

SCALE: 1"=1'-0"

**6**

**applicability:** EXTERIOR SHADES ARE A HIGHLY EFFECTIVE METHOD OF SUN CONTROL. MANY CAN PROVIDE SHADE FROM DIFFUSE AS WELL AS DIRECT SUNLIGHT. (THERE ARE MANY COMMERCIALLY AVAILABLE EXTERIOR SHADING DEVICES

**characteristics** | **variations**

MAY ALSO BE ARRANGED HORIZONTALLY TO PROVIDE FOR HIGH AND LOW VENTS. THE GREATER THE VERTICAL DIFFERENCE BETWEEN VENTS ON OPPOSITE SIDES OF THE BUILDING THE GREATER THE STACK EFFECT FOR INCREASING AIR FLOW.

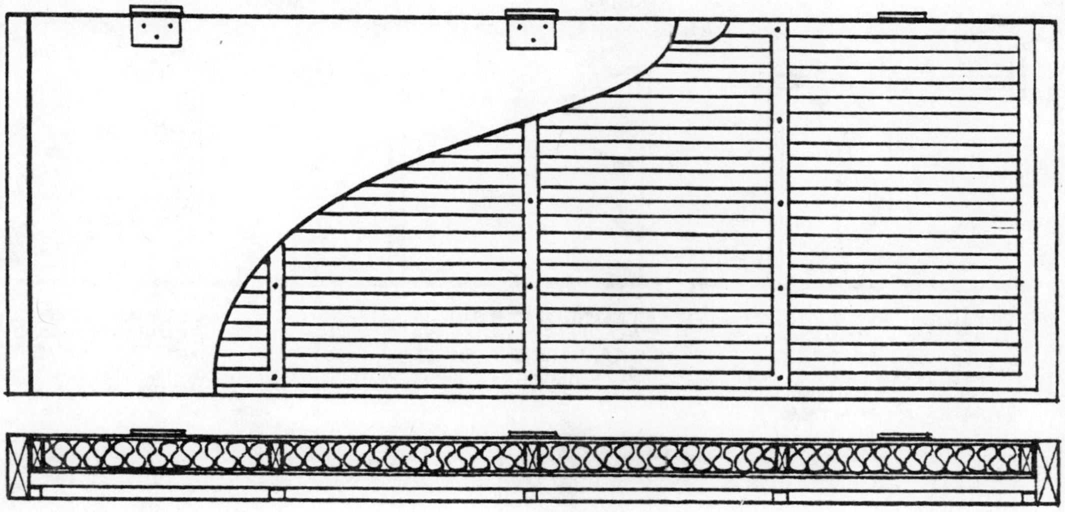

## SECURITY VENT

STANDARD GALVANIZED & SCREENED GABLE VENT

INSTALLED BETWEEN 16" O.C. STUDS WITH 1/8" MESH SCREEN FACING OUT

VENTS OVERLAPPED AND POP RIVITED

## INTERIOR INSULATING SHUTTER

1X2 PINE

1/8" DOORSKIN GLUED AND WIRE NAILED

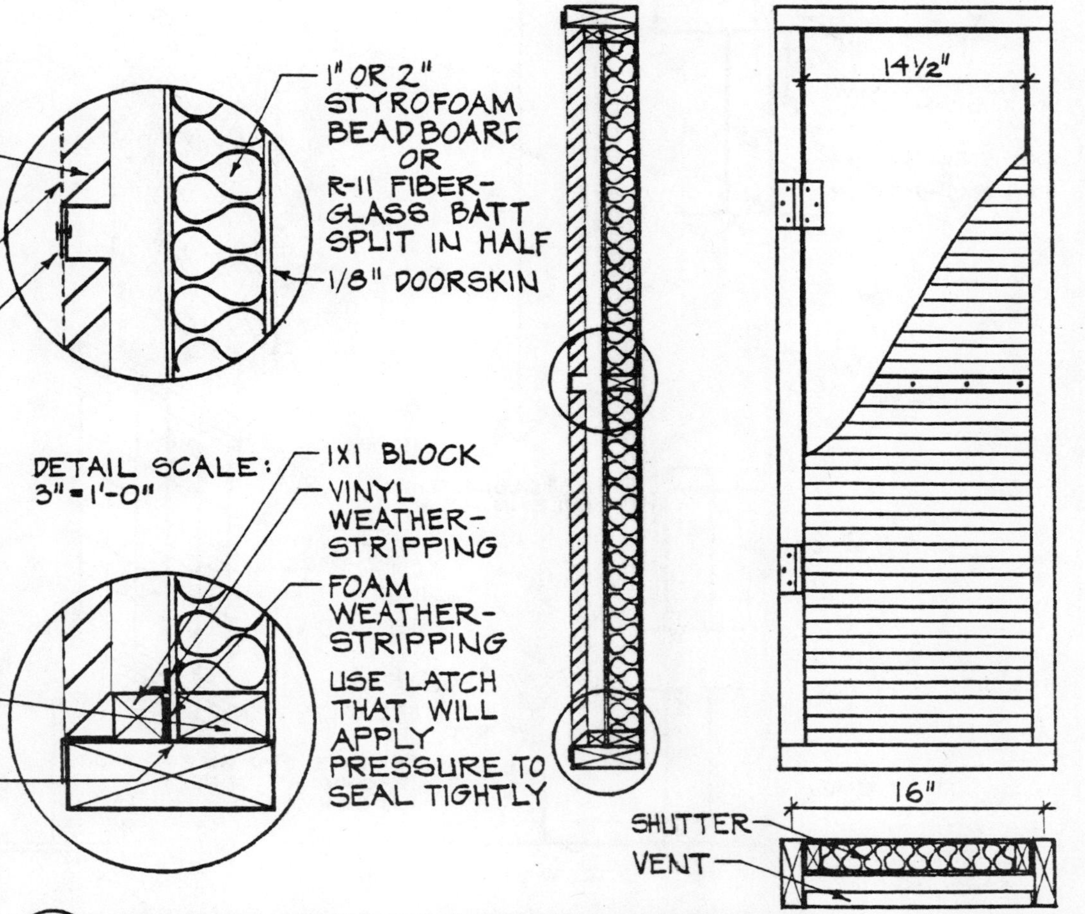


SECURITY VENT USING ATTIC GABLE VENTS STACKED VERTICALLY - SCALE: 1"=1'-0"

**7**

**applicability:** PASSIVE SYSTEMS EMPLOYING NIGHT VENTILATION REQUIRE VENTS THAT:

1. ALLOW FOR MAXIMUM AIR FLOW.
2. PROVIDE INTERNAL SECURITY FOR THE BUILDING.
3. ARE EASILY CLOSED BY THE OCCUPANTS.

## variations

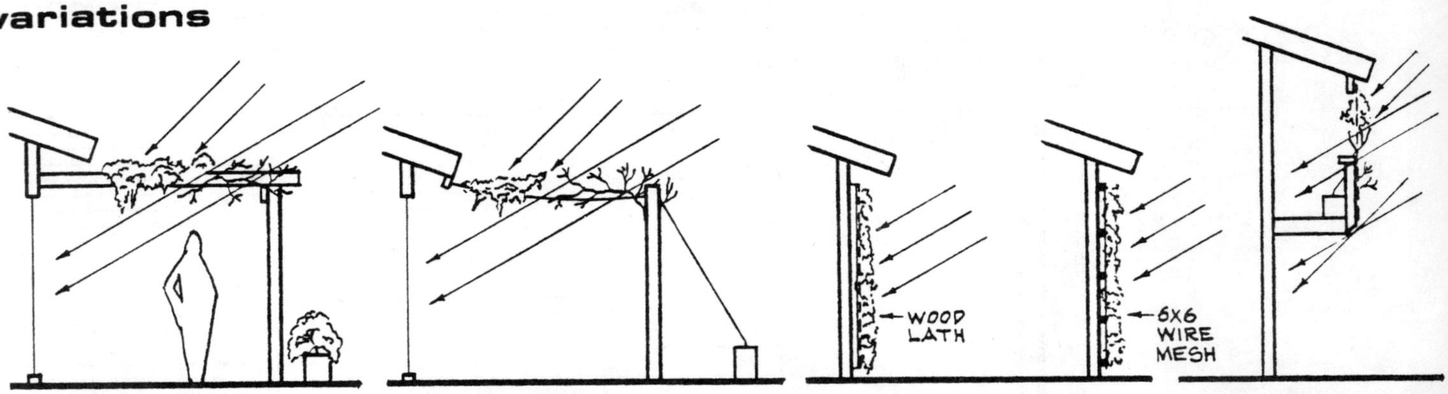


TRELLIS OVER SOUTH GLAZING— CARE MUST BE TAKEN SO THAT STRUCTURE DOES NOT CAUSE TOO MUCH SHADE IN WINTER.

TENSION TRELLIS ALLOWS MINIMUM SHADING OF SOUTH GLAZING.

VERTICAL TRELLIS PROVIDES GOOD SHADING TO PREVENT HEAT GAIN IN EAST AND WEST WALLS.

TRELLIS / BALCONY COMBINATION FOR SECOND STORY FACING EAST OR WEST.

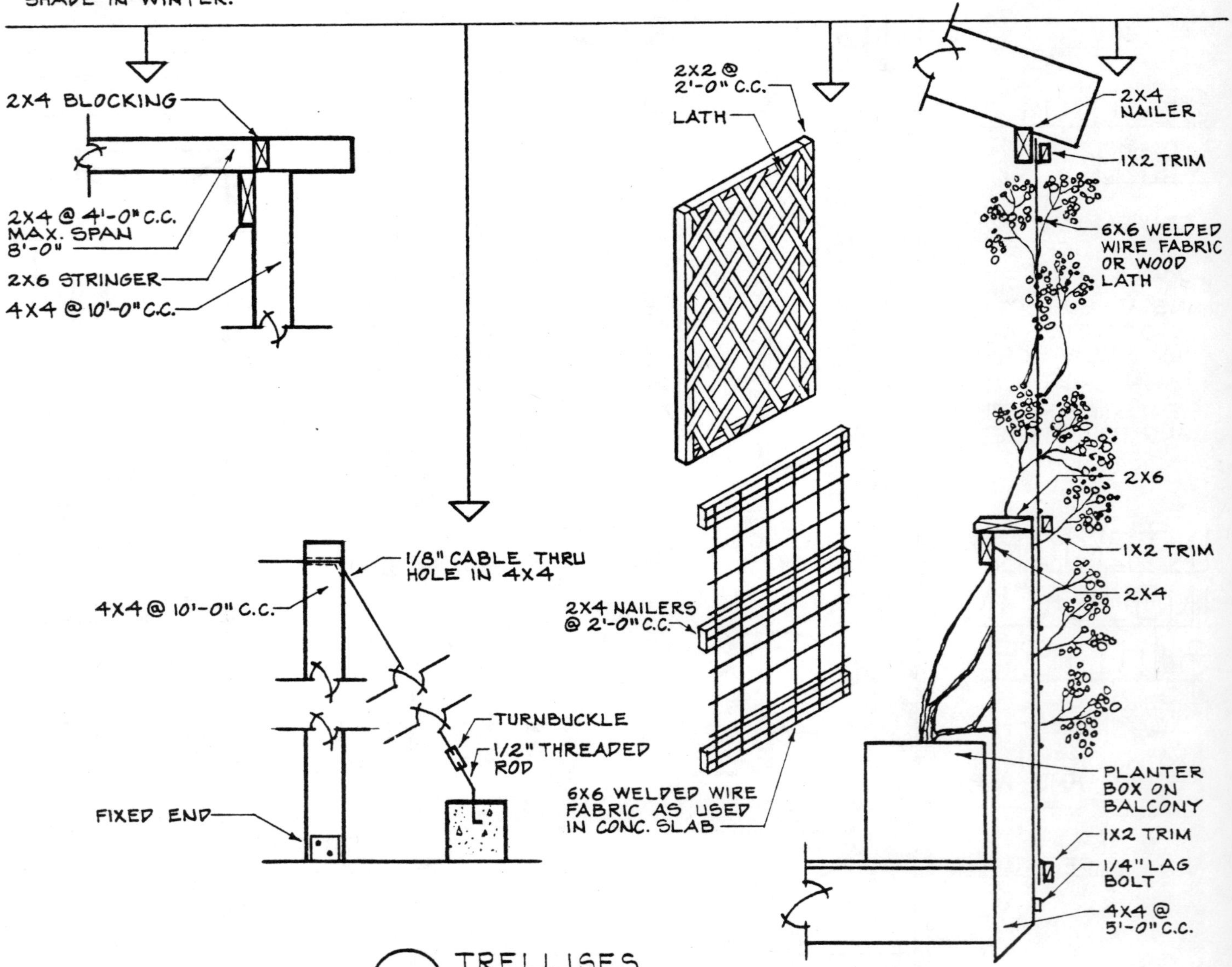


TRELLISES
SCALE: 3/4" = 1'-0"

| 8 | **applicability:** BESIDES BEING A VERY ATTRACTIVE METHOD OF SUN CONTROL DECIDUOUS PLANTS ON TRELLISES WILL AUTOMATICALLY ADJUST TO COOL SPRINGS OR HOT FALLS--PROVIDED THE PLANT MATERIAL IS CAREFULLY CHOSEN. |
|---|---|

## characteristics

THERMAL CAPACITY 3.3 BTU/□FT.°F

"U" VALUE ~ .25

COST MATERIALS 30¢

COST APPLICATION 10¢

WEIGHT 15 lb/□FT. (MEDIUM WEIGHT)

## variations

CAPABLE OF MANY PATTERNS SIMILAR TO BRICK.

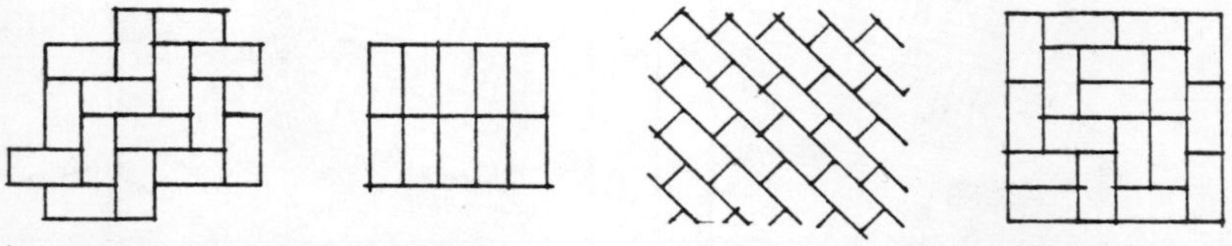


## relation to other components

STRUCTURAL CONSIDERATIONS

WOOD SUB-FLOOR

EACH SQ.FT. OF TILE ADDS 15 LBS. DEAD LOAD TO FLOOR.

FOR 4'-0" SPAN SUB-FLOOR 2" THICK T&G DECKING.

FOR 2'-0" SPAN SUB-FLOOR 3/4" PLYWOOD.

FOR 16" SPAN SUB-FLOOR 5/8" T&G DECKING.

# 2"x8"x16" CONCRETE BLOCK WALL CAPPING

1. LAID W/ TIGHT - DRY JOINTS IN CERAMIC TILE MASTIC.
2. 3/4" x 1 5/8" RW OR DF EXPANSION JOINTS. MAX 10' INTERVALS.
3. ENTIRE SURFACE WASH-GROUTED W/ THIN COAT OF TILE GROUT AND PREFERRED COLOR MIX.
4. FLOOR SEALED W/ SPECIFIED CHEMICAL PRODUCTS "COLORSEAL" OR EQUIV.

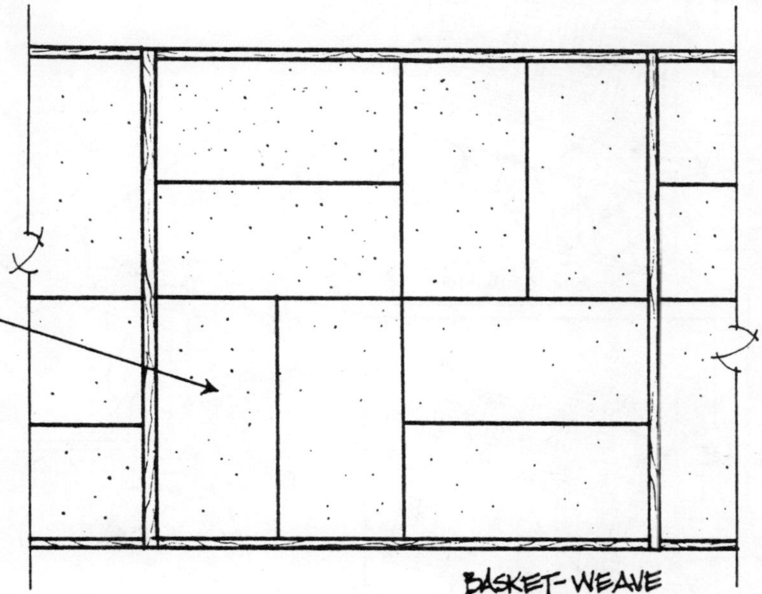


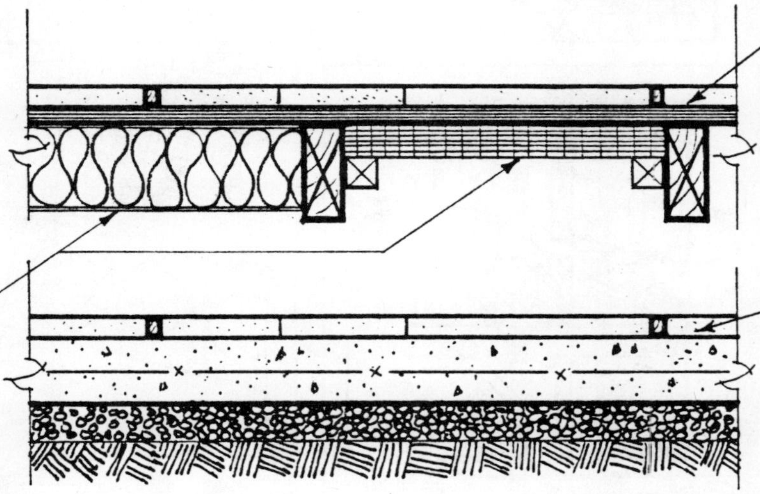


COLOR CONSIDERATIONS

HIGH THERMAL MASS FLOOR TILES

1" = 1'-0"

**applicability:** THESE FLOOR TILES ARE A SIMPLE WAY OF OBTAINING THERMAL MASS IN LIGHT WOOD FRAME CONSTRUCTION WHICH IS SO POPULAR IN CALIFORNIA. THEY ARE: 1. RELATIVELY INEXPENSIVE (AS LOW AS 40¢/□FT. INSTALLED - ABOUT 1/3 THE PRICE OF TILE.) 2. BEING 1 5/8" THICK THEY ARE THERMALLY FAR MORE EFFECTIVE THAN TILE. 3. THEY CAN BE EASILY APPLIED TO PLYWOOD SUBFLOORS AND DO NOT GREATLY MODIFY STANDARD WOOD FRAME TECHNIQUES.

**9**

**characteristics** WEIGHT OF SATURATED SOIL = 90 TO 100 LBS. PER CUBIC FOOT.

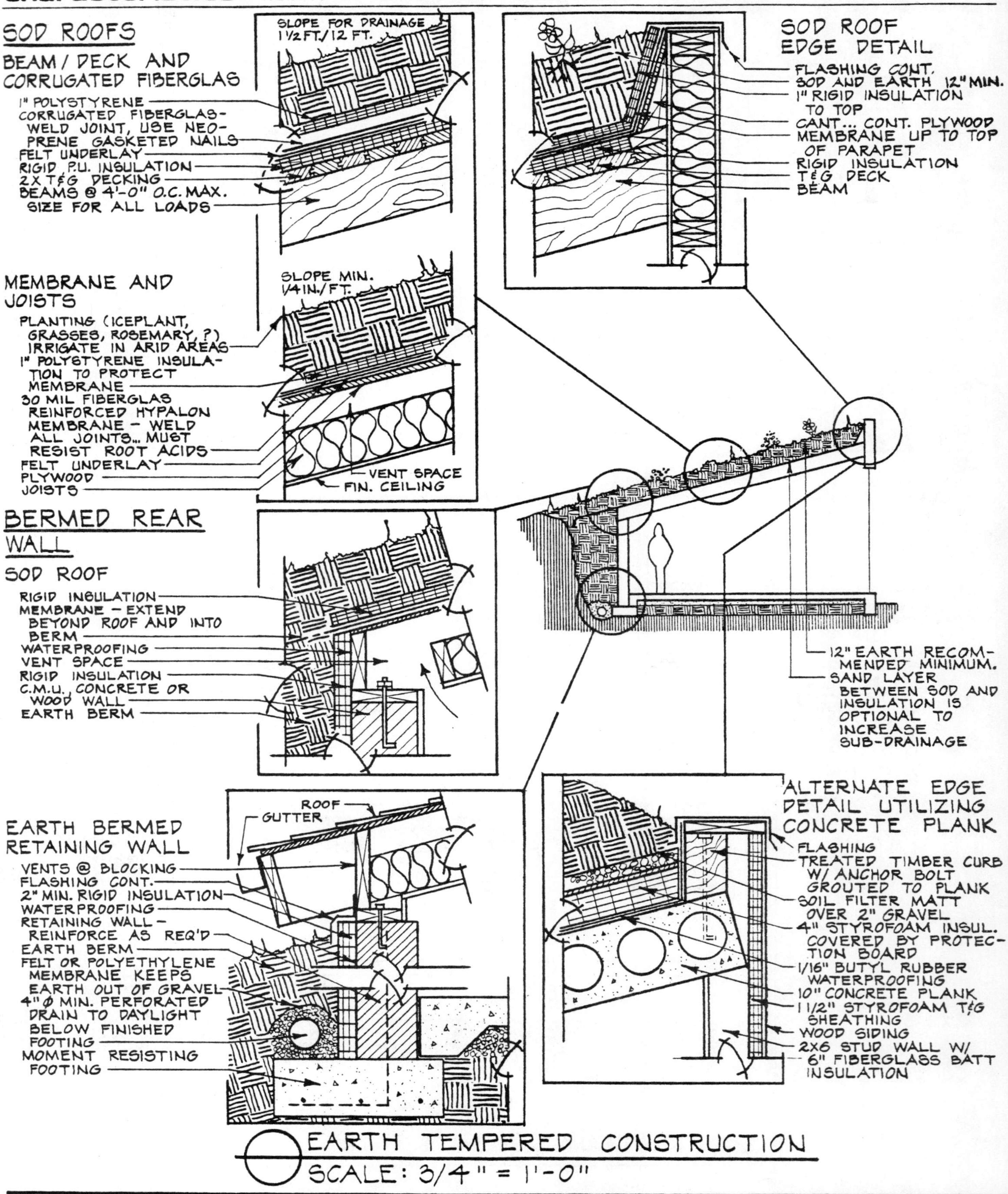


**applicability:** BESIDES HELPING TO VISUALLY INTEGRATE THE BUILDING WITH THE LANDSCAPE, BURYING PARTS OF THE STRUCTURE:

1. REDUCES AIR INFILTRATION BECAUSE OF WIND PROTECTION PROVIDED;
2. REDUCES CONVECTIVE LOSSES BECAUSE OF MORE STABLE EARTH TEMPERATURES;
3. EXTENDS LIFE OF BUILIDNG MATERIALS BY PROTECTING FROM FREEZE/THAW CYCLES.

**10**

## characteristics

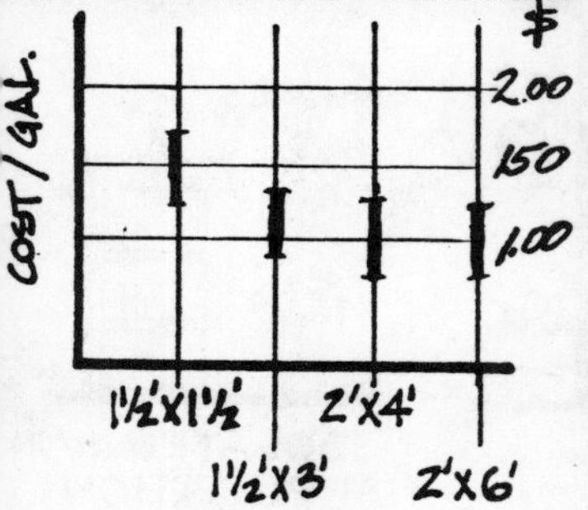


# BUILT-IN STEEL WATER TANK

SITE OR SHOP FABRICATED

1. 12 GAGE MILD STEEL USED FOR MODULE SHOWN
2. CUT AND WELD AS SHOWN
3. WELD SEAMS, ENDS AND BUNG. GRIND ROUGH EDGES
4. PRESSURE TEST BEFORE INSTALLATION
5. FOR STATIC (PASSIVE) WATER STORAGE...
   a. hard water, usually no treatment necessary.
   b. soft or corrosive water, paint with zinc-chromate or with lead primer... or insert magnesium anode to prevent corrosion!
6. FOR ACTIVE STORAGE. epoxy coat and magnesium anode to suit local conditions.
7. EXTERIOR TREATMENT sand or wire brush and paint w/ two coats primer.
8. FINISH
   a. texture & paint to match walls, or
   b. veneer w/ thin wood to match doors, or
   c. paint any color (at these low temps. any color of interior surface is O.K.

## variations

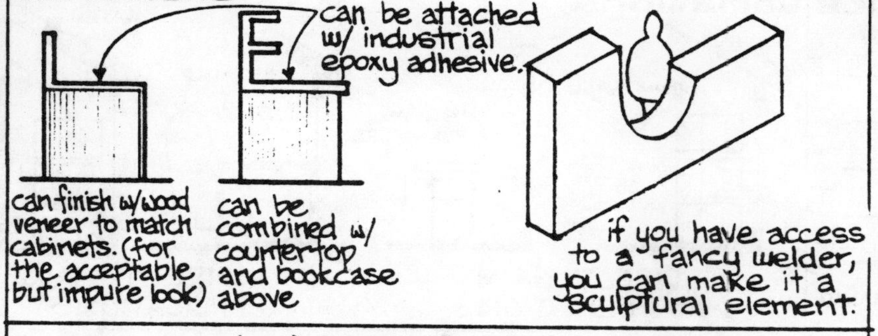


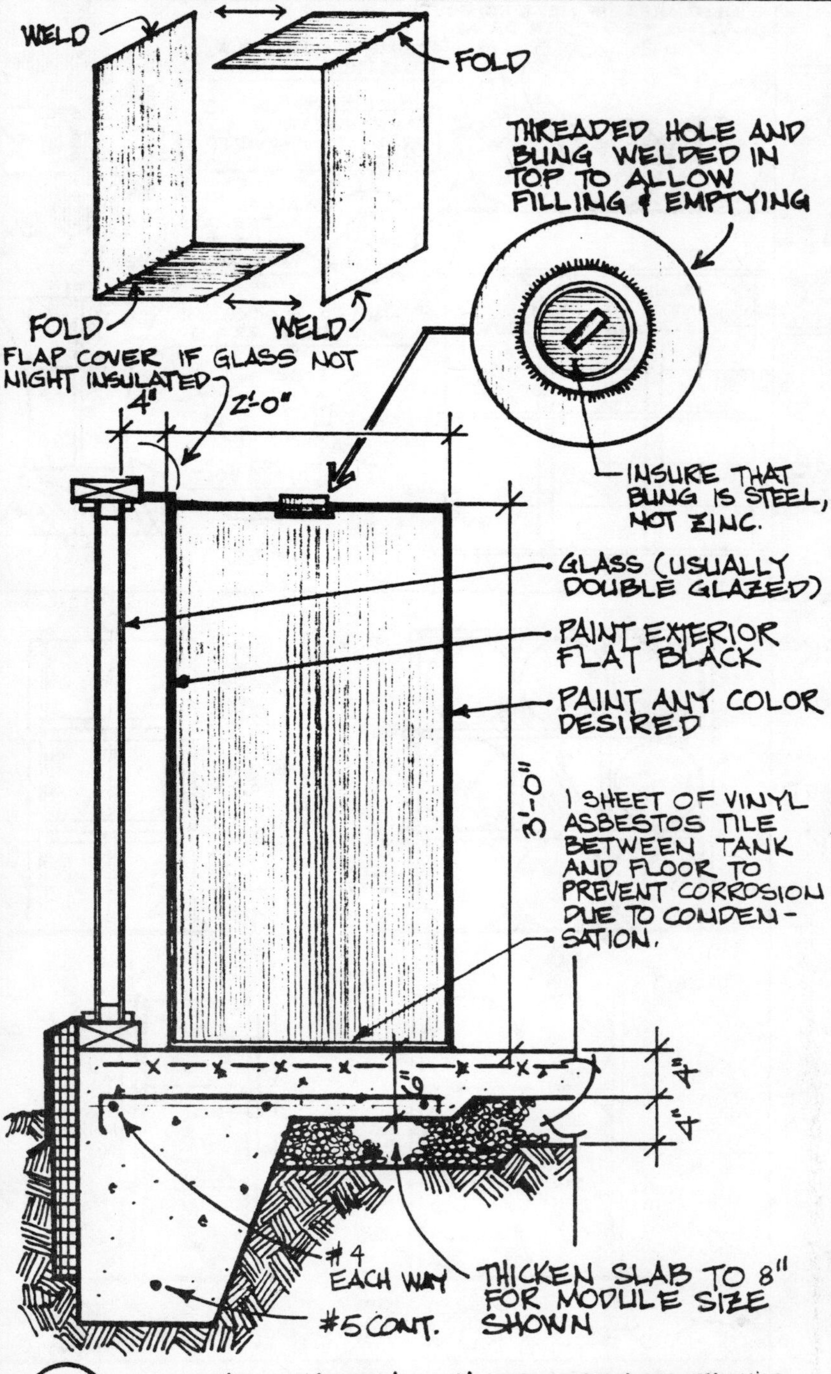


SITE MADE THERMAL WATER STORAGE TANKS
SCALE: 3/4" = 1'-0"

## relation to other components

STRUCTURAL CONSIDERATIONS

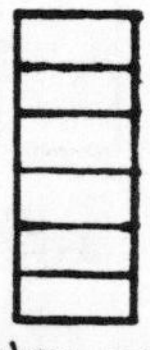

tall tanks can be internally braced w/ rebar spot-welded to each side

PLACEMENT CONSIDERATIONS

by south window

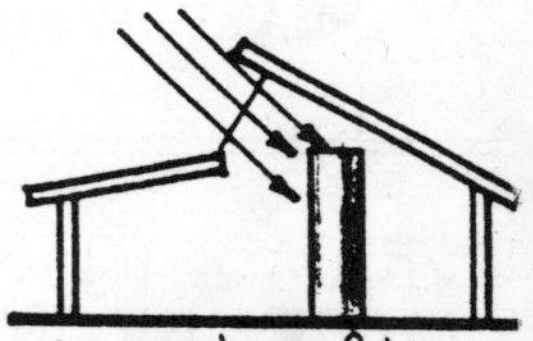

in center of house

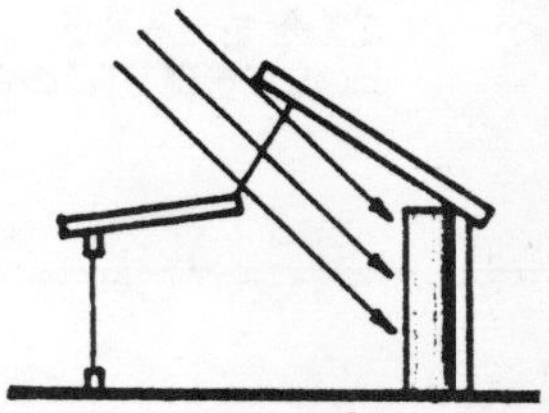

at rear of house

can be supplemented by a flat plate collector or wood stove for heating.... cool pool or tap water for cooling.

**applicability:** CUSTOM-MADE WATER TANKS CAN BE EVEN LESS EXPENSIVE THAN PRE-MADE TANKS — OFTEN AS LOW AS 50¢/GALLON. THEY HAVE THE ADDED ADVANTAGE OF POTENTIAL VISUAL AND ARCHITECTURAL INTEGRATION WITH THE BUILDING DESIGN.

11

**characteristics** | **variations**

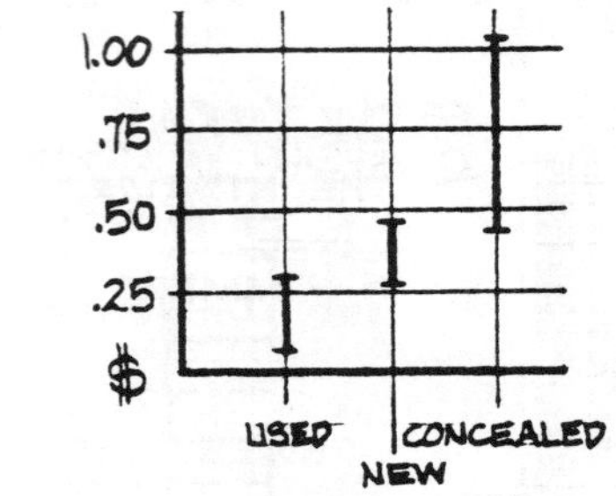


COST/GAL.

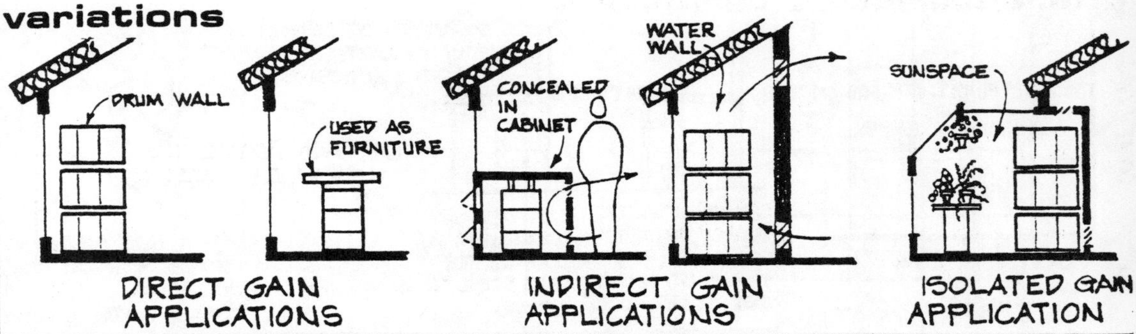


HORIZONTALLY STACKED DRUMS

CLEAN DRUM WITH TSP WASH

PRIME AND PAINT

TREAT WATER AS NECESSARY

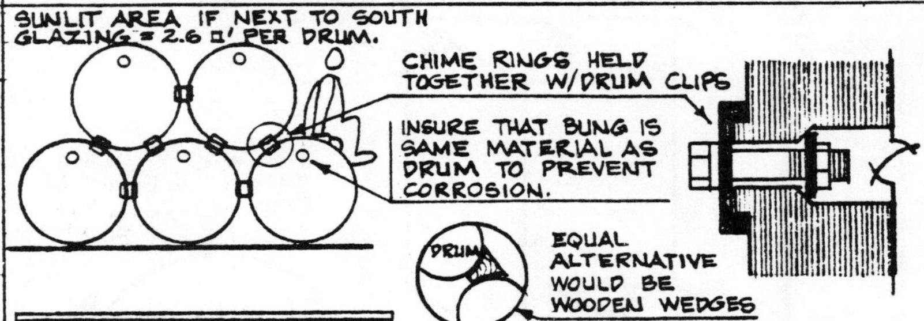


STRUCTURAL CONSIDERATIONS

FULL DRUMS CAN BE SAFELY STACKED 5 HIGH IN THE HEXAGONAL ARRANGEMENT SHOWN

VERTICALLY STACKED DRUMS

SET ON PLYWOOD PLATFORMS

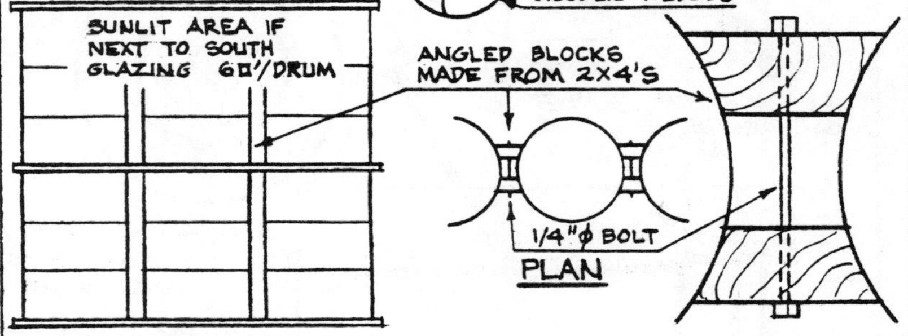


FULL DRUMS AT 500 LBS. EACH CAN SAFELY BE STACKED 3 HIGH IN THE ARRANGEMENT SHOWN

DRUMS IN RACK MADE FROM 2X4 LUMBER

2X4'S ARE BOLTED TOGETHER

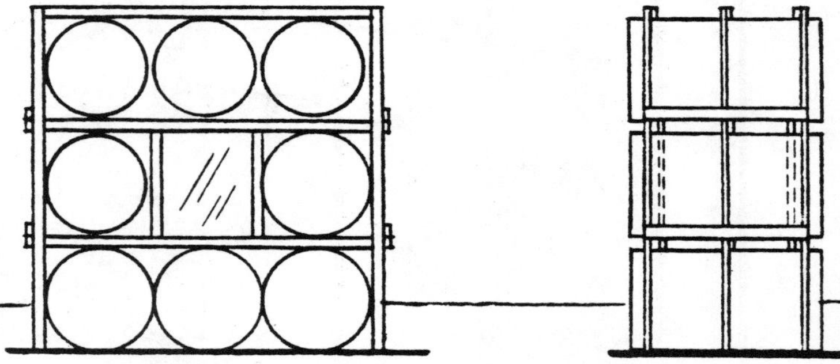

FULL DRUMS AT 500 LBS. EACH CAN SAFELY BE STACKED HIGH IN THE ARRANGEMENT SHOWN

DRUMS IN RACK MADE WITH STEEL CABLE AND PIPE

ANOTHER ALTERNATIVE IS A RACK MADE OF WELDED RECTANGULAR STEEL TUBING

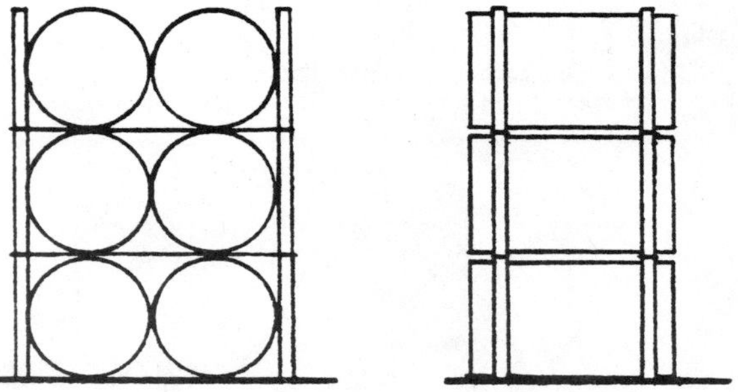

THERMAL TRANSFER CHARACTERISTICS

HEAT TRANSFER BY CONVECTION IF AIR CAN FLOW AROUND DRUMS =35 BTU/HR.°F

BY RADIATION =1 BTU/SQ.FT-°F-HR

STEEL DRUM WATER CONTAINER

NO SCALE

**applicability:** - ONE OF THE LEAST EXPENSIVE METHODS OF OBTAINING THERMAL MASS IS THE USE OF 55 GAL. DRUMS FILLED WITH WATER. THESE DRUMS ARE READILY AVAILABLE AND CAN BE USED IN A NUMBER OF WAYS THAT CAN BE AESTHETICALLY ACCEPTABLE.

**12**

## characteristics

| 12" WALL | 8" WALL | |
|---|---|---|
| .55 | .67 | U |
| 1.8 | 1.5 | R |
| 31 | 21 | STORAGE CAPACITY |

BTU/°F PER □' SURFACE

## variations

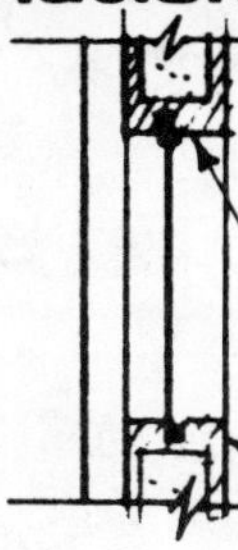


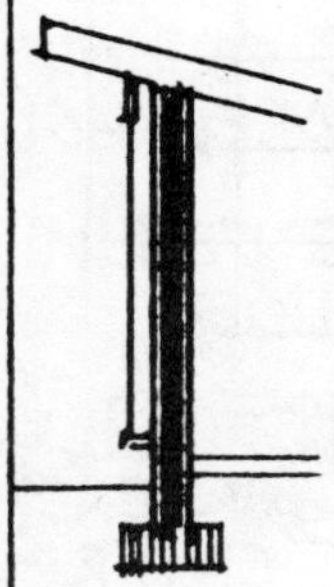

IN MILD CLIMATES THE THERMOSIPHON VENTS TO THE INTERIOR MAY BE ELIMINATED. IF SUMMERS ARE MODERATE AND ENOUGH SHADING OF THE WALL IS PROVIDED, EXTERIOR VENTS MAY BE ELIMINATED.

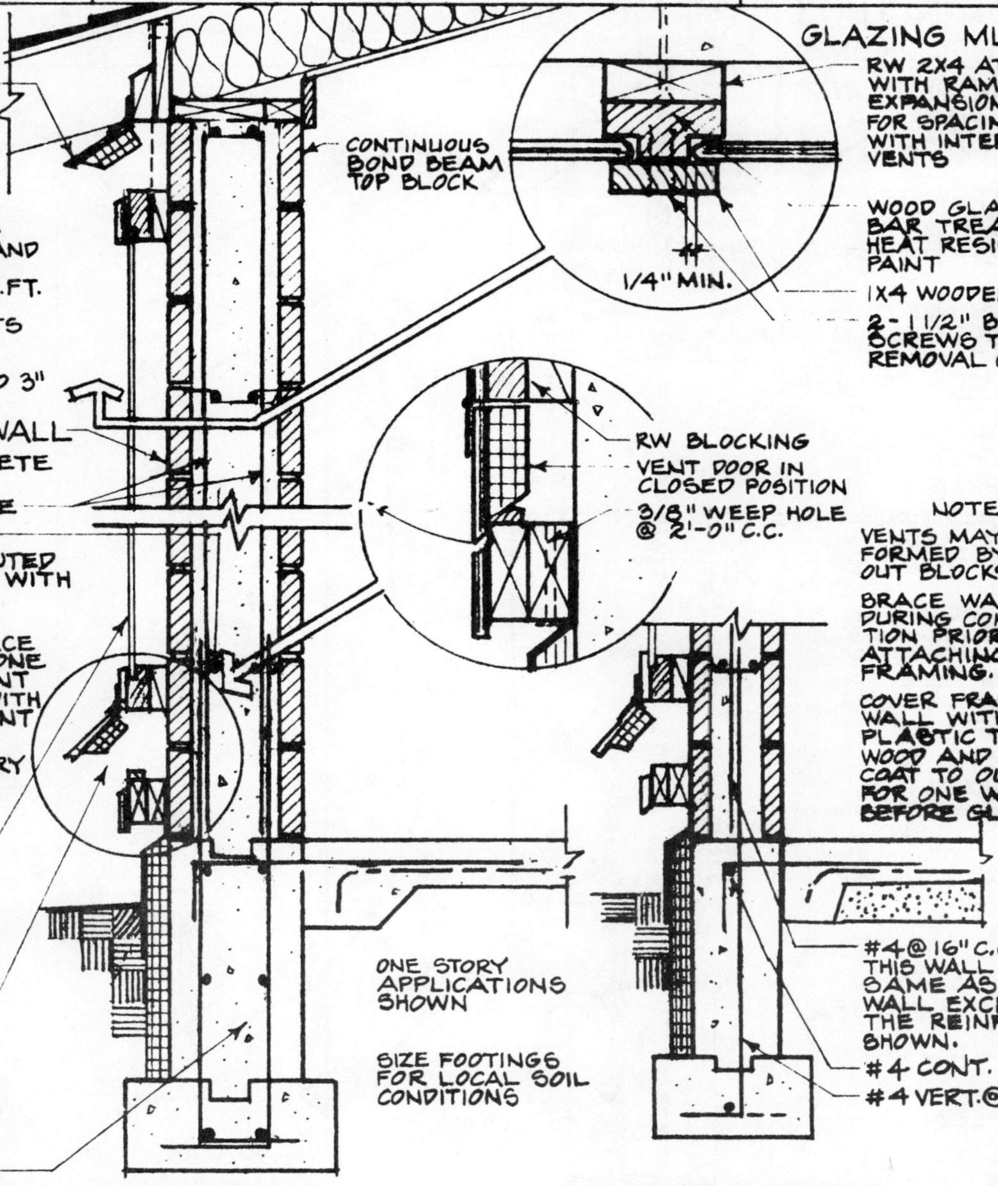


**CONCRETE BLOCK THERMAL WALLS**
SCALE: 3/4" = 1'-0"

**applicability:** GROUTED CONCRETE BLOCK WALLS AS A COMPONENT OF THERMAL WALLS ARE GENERALLY EASIER TO BUILD AND LESS EXPENSIVE THAN MONOLITHIC CONCRETE WALLS DUE TO THE ELIMINATION OF FORM WORK. HOWEVER, IN CALIFORNIA CAREFUL REINFORCING IS REQUIRED DUE TO SEISMIC CONDITIONS.

**13**

## variations

1. TO PROVIDE DEHUMIDIFICATION AND FRESH AIR FOR A WELL SEALED HOUSE.
2. TO PROVIDE DEHUMIDIFICATION AND FRESH AIR FOR A GREENHOUSE.
3. IN CLIMATES WHERE FREEZING OCCURS REGULARLY, A SOLAR AIR PREHEATER TO WARM OUTSIDE AIR BEFORE IT ENTERS THE HEAT EXCHANGER AND AID IN DEFROSTING THE EXCHANGER.

## relation to other components

EXCHANGER IS USUALLY PLACED ON BASEMENT FLOOR OR IN UTILITY ROOM NEAR AN OUTSIDE WALL AND CLOSE TO THE BATHROOMS AND KITCHENS.

# AIR-TO-AIR HEAT EXCHANGER

SHOP FABRICATED ORDINARILY; COULD BE BUILT ON SITE WITH TABLE OR RADIAL ARM SAW.

1. DEVICE IS BUILT IN A SANDWICH TYPE SEQUENCE.
2. CONSTRUCTION DETAILS ARE PROVIDED IN THE BOOKLET "AN AIR TO AIR HEAT EXCHANGER FOR RESIDENCES" BY R.W. BESANT, R.S. DUMONT AND D. VANEE.

AVAILABLE FROM EXTENSION DIVISION UNIVERSITY OF SASKATCHEWAN, SASKATOON, SASK. CANADA S7N OWO ($2.00)

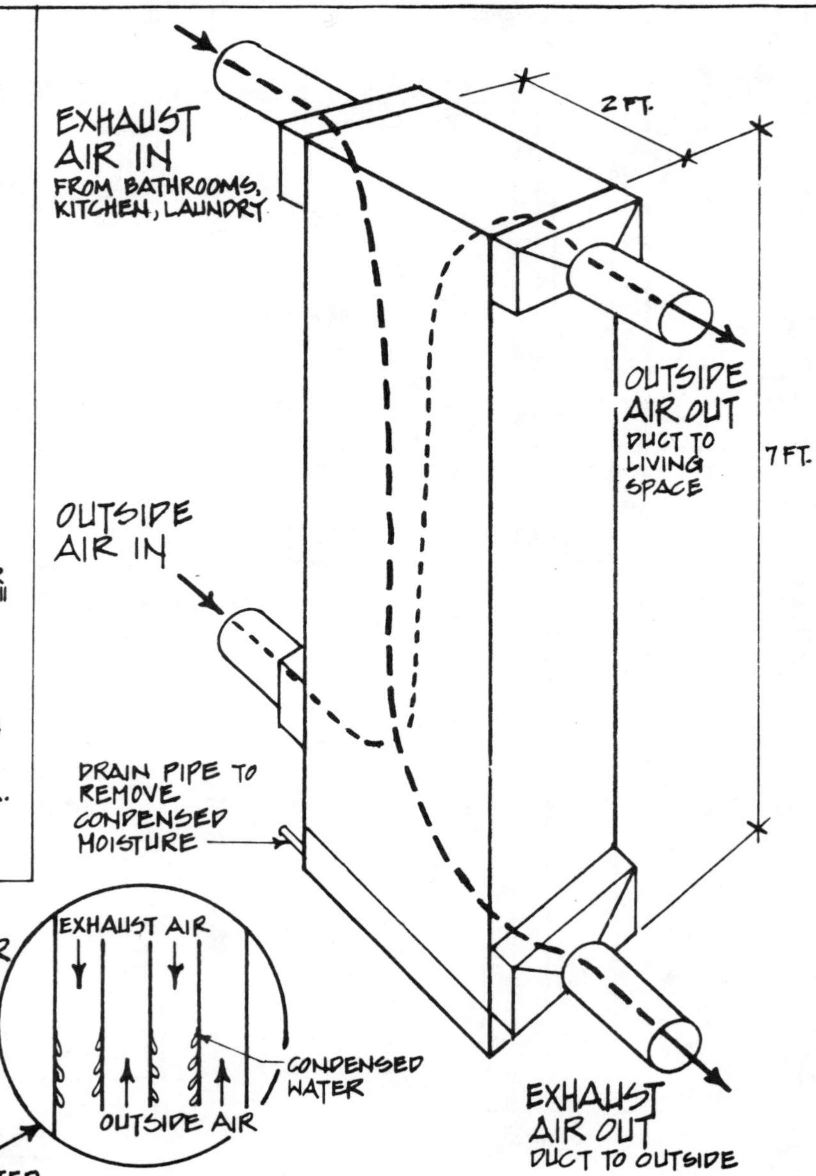


HEAT EXCHANGER PLATES ARE 6 MIL POLYETHYLENE SHEETS SPACED 1/2 INCH

NOTE: FANS OMITTED FROM SKETCH FOR CLARITY.

AIR-TO-AIR HEAT EXCHANGER
NO SCALE

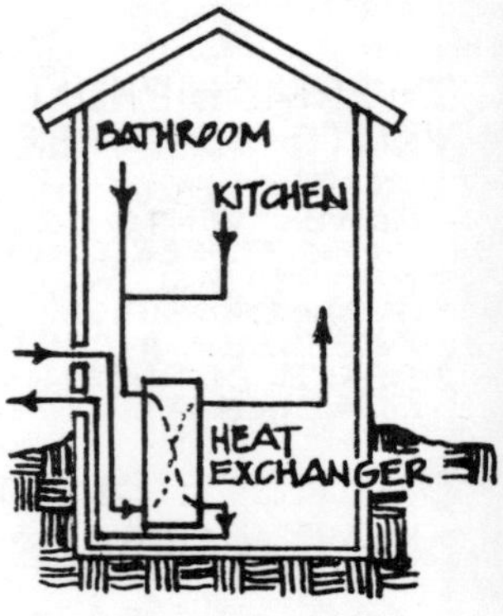


## FLOW RATES

A FLOW RATE OF ABOUT 70 TO 100 CUBIC FEET OF AIR PER MINUTE SHOULD BE ADEQUATE TO CONTROL HUMIDITY AND ODORS IN A NORMAL RESIDENCE. IF EXCESS HUMIDITY PERSISTS, MOISTURE SOURCES SHOULD BE REDUCED OR LARGER FANS USED.

**applicability:** IN WELL CONSTRUCTED HOUSES WITH GOOD VAPOR BARRIERS TO LIMIT AIR INFILTRATION ENERGY LOSSES, ONE CAN PROVIDE CONTROLLED FRESH AIR AT A LOW ENERGY COST USING THE HEAT EXCHANGER. IF EXCESS HUMIDITY IS NOT A PROBLEM, THEN YOUR DWELLING IS PROBABLY TOO LEAKY. SEAL IT FIRST.

**14**

# Chapter 7: Fourteen Examples

The fourteen examples presented in this chapter were chosen from the many thousands of passive buildings that are now functioning in America. These examples demonstrate passive design in both new and retrofit construction for both commercial and residential building. These examples cover the climatic extremes from Hawaii to Saskatchewan and include several projects that provide heating and cooling.

A close look at these retrofits and new buildings should help you understand the application of passive solar design to your own situation, whether you are building a new workshop or a large apartment building. As energy prices continue to skyrocket upward there will be little consolation for the people who had an opportunity to make their building a passive solar building and didn't. Take the time to determine what will work best for you and then do it.

The performance figures are difficult to verify and may not always represent the same thing. Far too little monitoring and evaluation has been done, not only of passive buildings but also of non-solar buildings. Where climate data was not available locally the nearest station was used. Cost data was supplied by the designer or builder.

When you build your passive building or retrofit your house for passive solar heating or cooling, please let us know. We like to keep track of activity in the field and may consider your project for use in the Third Passive Solar Catalog.

# Foothill Greenhouse Retrofit 1979

## Description:

This solar greenhouse retrofit was part of a remodel of a 1900's era log cabin. In addition to the 200 square foot greenhouse, a 60 sf. pantry/ laundry was added bringing the total house up to 1,200 sf. In addition, the fireplace was bricked up and replaced with an airtight stove.

Heating is provided by the double glazed south windows in the greenhouse. Thermal storage is provided by the slab and 5 gallon plastic drums full of water. The water is colored with food dye for better absorption. Summer cooling is by night ventilation and shading from an arbor. The end walls have screened sections covered with insulated panels in winter and left open in summer.

The solar greenhouse is used as a room. Heat is naturally covected into the house through double french doors that open into the kitchen. The owner is very pleased with the retrofit and will continue to further improve the house as time and money allow.

For more information:

Dave Norton
Habitat Construction
Moonshine Rd. #71
Camptonville, CA 95922

*Design: Dave Norton*

*Construction: Habitat Construction*

*Location: Nevada City, California*

*Latitude: 40° N*

*Altitude: 2520 ft.*

*Climate: 4900 HDD 378 CDD*

*Jan. Mean: 39° F July Mean: 70° F Yr.: 53° F*

*Solar Radiation: N.A.*

*System Data: 200 sq. ft. south glass, 5,000 lbs. water equivalent mass*

*Summer Design Temperature: N.A.*

*Winter Design Temperature: N.A.*

*Solar Heating: 60%*

*Cost: Approximately*

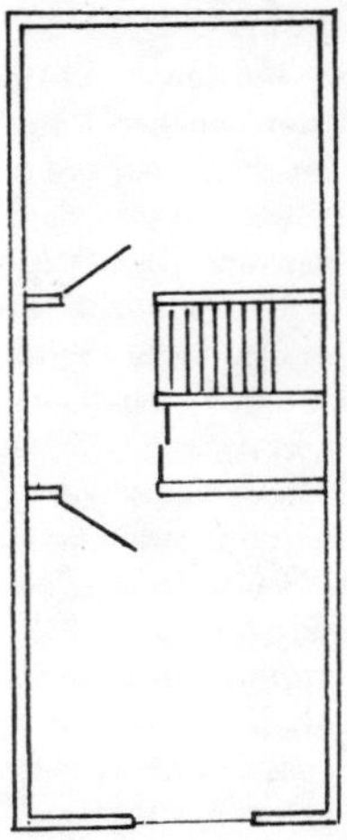

Second Floor

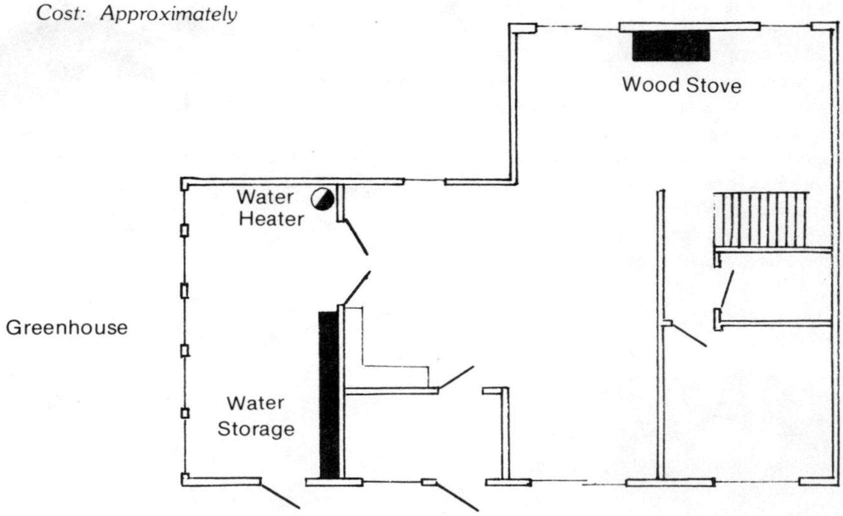


First Floor

# Passive Solar Retrofit, Farm House Addition 1973

## Description:

The Hammond retrofit is a classic example of a passive solar addition. It not only added 575 square feet to the house, a kitchen and dining room, but also included other floor plan revisions to make the house more livable. The house itself was insulated and resided, double pane windows added, and otherwise revamped.

The major element of the system is the on slab addition with thirtythree 55 gallon drums for heat storage in winter and coolth storage in summer. Heating is provided by 180 square feet of south windows in the addition and two 32 square foot skylights, one in the living room and one in the bathroom. The bathroom skylight shines on water filled aluminum culverts and the living room includes 5 gallon cans of water for added storage.

The solar heat rises by convection into the major part of the house which is 3' above the addition. Insulated shutters and drapes on all windows allow temperatures to reach equilibrium on cold winter nights.

In summer the large windows are shaded by a tension arbor with grapes. Insulated, reflective shutters are closed on the skylights and the house is ventilated at night.

I've spent much time in the house and can attest to how well it works. Internal heat gain and solar gains are usually sufficient for heating and summer cooling is very good.

For more information:

Jon Hammond
Living Systems
Rte 1, Box 170
Winters, CA 95674

*Design: Jon Hammond*

*Construction: Jon Hammond*

*Location: Winters, California*

*Latitude: 39° N*

*Altitude: 80 feet*

*Climate: 2819 HDD 1063 CDD*

*Jan. Mean: 45° F July Mean: 75° F Yr.: 60° F*

*Solar Radiation: 900 BTU/sf. south glass, 30,000 lbs. water equivalent mass*

*Summer Design Temperature: 100° F*

*Winter Design Temperature: 30° F*

*% Heating: 90-100% % Cooling: 100%*

*Cost: Approximately $10,000 for entire remodel, perhaps $2,000 going for the solar retrofit.*

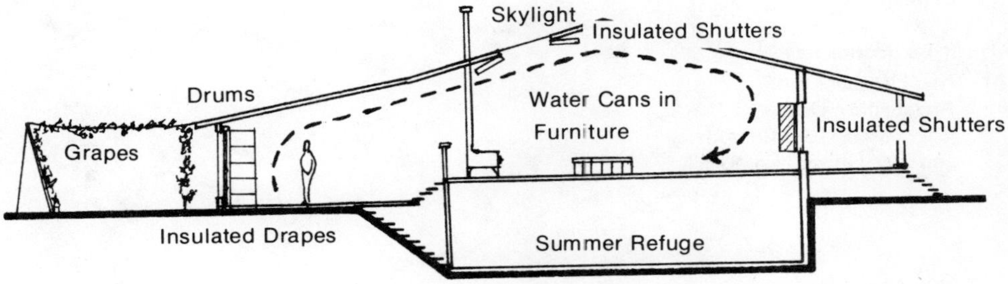

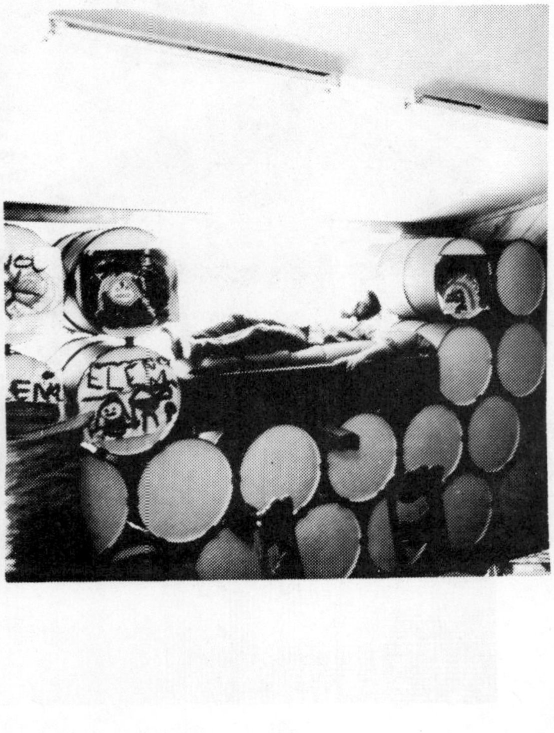

# Retrofitting A California Bungalow 1979

## Description:

This retrofit illustrates what can be done in an extensive retrofit. The 1930 bungalow was in fair shape when the remodel was begun but was too small at 900 sf., too cold in winter, and too dark. The wide overhangs and good ventilation provided reasonably good cooling.

The remodel added 300 square feet of space including a second bathroom, third bedroom, and dining room. The old ceiling was removed and a series of clerestory windows were added beneath an extended north roof. Mass for these is provided by pipes of water hung on the trusses. In addition, a bay window porch was added for a dining room. The thermal storage for the bay window is provided by a high mass floor built with masonry wall toppers used as tile. These are 1½" thick and are stained, set in grout, and varnished. Both bathrooms were designed with extra glass and tile for heat storage. Circulation is by natural convection.

Summer cooling is by good solar control and night ventilation. The thermal mass stores the coolth for use during the day.

For more information:


Ken Haggard
Architecture Dept.
Cal Poly San Luis Obispo
San Luis Obispo, Ca. 93407


*Design: Ken Haggard, Polly Cooper*

*Construction: Ken Haggard, Polly Cooper, and friends*

*Location: San Luis Obispo, California*

*Latitude: 35° N*

*Altitude: 315 ft.*

*Climate: 2472 HDD 246 CDD*

*Jan. Mean: 52° F July Mean: 65° F Yr.: 58° F*

*Solar Radiation: 1400 BTU/sf. Clear December day.*

*System Data: 200 square feet of south glass, 6,000 lbs. wate equivalent mass*

*Summer Design Temperature: 82° F ·*

*Winter Design Temperature: 37° F*

*Solar Heating: 80-90% Cooling: 100%*

*Cost: Approximately $15,000 for the entire project, about $5,000 can be allocated to the solar system.*

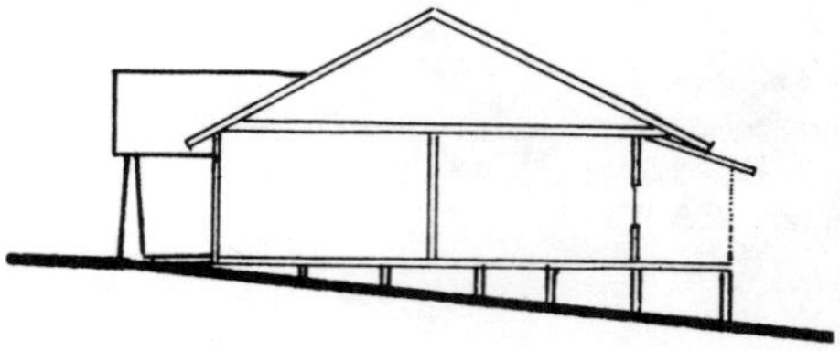

Before

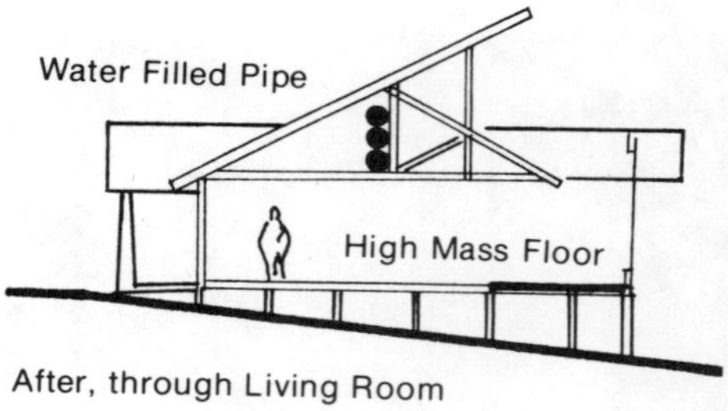


After, through Living Room

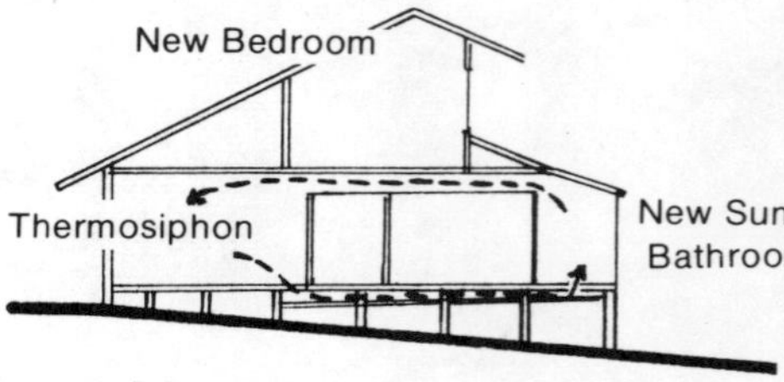


After, through Bedroom

# Atrium Retrofit 1979

## Description:

This retrofit is a fine example of what a turn-around a retrofit can make on a building that performs very poorly. This 2,300 square foot, flat roofed house had an open 20' x 25' atrium or central courtyard, surrounded by sliding glass doors and windows. It was causing overheating in the summer and made the house feel very cool in the winter. The extreme temperature fluctuations were also wreaking havoc on the collection of plants the owners had established in the courtyard.

The solution to the problem was a roof monitor with south and east windows. This well insulated structure radically changed the house performance and has made the atrium a real asset. Winter sun is collected by the south window and heats the atrium. Hot air from the peak is drawn off by a small fan and blown into the house. A ceiling fan also helps circulate air and helps move heat into the house when doors are open. An airtight stove was installed for better heating.

Summer comfort is also greatly improved by the retrofit. Overhangs provide good shading and nighttime natural ventilation pulls cool air into the house. When natural ventilation is insufficient a whole house fan is used. Terrazo tile (1½") is used for thermal storage. It was included in the original house.

A passive solar water heater was also included in the retrofit. It has 120 gallons capacity and 72 sq. ft. of double glazed Kalwall glazing.

For further information:

Marshall Hunt
2312 Elendil Ln.
Davis, CA 95616

*Design:* *Marshall Hunt*

*Construction:* *Bent Nail Construction, Marshall Hunt*

*Location:* *Davis, California*

*Latitude:* *38° 5' N*

*Altitude:* *30 feet*

*Climate:* *2814 HDD 1063 CDD*

*Jan. Mean:* *45° F* *July Mean:* *75° F* *Yr.:* *60° F*

*Solar Radiation:* *900 BTU/sq. ft. Dec. Day*

*System Data:* *155 sq. ft. south and southeast glass in remodel, 91 sq. ft. in existing wall.*

*Summer Design Temperature:* *100° F*

*Winter Design Temperature:* *30° F*

*Solar Heating:* *40-60%* *Cooling:* *70-80%*

*Cost:* *$24 sq. ft. including mechanical equipment, fans, stove, etc.*

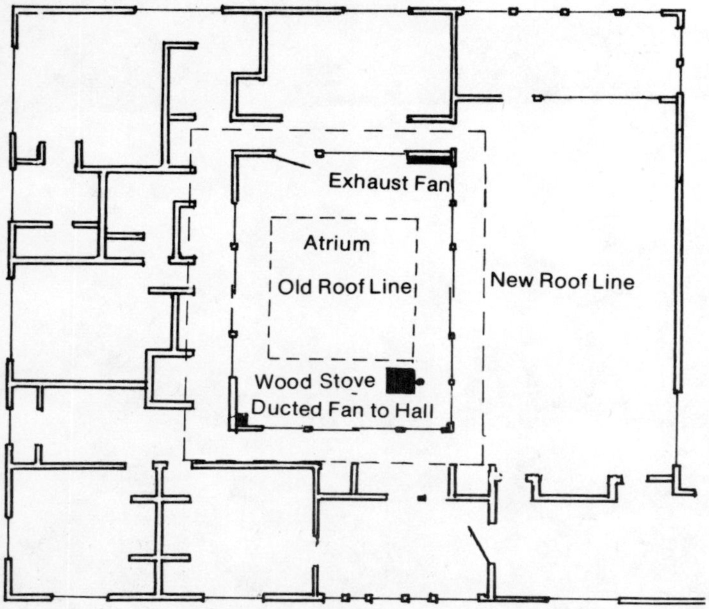


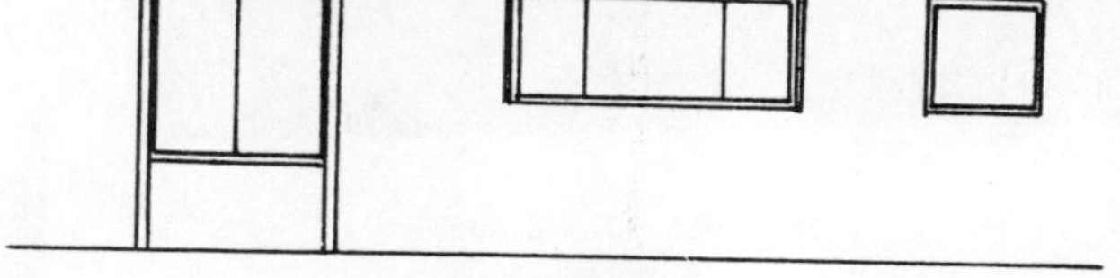


South Elevation

# A Mobile Home Retrofit 1979

## Description:

The retrofit suggests what can be done to improve the livability of a mobile home. The solar retrofit added considerable space to the 14' x 70' trailer as well as providing solar heating in the winter, cooling in the summer, and better light all year.

An addition of about 1,300 square feet was built on the south and west of the trailer. It includes about 200 feet of triple pane window facing south. Heat is stored in a 500 gallon jacuzzi. An active solar system is planned for additional solar heating, it will be coupled to the jacuzzi tank. Heat is conducted into trailer by existing door and windows. Exterior venetian blinds are used for summer shading and clerestory windows int he addition increase natural ventilation.

The owners are very pleased with the results and find the addition much warmer in winter and cooler in the summer.

For further information:

Dr. David Mitchell
P.O. Box 9
Glenboro, Manitoba
ROK OXO

*Design: David Mitchell, Solar Tech.*

*Construction: N.A.*

*Location: Glenboro, Manitoba*

*Latitude: 49° N*

*Altitude: 1200 ft.*

*Climate: 11,000 HDD 750 CDD*

*Jan. Mean: N.A. July Mean: N.A. Yr.: N.A.*

*Solar Radiation: 1,000 BTU/sq. ft. December day*

*System Data: 200 sf south glass, 4,000 lbs. water equivalent mass*

*Summer Design Temperature: 84° F*

*Winter Design Temperature: -26° F*

*Solar Heating: 30%*

*Construction Cost: Approximately $20 sq. ft. for the addition*

# An Affordable Passive Solar Design 1977

One of the biggest challenges in construction today is keeping the cost down. This simple passive solar house was designed to be affordable, easy to build, energy efficient, and space efficient. The designer's experience at the Minimum Cost Housing Group of McGill University enabled him to keep the cost down while still providing a very livable home.

The 2,107 square foot house is a two story cube with a 160 square foot greenhouse on the south. The two story design helps reduce heat loss and improve solar gain. Two thirds of the glazing faces south, south-east, or south-west for solar gain. A concrete slab is used for thermal storage. Walls and ceiling are R-20. Backup heating is with electricity.

The house was built in three months by the owner and his father with subcontractors doing the concrete, roofing, electricity, and plumbing.

For further information:
Witold Rybcynski
Minimum Cost Housing Group
McGill University
3480 University Avenue
Montreal, Quebec H3A 2A7

*Design: Witold Rybcynski*
*Construction: Florent Dumont*
*Location: Hudson, Quebec*
*Latitude: 460 NN*
*Altitude: 200 feet*
*Climate: 8200 HDD 800 CDD*
*Jan. Mean: 14° F July Mean: 70° F Year: 44° F*
*Solar Radiation: 1100 BTUsf clear December day*
*System Data: 330 sf south glass, 6000 pounds water equivalent mass*
*Summer Design Temperature: 79° F*
*Winter Design Temperature: -13° F*
*Solar Heating: 30 - 40%*
*Construction Cost: $15 sf*

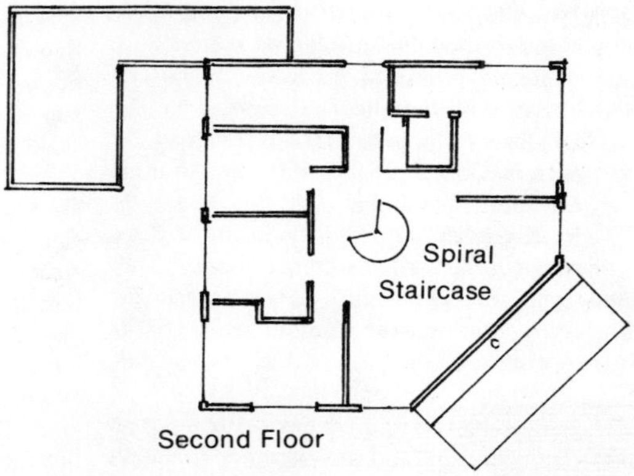


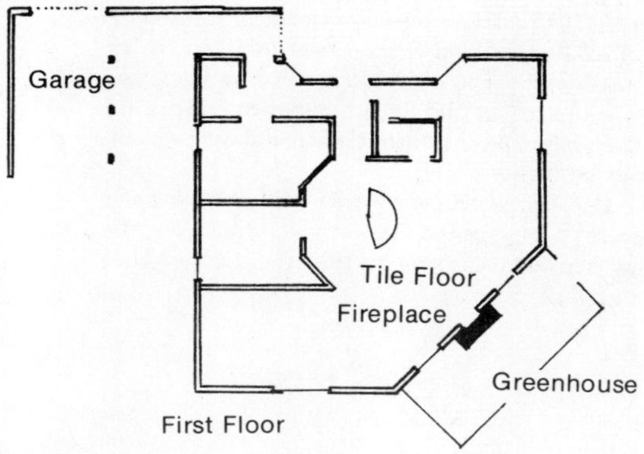

# The Saskatchewan House - Conservation/Solar Special 1977

## Description:

This large 2000 sq. ft., house is a good example of energy conservation/passive solar design in a harsh climate. It is a two story, boxy house of 3 + bedrooms and two baths. The house was designed to provide 100% solar heating: it used a very energy efficie..t passive solar house as the base and an efficient active solar system for backup. Water is also heate ' by the solar system.

The energy conservation efforts are worth particular attention. They include: R-41 wall insulation in an offset 2" x 6" wall, R-61 ceiling insulation, R-30 floor insulation, R-15 perimeter insulation, a carefully caulked vapor barrier (6 mil polyethelene with poly pans behind outlets), double and triple glazed windows, movable insulation systems for windows, insulated doors, a greywater heater reapture unit, and an air to air heat exchanger for make up air. Heat loss is reduced to 13,000 BTU/hr. at -29° F with shutters closed, 18,000 BTU/hr. shutters open.

The passive solar features include most glass on south and overhangs for summer shade. The extremely efficient house shell allows passive design (including interior heat gain) to meet 85% of the house heating load, 43.5% passive, 7% occupants, and 33.5% from lights and appliances.

The active solar system included 192 sq. ft. of vacuum tube collectors and was designed to supply the remainder of the heating. A water (60%) antifreeze (40%) mixture was used for heat transfer. A heat exchanger loop carried heat to a 2800 gallon storage tank. The $15,000 cost of this active system is unjustified and for a number of reasons the active system has been shut down and will probably never be repaired.

Backup heating is electrical with 2-3 kw water heaters. One is in the domestic water tank and the other is in the main waterline from the storage. They are used off peak.

The performance of all systems except the active solar has exceeded the designers expectations. They now feel a full passive solar house could be built in Regina, Saskatchewan by using more windows and more thermal mass. The house performance has been monitored carefully for almost two years now. It has been somewhat amibiguous because the house has been used as a demonstration house - with over 1,000 visitors per week. With a family in it, its performance would probably have been better.

In operation to date, electrical requirements for heating have been 5.1 G Joule per year. This is less than 5% of the typical home heating load in Regina. The first cost was higher than for a standard house - but the annual cost is probably less than for any kind of heating system except "neighborhood-scale seasonal-storage district heating."

It has served well as a demonstrator - and three firms are now building houses to similar standards. They include, Enercon in Regina; Concept Construction in Saskatoon; and Permanent Construction in Price Albert. Needless to say, they have improved on the aesthetics of the house. This was relatively easy to do once the active system was downplayed.

For further information:

Department of Mechanical Engineering
University of Saskatchewan
Saskatoon, Saskatchewan S7N OWO
Canada

*Design: Bob Besant and Rob Dumont*

*Construction: N.A.*

*Location: Regina, Saskatchewan*

*Latitude: 50° N*

*Altitude: 1884 ft.*

*Climate: 10,770 HDD 500 CDD*

*Jan. Mean: -6° F July Mean: 66° F Yr.: 25° F*

*Solar Radiation: Av. 614 BTU/sq. ft. day*

*System Data: 128 sq. ft. of south glass, 1072 pounds water equivalent mass.*

*Summer Design Temperature: 85° F*

*Winter Design Temperature: -20° F*

*Solar Heating: 84%*

*Cost: Construction to similar standards will cost approximately $2-3 more/sq. ft. than conventional construction for the passive and energy conservation features. Estimate cost to build it without active solar $50 sq. ft. (1979)*

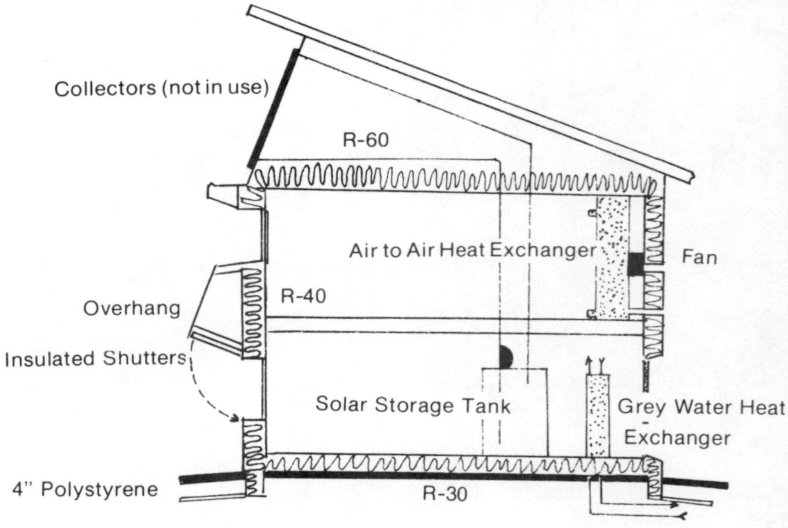

# A Naturally Cooled Residence In A Hot Humid Climate 1979

## Description:

This paper will discuss the design, construction, and performance of the Plad residence. This structure is a 900 sq. ft. vacation cabin on the island of Kauai in Hawaii. The site is a meadow near a creek which floods yearly. The living floor is raised ten feet above grade and is supported by poles. Rainfall at the site is eighty inches, the air humid, and temperatures average in the 70's the year around. Heating is not required and cooling the main design concern.

Plan orientation presents the closed corner of the living spaces, in a prow-like manner to the Kona storm winds. The open covered deck spaces lie on a lee-side toward the North, facing the best views. The center bay is open allowing a column of cool air from the shaded underside of the building to be drawn up through the structure by an open gable which faces the prevailing winds. The corrugated steel roof is hipped to provide rigidity and a low profile to strong winds. Eaves extend four feet beyond the walls, shading the Southeast and Southwest sides as well as the outdoor deck areas. The walls are constructed like giant louvers. Shingled planes three feet high overlap each other, with screened operable vents betwen, allowing breezes to freely flow through the building.

The architecture of this residence provides natural cooling by orientation, shading, transparency to breezes, cool central microclimate and venturi-like gable on hip roof which induces ventilation.

For further information:

Jonathan Stoumen, Architect
Box 87
Miranda, California 95553

*Design: Jonathan Stoumen*

*Construction: N.A.*

*Location: Kapaa Homesteads, Kauai, Hawaii*

*Latitude: 22° N*

*Altitude: 430 ft.*

*Climate: N.A. HDD N.A. CDD*

*Jan. Mean: 72 July Mean: 77 Yr.: 75*

*Solar Radiation: N.A.*

*System Data: Solar control and ventilation cooling*

*Summer Design Temperature: 82° F*

*Winter Design Temperature: 61° F*

*Solar Heating: None required. Cooling:*

*Cost: N.A.*

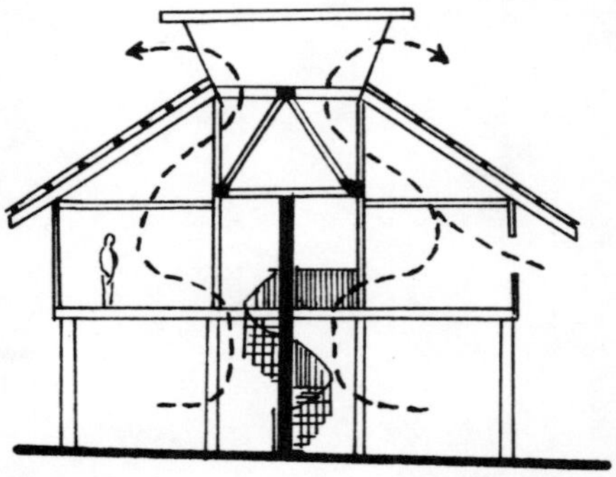

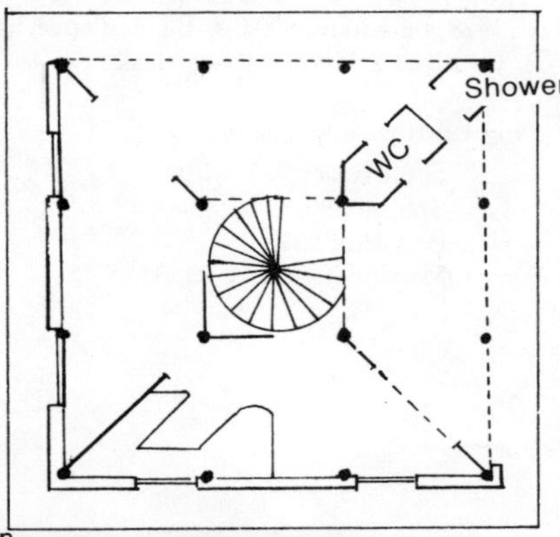

# Commercial Building Retrofit 1977

## Description:

This retrofit is an excellent demonstration of the potential for retrofitting high mass commercial buildings for passive solar heating. The building is an old brick factory building currently used by the Keller Products Company. Two thirds of the south wall was glazed with Kalwall's Sunwall Solar Window Panels ©. Before the glazing was installed the 12" thick masonry wall was painted black. The south wall provides the thermal storage for the system.

Existing windows provide direct gain in addition to the mass wall. The building was operational in two modes, unvented with windows closed and vented with windows open to set up a thermosiphon. Two large stairwells on the north provided a loop for the thermosiphon which was set up by opening the upper sash on the third floor and the lower sash of the second floor double hung windows.

The system has worked well and no detrimental effects have been detected. Indications are that the protection provided by the glazing may in fact prolong the life of the building. The system saved more than 2 billion BTU day during testing in early 1978, by adding 800,000 BTU of solar heat and preventing 1.2 MBTU of heat loss.

For further information:

Scott Keller
Kalwall
P.O. Box 237
Manchester, N.H. 03103

*Design: Kalwall Corporation, Scott Keller, Arthur Sedrick, et al*

*Construction: Kalwall*

*Location: Manchester, N.H.*

*Latitude: 43° N*

*Altitude: 250'*

*Climate: 7101 HDD 378 CDD*

*Jan. Mean: 20° F July Mean: 68° F Yr.: 45° F*

*Solar Radiation: N.A.*

*System Data: Windows 630 sq. ft., south wall glazing (incl. windows) 1970 sq. ft.*

*50,000 lbs. water equiv. mass in south wall alone.*

*Summer Design Temperature: 86° F*

*Winter Design Temperature: 1° F*

*Heating: N.A.*

*Cost: N.A.*

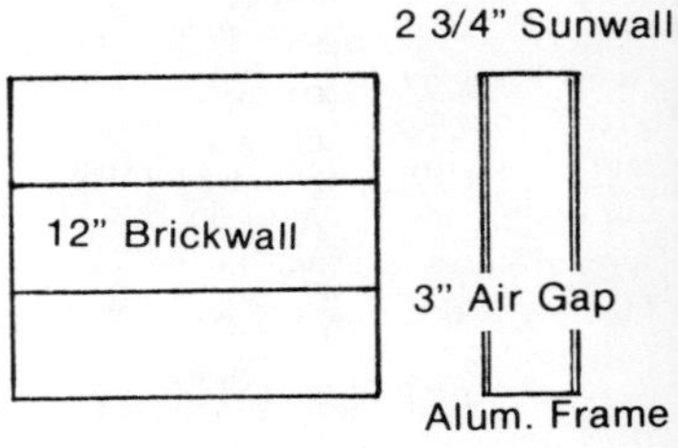


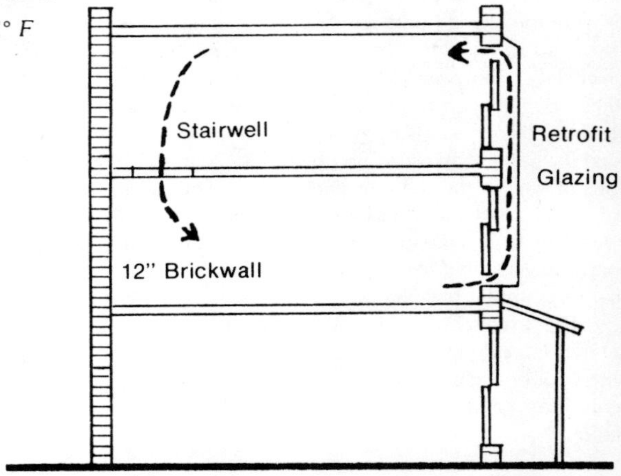

# Passive Solar Doctor's Office 1979

## Description:

This 1600 square foot commercial building was designed for a family practice doctor. The client, Dr. Robert Gentry, played an important role in design and was in large part responsible for including passive solar design features and for developing such a workable floor plan.

The building includes south glass for direct gain in winter. Storage is provided in the slab, tile floors, and water tanks. Cooling is by night ventilation. Cooling loads are minimized by overhangs, arbors, and absence of east and west windows.

The clerestory windows provide excellent daylighting which was an important consideration in the design. This not only reduces energy use for lighting and cooling but gives the building a much warmer feeling.

Dr. Gentry has been very pleased with the building and plans to retrofit his home with a solar greenhouse as time allows.

For more information:

David Bainbridge
PSI
P.O. Box 722
Davis, CA 95616

*Design: Ronald D. White, Architect; David Bainbridge, solar consultant*

*Location: Grants Pass, Oregon*

*Latitude: 42° N*

*Altitude: 925'*

*Climate: 5,000 HDD, 560 CDD*

*Jan. Mean: 37° F July Mean: 72° F Yr: 53° F*

*Solar Radiation: 838 BTU/sq. ft. Av. for heating season*

*System Data: 300 sf south glass, 8,000 lbs. water equiv. mass.*

*Summer Design Temperature: 89° F*

*Winter Design Temperature: 26° F*

*% Solar Heating: 50-60%*

*Cost: Approximately: $60 sf*

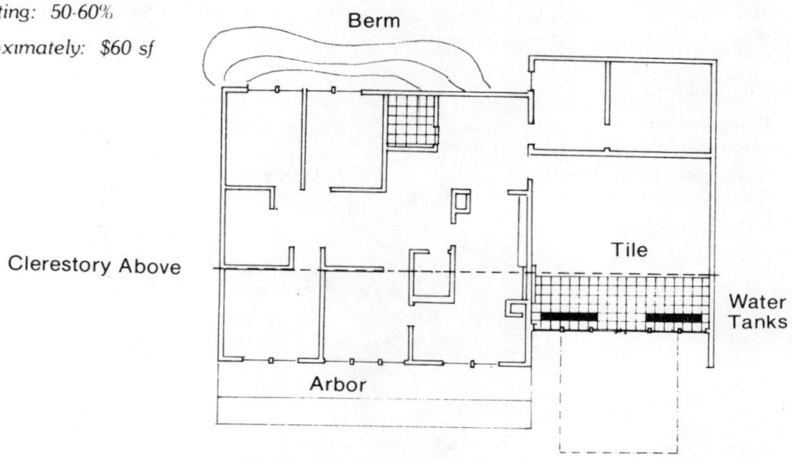

# Passive Solar Motel 1979

## Description:

The Best Western Motel in Osseo is an excellent demonstration of passive solar design used in a commercial project.

The motel uses fan assisted trombe walls to heat the rooms and a site built air collector to heat the public areas.

The project was financed and build using standard procedures. The mass wall was built by masons and glass company. The site built collector by the carpenters and glass company. The heating contractor handled ducts and control systems.

The project is a perfect demonstration of how solar should be done. Wise design and simple detailing allowed the project to be built with normal crews and standard materials. The reduction in standard heating and cooling equipment will usually offset the added material cost for the solar system as it did here. With tax credits it can be considerably less - and the annual savings even greater.

For futher information:

Rodney Wright
Hawkweed Group
4643 North Clark Street
Chicago, Illinois 60640

*Design: Hawkweed Design*

*Construction: N.A.*

*Location: Osseo, Wisconsin*

*Latitude: 45° N*

*Altitude: 800 ft.*

*Climate: 7,700 HDD 1,000 CDD*

*Jan. Mean: 10° F July Mean: 70° F Yr.: 43° F*

*Solar Radiation: 700 BTU/3F clear December day*

*System data: N.A.*

*Summer design temperature: 85° F*

*Winter design temperature: -11° F*

*Heating: 60%*

*Cost: 80¢ less psf than standard motel construction*

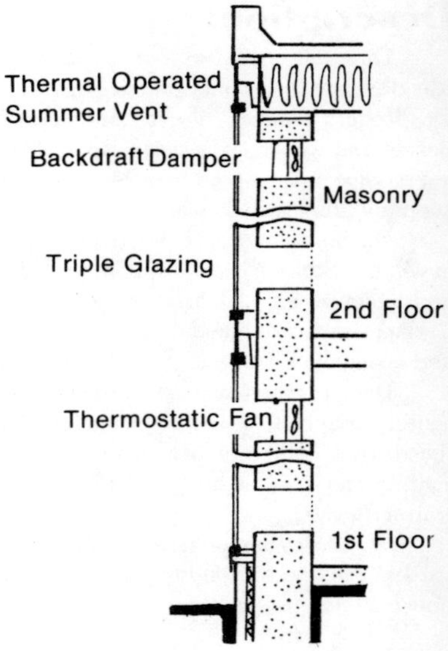


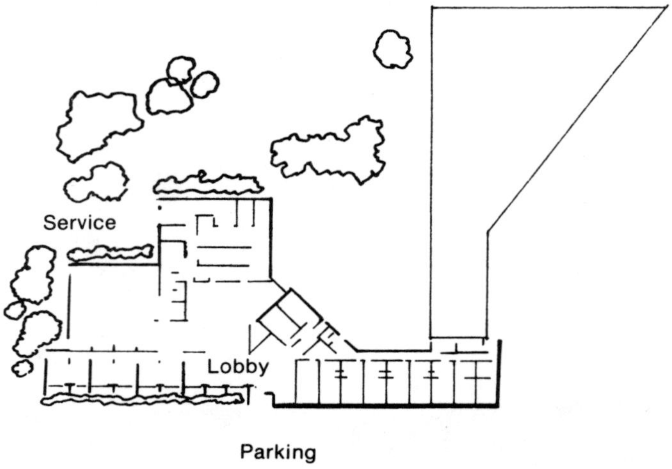

# A Passive Solar Apartment 1978

## Description:

The Kitsun Housing Cooperative is an eight unit townhouse that utilizes three passive solar systems for space and water heating. The units range in floor area from 600 to 1200 sq. ft. each. The seven ground floor units of the building utilize both Trombe walls and Direct Gain systems for 85% of their space heating on an annual basis. The eight second floor units utilize a larger direct gain system for about the same solar contribution. Solar domestic hot water is supplied to all units through the use of eight thermosyphon systems. These systems will supply 70% of the hot water needs on an annual basis.

The 12" thick Trombe walls are sixteen feet high to serve two floors. A large automatic insulating shade is used at night to reduce heat losses. The Trombe walls are painted dark shades of blue, green, brown and red as well as black to give individual identity to all units. The Trombe walls also provide excellent noice protection from a major street immediately to the south of the building.

The direct gain systems consist of 60° south facing skylights with skylid insulating shutters and concrete floor and walls. Concrete block walls running north and south were used as partitions between the units providing low cost effective thermal storage and noise separation.

This project is aimed primarily at providing quality housing for low income families. Most of the residents of the building live there because it is affordable housing and the solar aspects of the building are a secondary consideration. After living with these solar systems for a year, they have become enthusiastic supporters of solar energy.

For further information, contact:
Solar Application & Research Ltd.
3683 West 4th Ave.
Vancouver, B.C. Canada, V6R 1P2

*Type:* *Direct Gain, Trombe Wall, Thermosyphon Water Heater*

*Design:* *Solar Applications & Research Ltd.*

*Construction:* *Lewis Construction*

*Location:* *Vancouver, British Columbia, Canada*

*Altitude:* *100 feet*

*Climate Data:* *5500 Degrees Days*

*Solar Radiation:* *1924 hours/year bright sun*

*January Vertical South facing direct and diffuse Isolation:* *440 btu/sq. ft./day*

*Winter Design Temperature:* *19°F*

*System Data Direct Gain:* *372 sq. ft. total south facing glazed area/ 349,997 lbs. concrete heat storage; R16 insulated shades on all windows; R7 insulated shutters on all skylights; R35 ceiling insulation; R20 wall insulation*

*System Data Trombe Walls:* *1015 sq. ft. total glazed area (145 sq. ft. per unit); 12" concrete walls; 168,000 lbs. of concrete (24,000 lbs/unit)*

*R16 insulating shade*

*System Data Thermosyphon Water Heater:* *632 sq. ft. of solar collectors (93 sq. ft. per unit for 7 units and 62 sq. ft. per one unit)*
*storage tank capacity 706 gallons*
*(100 gallons per unit for 7 units and 60 gallons per one unit)*
*Freeze protection - propylene glycol solution*

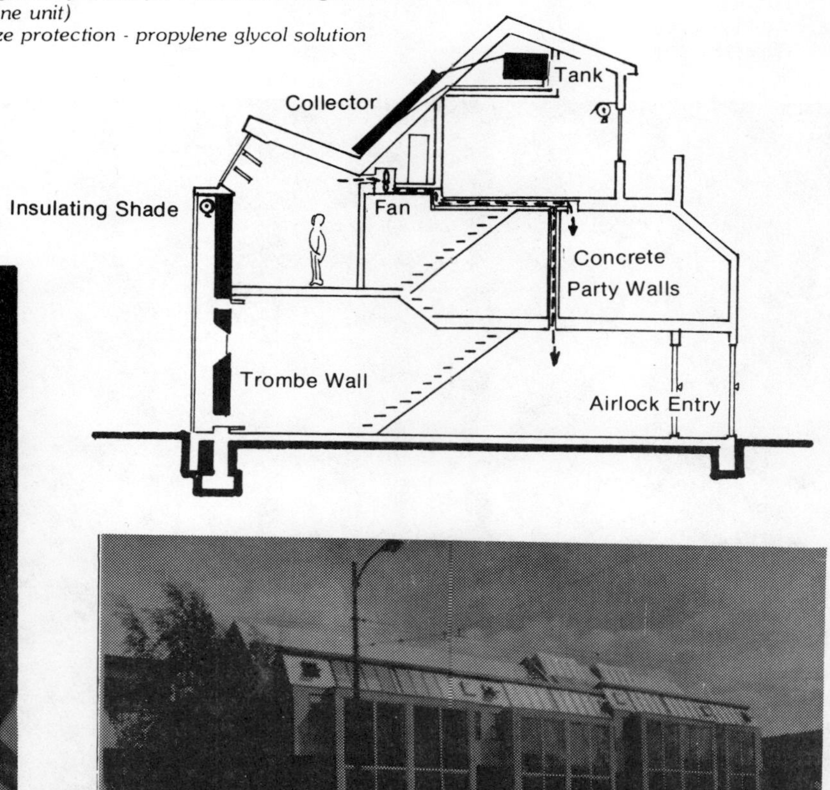

Insulating Shutters

# A Passive Solar Savings & Loan Building 1979

## Description:

This 4,000 square foot branch bank utilizes a 625 square foot trombe wall on the south, as well as a skylight lobby area, to provide approximately 60% of the annual heating requirements for the building. Thermal storage is provided by the masonry walls and slate and concrete floor. Movable shutters called skylids screen the skylight in summer, to prevent heat gain, and at night or on overcast days in winter, to prevent heat loss. Backup for the system is provided by an energy conserving fireplace (Hydrohearth), containing a grate which circulates water through the fireplace when the fire is lit. Heated water from the grate is then either distributed throughout the building immediately or transferred to a 450 gallon heat storage tank for use later. If added backup is needed a 100 gallon natural gas water heat goes on.

Cooling is provided by a conventional air conditioner. Cooling loads are expected to be much lower than average because of the mass and ventilation.

"Solar energy will have to be the source of the future", R. E. Knapp, President Friendship Federal Savings and Loan.

For further information:

Burt, Hill, Kosar, Rittleman Associates
400 Morgan Center
Butler, Pa. 16001

*Design: Burt, Hill, Kosar and Rittleman Associates*

*Construction: N.A.*

*Location: Butler, Pa.*

*Latitude: 40° N*

*Altitude: 1100 ft.*

*Climate: 6,000 HDD 800 CDD*

*Jan. Mean: 26° F July Mean: 70° F Yr. 49° F*

*Solar Radiation: N.A.*

*System Data: Trombe wall 625 sq. ft., direct gain 150 square feet; 15,000 lbs. water equiv. massive trombe, 60,000 lbs. in building*

*Summer design temperature: 89° F*

*Winter design temperature: 4° F*

*Solar Heating: 60% Cooling: 10-20%*

*Cost: About $3 square foot more than standard construction.*

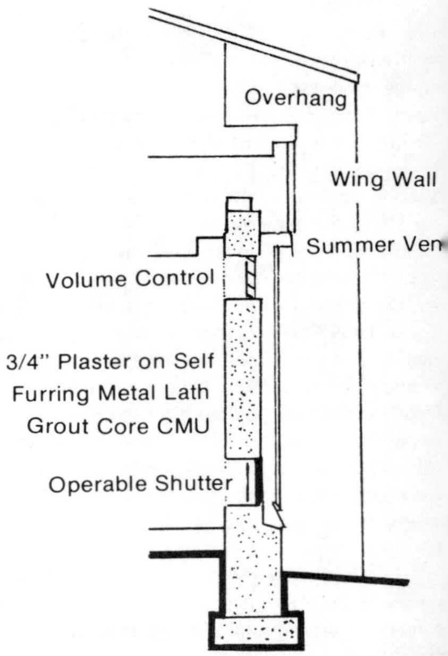

# Suntrees - Passive/Active Solar Apartments 1979

## Description:

Tandem Properties SUN TREES townhouse and apartment complex is on the leading edge of solar development in America. The 95 unit complex includes: solar heating and cooling and solar hot water and promises to keep tenants utility bills below $10 month. It includes 34 four bedroom apartments (1559 sq. ft.), 48 three-bedroom apartments (1377 sq. ft.) and 13 two-bedroom apartments (1070 sq. ft.).

Developer John Whitcombe feels that the tenants who live in a comfortable, economical, and reliable solar apartment will be more likely to ask for solar in their next home.

The system is an outgrowth of Tandem's earlier passive apartment design work, (see the First Passive Solar Catalog for a review of Chatauqua Apartments) and their ongoing research with solar heating/cooling systems. Rather than relying on expensive absorbtion cycle cooling, Tandem's solar expert Blaine Juschaw has used an IR transmissive glazing on the collectors to allow effective night sky cooling to occur.

The solar apartments were designed, from the ground up for solar. They begin with good passive solar energy conserving shell design, facing south with most windows on the south, well protected with overhangs and wing walls. The floor slab is used for much of the heat storage for both passive and active solar systems. Water storage tanks are used for additional storage and for hot water. The active system uses polybutylene pipe in the slab for transferring heat to the slab during the winter, and carrying heat away to the collectors for dissipation on summer nights.

The control system for pump operation and system operation was custom engineered for the system by Tandem Associates and Welco Electric of Sacramento. The microprocessor regulates pumps, flow direction, valves, and the backup water heaters (which will rarely be needed). The microprocessor includes energy use display as well as a thermostat.

The apartments are attractive and make a strong statement for solar in the multi-unit market. They also will prove very comfortable and pleasant to live in. The passive features (and radiant backup with active) all but eliminate noise and vibration, cut down on dust, and are quieter (thanks to double pane windows and better insulation). The radiant heating and cooling allow air temperatures to fluctuate in a wider range - with equally good comfort.

For further information:

Tandem Associates
709 Barcelona
Davis, CA 95616

*Design: Tandem Properties, solar design: Blaine Jushaw, Architect: Bill Rowe*

*Construction: Tandem Properties*

*Location: Davis, CA*

*Latitude: 38.50° N*

*Altitude: 30 feet*

*Climate: 2414 HDD 1063 CDD*

*Jan. Mean: 45° F July Mean: 75° F Yr.: 60° F*

*Solar Radiation: 900 BTU/sf. December day, 1500 BTU/sf. Av. day.*

*System Data: Area South Glass Unit: 60sf (varies), Collector: 5 panels per unit, Equivalent water mass per unit: 2,400 lbs.*

*Summer Design Day: 100° F*

*Winter Design Day: 30° F*

*% Solar Heating: 80-90% Cooling: 100% Hot Water: 70%*

*Cost: Solar added $2 sf. (no including tax credit).*

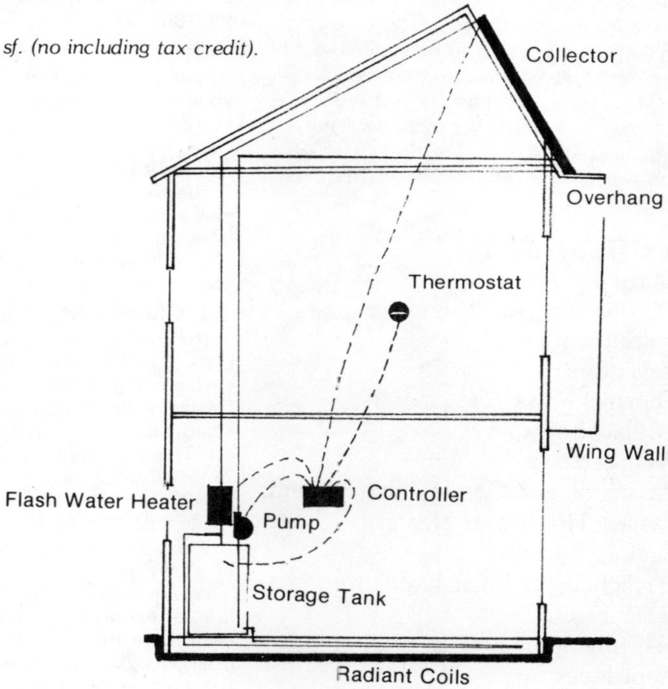

# Chapter 8: Passive Solar Products

This chapter includes listings of passive solar components divided in 17 different categories.

This list has been made as complete as time allowed; however, some products are probably still not listed. Please let us know if we missed you—or if you know of anyone making a product we should list. Forms are free for SASE.

A listing here in no way constitutes a recommendating or warranty and you should evaluate product claims and specifications carefully to make sure they will work in your system.

Many of these companies are small and struggling to get established. Therefore, when asking for information send at least a self-addressed, stamped envelope (8½" x 11") and/or a check for $1 to $5. And only ask for information if you are serious about it. The Whole Earth Catalog inadvertently put several companies out of business with requests for information that were never followed with orders. So when you see a product that you can/should use don't just think about it—order it. And be patient if it takes a while to get there.

If you have the means to do so, consider becoming a dealer or distributor for passive solar products. Many of these products need regional representatives—and will pay handsome returns to those who invest now.

Tell them you learned about them in the Passive Solar Catalog!

## Classifications

1. Glazing
2. Glazing Control
3. Solar Control
4. Reflectors
5. Thermal Mass
6. Insulations
7. Weatherizing
8. Passive Space Conditioning Systems
9. Passive Hot Water Heaters
10. Backup Systems
11. Daylighting & Solar Exposure
12. Solar Electric
13. Instruments
14. Appliances
15. The Good Life
16. Simulation
17. Catalogs and Sources

## MAJOR SUPPLIERS — order their catalogs

Peoples Solar Sourcebook $5
450 E. Tiffin
Bascom, Ohio 44809
419-937-2226

Energy House $3
P.O. Box 5288
Salem, Oregon 97304

Solar Hardware Supply Company $1.00 (refunded with first purchase)
2160 Clay Street
Denver, Colorado 80211

Energy Shack N/C
P.O. Box 7305
Flint, Michigan 48507

Kalwall Solar N/C
P.O. Box 237
Manchester, N.H. 03105

Zomeworks (send 9" × 12" SASE)
P.O. Box 712
Albuquerque, New Mexico 87103

You may also find many products you will need at your local hardware store or through Sears or Wards.

## 1. Glazing

The choice and placement of your glazing will play a dominant role in how your house works in both summer and winter. If glazing is poorly placed in your existing home you can rectify the error using solar control (3) and glazing control (2).

### 1.1 Glazing Materials

The best glazing material for most purposes is glass. It is the most durable, attractive, and is recyclable. For added strength use tempered glass. Design your structure for existing glass size as custom tempered costs twice as much. The following sizes are usually available locally: 28 × 54, 28 × 76, 34×76, 46×76, 22×76, 33×76, 34×74, 34×78, 34×79, 34×92, 45×76, 46×77, 46×78, 58×76, 46×79, 46 × 92, 58 × 92.

You can often reduce the cost by buying seconds or used glass.

Where impact strength, lightness, ease of handling, or economy are important you may wish to use one of the plastics. The specially formulated fiberglass glazings are easy to work with and will last 15-20 years or more. For maximum strength and clarity polycarbonate may be the answer. Acrylic is cheaper and not quite as strong. For interior double glazing, Tedlar may be used. And for ultimate low cost polyethylene may be used. The chicken wire polyethylene combination is stronger and good for cold frames, etc.

For storm windows see 7.3.

SolaRoll
Bio-Energy Systems, Inc.
Box 87
Ellenville, NY 12428
A. Mail order or Dealer
SolaRoll passive components called framing strips, are low-cost extrusions of EPDM for attaching glazing to passive and active solar installations. SolaRoll framing strips lock glazin securely in place while acting as weatherproof, watertight seals and flashing. Flexible, durable SolaRoll molds readily to any countour and simply glues on to framing lumber.

AFG Industries
P.O. Box 929
Kingsport, TN 37662
615/245-0211
Ralph W. Carter
N/C
Will quote on inquiry
A clear glazing material
Windows and openings where vision and sunlight are required
Request by mail

Insulated Double Pane Glass
Good Earth Glass & Sun
729 Heinz Ave. #10
Berkeley, CA
415/843-3019
Bob Daley
$1 - Dealer only
⅝" overall thickness tempered glass
Aluminum spacer filled with dessicant
Sealed with Polysulfide
New $3 sq. ft./Seconds $2 sq. ft.
Depends on construction methods, labor and application
Insulated Tempered Double Pane standard patio door units
Greenhouse, Solarium, Skylight, general Solar wall applications
Tempered Doublepane Insulated Units

| | | |
|---|---|---|
| New | 28" × 76 | $49 |
| | 34" × 76" | $56 |
| | 46" × 76" | $72 |
| Seconds | 28" × 76" | $30 |
| | 34" × 76" | $36 |
| | 46" × 76" | $49 |

Tempered Glass Factory Seconds
28" × 76" $12. 34" × 76" $15. 46" × 76" $20.
Glass & Supplies for Greenhouses, Solariums, Skylights, Solar & General Building

Glass Factory
Box 1372
Taos, New Mexico 87571
505/758-9344
Tempered insulating glass. Factory direct prices.

PPG - Tech. Service
One Gateway Center
Pittsburgh, PA 15222
Full range of glass, also exc. info on daylighting
in Canada PPG
834 Caledonia Rd.
Toronto, Ontario, Central M6B 3X9

Kalwall Corp./Solar Division
P.O. Box 237
Manchester, NH 03105
(603) 668-8186
sales department
N/C

Mail Order and stock distributors
.040" thick; widths of 24", 36", 48", 49½", 60", lengths up to 50' standard.
$ .50 - 1.00 per sq. ft.
$1.00 - 2.00 per sq. ft.
Sun-Lite® is a high performance glass fibre reinforced polymer material with 85-90% solar transmittance.
Solar collector cover for water walls, trombe walls, solar greenhouses
Solar windows with 2, 3, or 4 glazin g layers, Solar Storage Tubes, Solar Furnace, Solar Attic

Glasteel Tennessee, Inc.
1727 Buena Vista
Duarte, CA 91010
(213) 357-3321
Thomas C. Wright
N/C
Mail Order
Fiberglass translucent panels with extra clear resin offered with or without Tedlar in weights from 4 oz. per square foot to 12 oz. per square foot.
Face place for collectors

Reichold-RPD
P.O. 81110
Cleveland, OH 44181
Alsynite-structoglas fiberglass glazing material

LASCO
12221 A Carlsbad
Houston TX 77085
(713) 723-3250
Lascolite cyrstalite fiberglass greenhouse panels. 48" coverage width, length as required for commercial applic. Home owner 26" by 8', 10', 12' lengths.

Filon Div.
Vistron Corp.
12333 South Van Ness
Hawthorne, CA 90250
Fiberglass panels. Send out plans for a lean-to green house for 35¢.

Collier Wharehouse
1485 Bayshore Blvd.
S.F., CA 94124
(415) 467-9235
Kalwall, plastics and skylights

Park Energy Company
Star Route Box 9
Jackson, Wyoming 83001
(307) 733-4950
Richard C. Greig
N/C
Mail Order or Phone
Resin: UV inhibited polycarbonate
Treatment: modified acrylic UV screen applied to outer surface.
Normal Transmittance: 78%, transparent
Flammability: Self-extinguishing
Retail: \$3.43/ft² cover
\$ .50/ft² brackets
Dependent on window size and edging costs.
*Double glazed (1" space) polycarbonate extruded is available in 1 ft. widths of various lengths. Mounted from its underside using special brackets, it results in a monolithic glazing of nearly any desired length and width. Features: high impact strength, impervious to thermal shock, aesthetic appearance.
All glazing applications.
*Guarantee/warranty: 5 years.

Sheffield Plastics, Inc.
P.O. Box 397, Salisbury Road
Sheffield, MA 01257
(413) 229-8711
William R. Preston, Sls. Mgr.
N/C
Dealer Only
extruded polycarbonate sheet (Poly-Glaz)
extruded acrylic sheet (Acry-Pane) for glazing

Chemplact
150 Dey Road
Wayne, NJ 07470
(201) 696-4700
Plastic films for collector glazing. Including "Llumar" clear polyester, 5 and 7 mil. Also Teflon FEP in 1 mil, at 40¢ sf.

CY/RO Industries
West Main St.
Bound Brook, NJ 08805
(201) 356-2000
John N. Harvey/Market Development
N/C
Dealer Only
Exolite acrylic double skinned sheet and CY/RO glazing systems (also available in polycarbonate plastic).
\$2.50/sq. ft., basic
\$5.00/sq. ft., installed
Primary glazing, high solar transmittance (83%). Good insulation value (U= .57) Stiff and light weight. Low in-place cost for greenhouse, skylights. Trombe walls, atriums, etc.
Standard acrylic sheet.

Chemacryl Plastics
73 Richmond St. West
Toronto, Ontario M5H 2A2
(416) 869-0013
N/C
distributor or direct
Exolite double skinned acrylic and polycarbonate sheets, R-1.8. Integral ribs provide high strength with light weight. Maximum service temp: 60°F for acrylic, 240°F for polycarbonate. 47¼" wide, 8', 10', 12' and 16' long.

Thermo-Insul Wall
All American Mktg. Systems
26 Broadway
NY, NY 10004
Polycarbonate double glazed

American Acrylic
173 Marine St.
Farmingdale, NY
(516) 249-1129
Fiberglas Acrylic sheet, good only for low temperature uses.

Dupont-Tedlar PVF
1007 Market St.
Wilmington, Delaware 19898
An interior glazing for economical double glazing: Lifetime 5-10 years

SEARS
wire reinforced plastic film

## 1.2 Windows

South facing windows are usually the best passive solar collector. Don't overdo it, though. Provide summer shade with an overhang or see solar control, section 3 and add insulated interior shutters with drapes, see section 2.

Anderson Windowwalls
Box 12
**Bayport, MN 55003**
Woodwindows, double pane. Very high quality and commensurate price. Various types and sizes.

Marvin Windows
Warroad, MN 56763
(218) 386-1430
Wood windows, double and triple glazed, many styles.

Clearline Inc.
P.O. 1368
North Wales, PA 19454
215-699-4861
Eugene Storek

Clerestory window operating systems for remote operation. Can be manual or motorized. Surface or concealed. *"Looks good."* ed.

ROW
1365 Academy Ave.
Ferndale, Mich. 48220
Woodwindows, double pane.

Caradco Corp.
P.O. 920
Rantoul, IL 61866
Woodwindows, double and triple glazed.

Season All
Indiana, PA 15701
Window replacements. Various types.

General Aluminum Corp.
P.O. 8987
Boise, Idaho 83707
(208) 376-4531
N/C
Aluminum frame windows, many thermal glazed. Includes Sun Garden mini-greenhouse window.

## 1.3 Skylights

Skylights can be very effective in a solar system. However, they are more likely to leak and become overheated in summer. Install them carefully, shade them in summer, and provide interior insulated cover of some kind.

Kennedy Sky-Lites, Inc.
3647 All American Blvd.
Orlando, Florida 32810
(305) 293-3880
Jim Childers or David Ruddell
N/C
Dealer Only
International Congress of Building Officials — Compliance Numer 79-2
Underwriters Laboratories 972 Listed — BP 3377
Energy-efficient Double, Triple and Quadra-Therm™ Domed Skylights.
Also manufacture Caulking

Ventarama
75 Channel Drive
Port Washington, NY 11050
(516) 883-5000
N/C

Roto-Roof Window
Box 758
Essex, Conn. 06426
(203) 767-2015
A European skylight. Opens like a window. Can be cleaned easily. Insulated glass and supertight weatherstrip.

O'Keefes, Inc.
75 Williams Ave.
**San Francisco, CA 94124**
Skylights of many types and sizes.

Pfeiffer Skylight Corp.
14692 Wicks Blvd.
San Leandro, CA 94577
(415) 357-1400

Rollamatic Roofs
1400 Yosemite
San Francisco, CA 94124
(415) 822-5655
More for commercial applications. These are skylight roofs that open. Nicely done. Up to 200 ft. unsupported span.

Wasco Products
P.O. 351
Sanford, ME 04073

## 1.4 Solar Greenhouses

A solar greenhouse is one of the best passive solar retrofits and also good on a new house. Added mass and movable insulation are desirable.

Lord and Burnham
Irvington, NY 10533
(914) 591-8800
Excellent quality greenhouses. In business since 1856.

Peter Reimuller
980 17th
Santa Cruz, CA 95062
Redwood frame with fiberglass glazing. Various styles and shapes.

Pacific Coast Greenhouse Mfg. Co.
8360 Industrial Ave.
Cotati, CA 94928
(707) 795-2164
William R. Snow, President
N/C
Mail Order
factory direct, retail outlet
Prefabricated redwood and glass and/or fiberglass greenhouses, classic freestanding or solar in configuration. Solariums prefabricated and in full compliance with the guidelines and criteria for the solar energy tax credit (California).
$10-15/sq. ft.
greenhouses: c. $18/sq.ft. solariums: c. $20/sq.ft. (minus approx 30% when tax credit figured in.)
Also carry glazing materials, full line of various products including caulking. Sell silicone tubes $3.75 each

J.A. Nearing Co., Inc.
9390 Davis Ave.
Laurel, Maryland 20810
(301) 498-5700
Home Office
$1.50
Mail Order
Representatives in most states
Greenhouses

Vegetable Factory, Inc.
100 Court St.
Copiague, NY 11726
NY State: (516) 842-9300; (800) 221-2550
John Stevens, Sales Manager
$ .50
Mail Order
Solar-Panel greenhouses, lean-to and free-standing models. Patented, rigid double-wall construction, aluminum framework. Five-year warranty. Panels-only available for retro-fit, heat-gathering solar gain application
$4.00/sq. ft. and up
Two rigid layers of flat acrylic/fiberglass bonded to an aluminum I-Beam supporting grid core. Also available in GE Lexan. for Porch covers, Greenhouses, Solar Collectors, Swimming pool covers.

Santa Barbara Greenhouses
390 Dawson Drive
Camarillo, CA 93010
(805) 482-3765
Robert Solakian
N/C
Mail Order
Professional greenhouse coverings of fiberglass. All clear redwood with all joints bolted.
$4.50/sq. ft.
Free standing and lean-to greenhouses. 7 × 4 to 14 × 20 in size.
Home gardener or solar collectors
one 7 × 20 in use at Univ. of Cal. Davis, Cal.

Sturdi-Build Mfg. Co.
11304 S.W. Boones Fy. Rd.
Portland, Oregon 97219
(503) 244-4100
$1.00
Mail Order
Redwood and Glass Residential Greenhouses. Custom units available. Double-glazed. Freestanding and lean-to greenhouses.

McGregor Greenhouses
Box 36
Santa Cruz, CA 95063
(408) 476-5390
Corrugated filon on redwood frame.
Attached and freestanding designs.

Four Seasons Solar Products Corp.
672 Sunrise Highway
West Babylon, NY 11704
(516) 422-1300
Lexan or glass, double and triple wall solar greenhouses and other greenhouse products. Many styles $25+ sf.

Solar Greenhouse Inc.
2412 Elendil Lane
Davis, CA 95616
(916) 758-4534
Susan & Ted Van Syckel
None
Dealer only
Custom Design
$20.00 to $30.00/sq. ft.
Construction of custom designed solar greenhouses. Also sell Phase change storage rods.

Solar Technology Corporation
2160 Clay Street
Denver, Colorado 80211
(303) 455-3309
Carl Lehrburger
A) Free brochure B) $1.00 for catalog
Dealer Only
SOLTEC GreenRoom™
Dimensions: 12′ × 12′; length expandable by 4 sections. Construction: prefabricated modules. Framing: Redwood; roof and solar walls, Fir; sealed sidewalls. Trim: Redwood. Insulation: 3½″ polystyrene foam. Glazing: Acrylite SDP; tempered thermopane glass. A: Manufacturer B: Distributor C: R&D
A 1.4 Greenhouses (GLAZING)
A 8.2 Heating (PASSIVE SPACE CONDITIONING SYSTEMS)
GLAZING
B 1.0 Plans
B 1.1 Glazing materials
B 1.2 Windows
B 1.3 Skylights
GLAZING CONTROL
B 2.0 Plans
B 2.2 Insulated drapes
B 2.3 Insulated shutters
SOLAR CONTROL
B&C 3.2 Movable shades (see 2.3)
5. THERMAL MASS
B 5.1 Water tanks and containers
6. INSULATION
B 6.1 Building
B 6.2 Pipes
7. WEATHERIZING
B 7.1 Caulking
B 7.2 Weatherstripping
9. PASSIVE HOT WATER HEATERS
B 9.3 Systems
10. BACKUP SYSTEMS
B 10.1 Fans and Blowers
B 10.2 Ceiling Fans
B 10.3 Ventilation/Heat Exchangers
B 10.4 Vents
B&C 10.5 Space Heating
B 10.7 Water Heating
A Passive Solar Modular Building System for use as a solar greenhouse, room addition, supplemental solar heaters, hot tub enclosure, etc.
WINDOW QUILT™: Acrylite SDP; SOLTEC Hot Water Syste; a complete line of Energy Conservation Products.

Dalen
201 Shelake Drive
Knoxville, TN 37922
(615) 690-0050
Automatic vent for greenhouses. Operated by thermal expansion of working material.

Heat Motors, Inc.
635 W. Grandview Ave.
Sierra Madre, CA 91024
A series of heat motors. Operate with thermal expansion of working fluid. Different strength and temp. available.

John W. Stewart
Solarama
220 N. Richards St.
Colo. City, AZ 86021
(602) 872-2368
John W. Stewart
Mail Order
Opens and closes auto. with passive heat. To be used in greenhouses and windows
$29.95 each
Solar Vent
Greenhouse Vents
Also sells House Plans

Solar Resources, Inc.
P.O. Box 1848
Taos, NM
(505) 758-9344
Norman Brooks
Free
Mail Order
Dealer Only
Twin-Skin Solar Room. 6 mil film, translucent for privacy. Perimeter frame. Kiln dried, clear heart straight grained redwood (weather and rot resistant). Twin-Skin fastening system. A patented, two part extrusion, aluminum and rubber, locks the glazing in place.
Struts. Non-rusting, galvanized ¾" steel tubing supports Twin-Skin glazing.
$4.50/sq. ft. (24' unit)
Product is self installed
Solar Room Dimensions: A Solar Room is an optional 8' or 9' high at the house. It is always 7' deep and comes in lengths 9' to 39' and is available at any 3' interval.
The Solar Room has 35 ft² of collector area for each 3' section. Thus, a 30' Solar Room (10 three-foot sections) has approximately 350 ft² of collector area. Retrofit passive collector buildings with south-facing wall orientation.
None

Rodale Press, Inc.
Plans Department
33 E. Minor St.
Emmaus, PA 18049
(215) 967-5171
Ray Wolf
Mail Order
Retail book stores
Rodale's Solar Growing Frame
Includes 80 page book and 10 pages of blueprints
$14.95
App. $225.00
Book and blueprints to explain exactly how to build and garden a solar growing frame. Grow frames perform almost as well as solar greenhouses, at a fraction of the cost. One 4×8 ft. frame will supply more than 200 servings of salad every winter.
Also sell insulating window shade plans.

## 2. Glazing Control

Glazing control covers the movable insulation used to reduce heat loss in winter and to a lesser extent heat gain in summer. Most glazing control is done inside because it is cheaper and easier than dealing with exterior panels or drapes which are at the mercy of ice and wind. You can build your own or buy them ready made. In many areas you will have to install them yourself, or hire someone to do it for you.

### 2.1 Glazing Control Materials

You can make your own insulated drapes or shutters using these materials and the plans published in such magazines as *Organic Gardening*, Jan. 90, NMSEA, or *Alternative Sources of Energy*. The two books on glazing control soon to be available are:

William Shurchiff (1980) *Thermal Shutters and Shades*, Brick House Publishing, 3 Main Street, Andover, MA 01810
William Langdon (1980) *Insulated Shutters and Drapes*, Rodale Press, Organic Park, Emmaus PA 18049

I generally prefer shutters because you can get a better seal and higher R-value, but drapes and shades do have their uses. You may find the drapery pattern from Randa, 7685 Sycamore Drive, Citrus Heights, CA 95610 helpful. It helps you lay out drapes simply. You'll need to add weights to bottom, valance to top, insulation to drape, and side seals to make a thermal drape.

3M
3M Center 223-6SN
St. Paul, MN 55101
(612) 733-9139
C.H. Fridley
Thinsulate Thermal Insulation for manufacturers of Insulated Drapes
*Types* C 1.9 CLO/Cm, 70 CM 3/g, 35% Polyester, 65% Polyolefin, Compressible
*Type M* 1.9 CLO/Cm, 25 CM 3/g, 100% Polyolefin, resists-compression
$ .80 to $1.86 per yard
Organic microfiber non-woven batting for Insulated drapes Insulated apparel.

Dacron Holofil II
E.I. DuPont DeNemours & Co.
Wilmington, DE 19898

Polarguard
Fiber Industries, Inc.
Barclay Downs Drive
Charlotte, NC 28209

These materials are also available from backpacking supply houses, see 6.3, REI, EMS, others.

Duracote Corporation
350 North Diamond Street
Ravenna, Ohio 44266
(216) 296-3486
N/C
Direct Mfg. Sales
FOYLON 7001
Weight (5041 oz./sq. yd. ± 10%) 3.4
Shading Coefficient (ASHRAE) 0.29
Heat Reflectivity (ASHRAE) 85%
Heat Transfer Coefficient 0.4 to 0.8 BTU's/hr/ft/°F
$1.63 /lin. yd., 54"
FOYLON 7001 is a foil-fabric hybrid. Produced by a patented process, the foil is embedded into the fabric.
Drapery liners, window shades Reflective Mulch and Horticultural thermal blankets.

Space Blanket
Thermos Division
King Seeley
Norwich, Connecticut 06360

Rodale's Insulating Window Shade
Rodale's Plans Department
33 E Minor St.
Emmaus, PA 18049
(215) 967-5171
Ray Wolf
Mail Order
Retail book stores
Book and blueprints to build insulating window shades
$2.00/sq. ft.
An insulating roll-up window shade with an R-value of 5. The shade is designed to be built by the homeowner. Complete plans and descriptions for fitting the shade to individual window sizes.

### 2.2 Glazing Control Drapes and Shades

Several new manufacturers entered the field this year joining the older more established firms. These are generally very well built, effective, and worth buying. Finally a wider choice is available and I'd urge you to try those you like best.

Appropriate Technology Corporation
14 Green St.
Brattleboro, VT 15301
(802) 257-4501
John McPherson
$2.00
Mail Order
Dealer Only
$3.95/sq. ft.
WINDOW QUILT™
Installed cost varies w/dealers.
Insulated roll down shade. Five layers. Edgetracks and valance included. R 5. 15 with double pane
For windows, patio doors, etc.

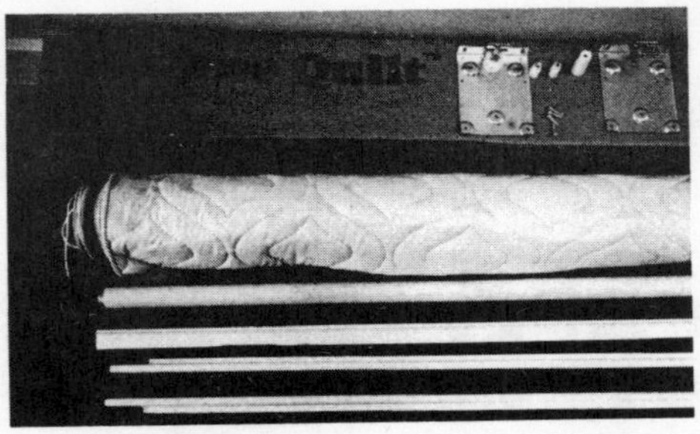

Simply assemble your
Window Quilt according to our directions.

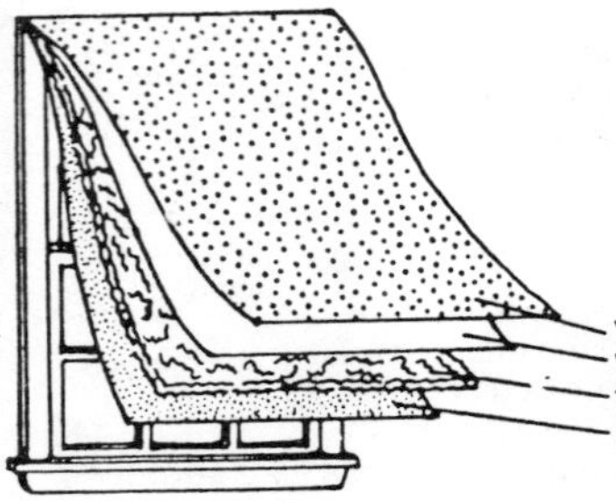


CREATIVE ENERGY PRODUCTS
1053 Williamson St.
Madison, WI 53703
(608) 256-7696
Randy Korda
Mail Order
WINDOW WARMER KITS, also Custom Work — Madison WI area
This product is a modified Roman shade with four layers and wood hinged side clamps to form a seal. It has an R Value of about 4
$1.95/sq. ft.
$5 to $6/sq. ft.

SUN QUILT
Box 374
Newport, NH 03773
(603) 863-2243
A quilted drape system designed to open and close automatically.

Insulating Shade Co.
P.O. 282
Branford, Conn. 06405
N/C
A multilayer insulating shade with hi R value, up to 15 with insulated window. Available for a variety of window types and applications. Very airtight. About $3.75 sf. with frame.

WINDOW BLANKET
Rt. 1, Box 83
Lensin City, IN 37771
A quilted curtain for patio doors. Needs a valence, bottom weights and attachment at center connection for good performance. About $2.00 sf.

SECI Solar
212 Welsh Pool Road
Lionville, PA 19353
An airtight shade built with slats (like and old roll type desk). Low R-Value and requires large roll. $6-sf.

Solar Power West
709 Spruce St.
Aspen, Colo. 81611
A plastic bamboo type shade with plastic film on outside. Moderately airtight. Fairly low R.

Wind-N-Sun Shield, Inc.
P.O. Box 1434
Melbourne, FL 32935
(305) 259-6988
Carol Boyle
N/C
Mail Order
$ .75/sq. ft.
Liner for behind regular curtains or rollershades. Metalized polyester.

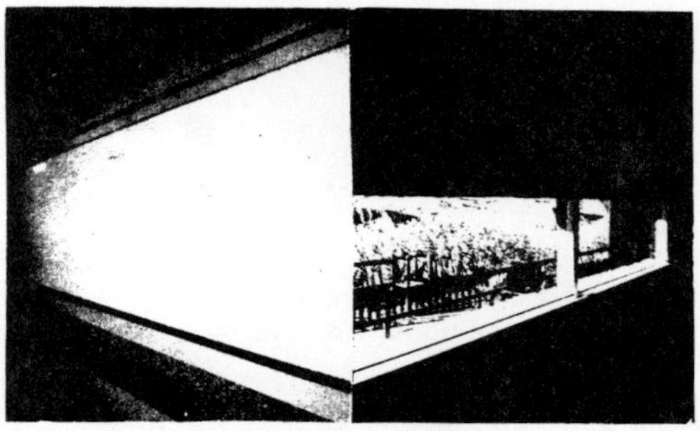

Thermal Technology Corp.
P.O. 130
Snow Mass, CO 81654
(303) 963-3185
Staff
N/C
Multilayer roll down shade made of metalized reflective fabric. Motor control. Works automatically. Insulating curtain wall™ is available in sizes for trombe wall and large solar windows — up to 20′ wide by 13′ high standard. Custom oversize. R-9. Good to 220°F.

## 2.3 Insulated Shutters

I like the durability, tight seals and high R value of shutters. Some very good ones entered the field this year. Erickson and Wallrich are worth a very close look.

**Folding Insulated Window Shutters**

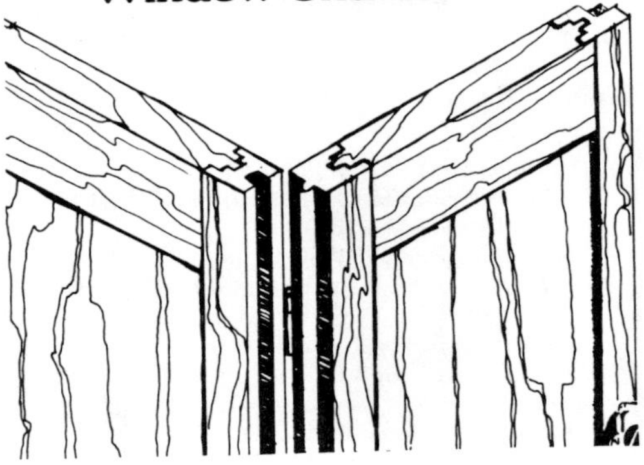

InsulShutter, Inc.
401 Main Street
Silt, Colorado 81652
(303) 876-2743
George Eriksen/Peggy Hart
N/C
Mail Order if no dealer in area.
$9.10-13.50/sq. ft.
Woodframe, wood skin insulated shutter. Nice detailing for airtight seal. New construction/retrofit.

Wallrich, Inc.
2601 E. Missouri
El Paso, Texas 79903
(915) 566-9426
Wally Sheid
N/C
Mail Order
THERMA-SHUTTER
1/8′ Birch plywood laminated to polystyrene core to give a calculated "R" value of 3.44 and a tested "R" value of 5.
Unfinished birch panels from 6″ to 16″ in width and 24″ to 82″ in height in 2″ increments. Special sizes upon request.
$7.00/sq. ft.
Self installed by Purchaser
To the inside of windows and patio doors.

Zomeworks Corp.
P.O. Box 712, Albuquerque, NM 87103
(505) 242-5354
Staff
N/C
Mail Order
**SKYLID® Insulating louvres.**
**Standard units consist of either two or three louvres mounted in a wood frame ready for installation. Standard sizes range from approx. 4′×4′ to approx. 6′×10′. Special sizes available on request. Installed R value 3-5 (does not include skylight glazing).**
Self-operating aluminum-faced insulating louvers which pivot in wood frames beneath skylights. They open automatically when the sun comes out, close when it disappears, and can be manually overridden and tied in position. For control of heat-loss/heat gain beneath skylights.
$12.50-$25.00/ft²
Two man hours installation time. 4 lbs/ft² shipping weight
Also make BEADWALL® self-insulating window panels, NIGHTWALL™ magnetic clips for using insulation board on windows; water heating systems; plans; sun tracking photovoltaic arrays.

SUNFLAKE
P.O. 676
Ignacio, Colorado 81137
(303) 563-4597
Integral shutter/window system. Shutter slides into wall pocket. R-14 when shut. Excellent for new construction but can also be retrofit.

SHUTTER SHIELD
C. D. Davidson & Assoc.
1014 Baldwin Ave.
Pontiac, MI
Sliding shutters for windows and patio doors. 3/4″ of insulation. Seal tightly and lock.

Homesworth Corp.
54 Cumberland St.
Brunswick, ME 04011
An insulated shutter kit. Cut to fit, make frame, and install. Good instructions. Reasonable R value, airtightness good. About $2.00/sq. ft.

Shutters, Inc.
110 E. 5th Street
Hastings, MN 55033
(612) 437-2566
Paul Swanstrom
N/C with self addressed stamped envelope
Mail Order
Seeking Dealers
Thermafold Shutters for patio doors
Folding, insulated shutters operated from inside by a crank on each side. Fully weatherstripped to stop drafts, frost, and condensation. Cut heat loss 500%, block out sun in summer, and reduce vandalism. Designed for retrofit. Six foot patio door is the only size shutter now available.

| | |
|---|---|
| Double glazed patio door | R 1.9 |
| Thermafold shutters | 7.1 |
| Air space | .9 |
| Total R value (closed) | 9.9 |

$259.00 for six foot patio doors,
$379.00 installed, about
$400 including shipping at 1000 mi.

## 3. Solar Control

Keeping the sun out in summer can be as important as keeping the heat in during the winter. A variety of materials and techniques can be used for solar control. Building your own is relatively easy but you can also have them installed easily in most areas.

### 3.1 Solar Control Materials

The most commonly used materials for solar control are overhangs and wingwalls. These can be built on site using wood. Or make awnings with canvas or using shade cloth. Alternatively, shade screen can be used. This is available in fiberglass, aluminum, and bronze. Fiberglass is tougher but doesn't last as long. Metal is more expensive, lasts longer, but is more easily damaged.

Reducing heat gain through the roof is also important. Many white paints will help but Plasticoat is specially designed for solar control.

Zip Dee
96 Crossen Ave.
Elk Grove Village, IL 60007
Awning material and manufactured awnings

Canvas Center
Box 11357
Tampa, FL 33680
Trailer awnings, stake, frames, and automatic. Many possible applications.

Sunkit Awnings
12937 Venice Blvd.
Mar Vista, CA 90066
(213) 391-3560
Synthetic fabrics for awnings as well as all hardware and manual for making them.

Sidewinder Inc.
16730 Gridley Road
Cerritos, CA 90701
(213) 865-0401
Self storing compact awning for campers and trailers. May be useful on that patio door that faces SE or?

Weblon, Inc.
4 Westchester Plaza
Elmstad, NY 10523
Vinyl canvas for awnings.

Golden Earth
361-N Oak Place
Brea, CA 92621
(714) 990-0681
$1.00 — Refundable with order
Mail Order postpaid
Shadecloth

| *Color* | *Shade Density* | *Price per linear foot* 4' width | 6' width | 7' width |
|---|---|---|---|---|
| Bone | 73% | | | $1.04 |
| White | 73% | .80 | | |
| | | | 1.05 | |
| Charcoal | 73% | | 1.20 | |
| Green | 75% | | 1.20 | |
| Bronze | 75% | | 1.20 | |

$ .20/sq. ft.
Environmental Protection Fabric can be cut to any shape and length with household shears. *Will not unravel.*
Durable fibreglass bonded weave to give many years of protection from weather extremes from: High winds, hail, heavy rainfall, blistering sun.

The Screen Manufacturers Association
410 North Michigan Ave.
Chicago, IL 60601
Booklet on shadescreen installation, new and retrofit. Free.

Reynolds Metals
P.O. 27003
Richmond, VA 23261
Aluminum shadescreen.

J.P. Stevens and Co.
P.O. 1138
Waterboro, SC 29488
(803) 538-8048
Fibreglass shadescreen available in 4 meshes and 5 colors.

KOOLSHADE CORPORATION
P.O. Box 210
Solana Beach, CA 92075
(714) 755-5126
R. Scott Townsend
N/C
Dealer only.
I. External Bronze Louvered Sun Screens
A. Standard (17 louvers per vertical inch)
B. Low Sun Angle (23 louvers per vertical inch)
Jobs individually priced by local KoolShade Distributor
KoolShade®-Machine woven, bronze, louvered, eternal solar screens
Custom made for exterior application to windows.

**ALL YOU NEED IS A SCREWDRIVER**

VIMCO SOLAR SHIELD
Virginia Iron & Metal Co.
P.O. Box 8229
Richmond, VA 23226
(804) 266-9638
Sidney Burnstein, Jr.
N/C
Mail Order
Service Dealers and Retail Stores
Specially woven fiberglass screening held in tension over sun-exposed openings with aluminum rails that attach to the head and sill of openings.
$1.00/sq. ft.
Simple do-it-yourself attachment requires only a screwdriver.
Vimco Solar Shield, the new energy saver that attaches to your windows to save money and keep you comfortable. Keeps your home cooler in summer. Saves energy dollars all year round. Easy to attach to your windows. The screen mesh stops about 75% of the sun's heat outside the window while allowing fresh air in, vision out, and providing daytime privacy.

**Fig. 1** Remove old retaining spline. If necessary, use tool to lift out spline from channel. Discard old screening. If old spline is soft and pliable, it can be reused.
**Fig. 2.** Place SunScreen® over frame, overlapping all sides 1/2 to 1". Starting at any corner, roll screening and spline into channel with concave roller. Continue in same direction until all four sides are completed. Screening should be taut.
**Fig. 3.** Trim excess screening from frame using a sharp utility tool. Cut at 90° angle, just above top of spline.
**Fig. 4.** The result, an energy saving SunScreen®.

Phifer Western
14408 E. Nelson
City of Industry, CA 91744
(213) 968-0587
David Ballard
N/C
Dealer only
Wholesaler
53 × 16 mesh vinyl coated Phiferglass Sun Screen
$ .30/sq. ft.
$2.50/sq. ft. of glass area installed
Phiferglass SunScreen Energy saving solar screening. Excellent exterior shading device. Blocks 70% of sun's heat and glare. Write or call for dealers in your area.
Vinyl coated Phiferglass SunScreen for shading windows and doors against unwanted sun's heat and glare.
Installed on windows with regular screen frame.

PLASTICOOL
Coating Laboratories, Inc.
3133 East Admiral Place
Tulsa, OK 74110
E. B. Kaplan
N/C
Through distributors or direct from factory.
$15 per gallon
$ .25 to $ .40 per sq. ft.
Solar Heat Control with Plasticool
High-heat solar energy. 90% high-heat energy rejected. 10% high-heat absorbed. Plasicool is a development of Coating Laboratories' 25 years of research and testing. By forming a reflective shield Plasticool bounces most of the sun's heat away.

## 3.2 Movable Shutters

For solar control, a shutter or shade should be on the outside of the window. Several systems are available. You might also consider some of those in 2.3.

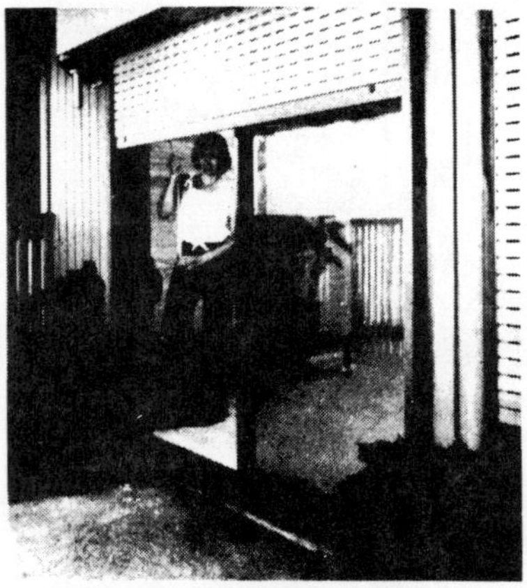

Serrande Exterior Shutters
Serrande, Inc.
P.O. Box 1034
West Sacramento, CA 95691
(916) 371-6960
Riccardo Addobati, President
N/C
Mail Order
$7.50/sq. ft.
$10.00/sq. ft.
$10.00/sq. ft.
Roll up shutters for exterior or between panes. Manual or electric

Pease
EVER-STRAIT DIVISION - Rolling Shutters
2001 Troy Ave.
NewCastle IN 47362
(317) 529-1700
Daniel J. Antaya, Mktg.
Exterior shutters. PVC slats with optional motor operation. R value to 2.47

Therma Roll
512 Orchard St.
Golden, CO 80401
Exterior plastic shutters. Provide sun control and wind protection.

## 3.3 Awnings

You can buy metal awnings from SEARS, WARDS, or many other outlets. We have not listed any other for that reason.

## 3.4 Other

Solar control can also be provided with reflective materials or reflective glass. This is best where winter heating is not a concern because the films and tints will reduce winter heat gain. In hotter areas you may want full shade in summer plus control films.

MADICO
64 Industrial Parkway
Woburn, Mass 01801
(617) 935-7805
Reflective, transparent, window insulating film. Best for solar control although reduces single pane heat loss 14%.

3M
1121 Grandview Dr.
S. San Francisco, CA 94080
(415) 761-7611
Bernie Bramante
N/C
Mail Order
$2.00-$3.00/sq. ft.
"SCOTCHTINT" Sun Control Film
A polyester film, approx, $\frac{1}{1000}$" thick, designed to reduce heat, glare and fading, also some insulating qualities.
Water activated/pressure sensitive
3M adhesive Weather Strip; 3M Insulating Window

Ford Glass Division
Box 43343
Detroit, MI 48243
(313) 568-2350
Sunglass reduces heat gain yet is lighter than bronze glass.

Window Shade Manuf. Assoc.
1211 W. 22nd
Oak Brook, Ill 60521

# 4. Reflectors

A reflector can be used to provide added gain to a passive system. Where glare might be a problem white paint is best. For better reflectivity use aluminum foil or one of the reflective materials available from:
Edmund Scientific
7780 Edscorp Bldg.
Barrington, NJ 08007

# 5. Thermal Mass

Adding thermal mass will increase your structure's ability to store heat in winter, coolth in summer. Add mass only after you've improved the shell or it may work against you. Water is usually best, although masonry and phase change materials have their uses.

## 5.0 Thermal Mass Plans

Alex Wilson (1979) *Thermal Storage Wall Design Manual*, NMSEA, PO Box 2004, Santa Fe, NM. $4.

D.A. Bainbridge (1979) *How to Build a Water Wall*, PSI $3.

## 5.1 Water Storage

Water is more than three times as effective at storing heat as concrete. Water can also be emptied out so mass tanks can be moved for remodeling. You can have tanks custom built or use existing materials. Plastic drums and steel cans have also been used successfully.

One Design, Inc.
Mt. Falls Route
Winchester, Virginia 22601
(703) 877-2172
Tim Maloney
Mail Order
Waterwall Modules
Part Name: IBB (Isopthalic Black Beauty). Overall length: 95 inches. Price: $200. Overall Width: $16\frac{1}{2}$ inches. Maximum Water Pressure: 1 p.s.i. Contents: 12.6 feet$^3$, 94 gal. 916 lbs. $1/sq. ft.
$15/sq. ft. (includes everything)
This sculptural thermal mass module is a snap to install: simply stack them in place. Built-in water connections facilitate filling and draining via your garden hose. The modules have been tested for strength and reliability. Residential homes, commercial and industrial buildings, greenhouses. Also make rigid, external movable insulation system; passive solar distillery.

Kalwall Corporation
Solar Components Division
Box 237
Manchester, NH 03103
A full range of fiberglass tubes for water walls. Not as strong as culvert. Can be used where light through wall is desired.

Tabline Company (custom only)
Box 1135
Berkeley, CA 94701
A modular thermal mass tank.

Reikes
1818 Leavenworth
Omaha, NE 68102
A wide range of bottles and drums, glass and plastic to 57 gallons.

ARMCO STEEL
METAL PRODUCTS DIV.
925 Village Center
Lafayette, CA 94549
(916) 444-0230
Contact: Jay Repors
Helical welded steel culverts. Galvanized. Used in water wall applications. Stock lengths 14', 20', and 24'.
Cost: 12" diam. $5.00 ft
15" diam. $6.50
18" diam. $7.75
24" diam. $9.50

Tanks-A-Lot
Rt 1 Box 158
Woodland, CA 95695
916-666-5695
Denny Long

Custom made steel tanks and culverts for new and retrofit in Sacramento area, $1.

D.S.A.
2655 Portaze Bay Suite #1
Davis, CA 95616

Gearing up for retrofit waterwall modules, ~ 90¢ gallon, SASE.

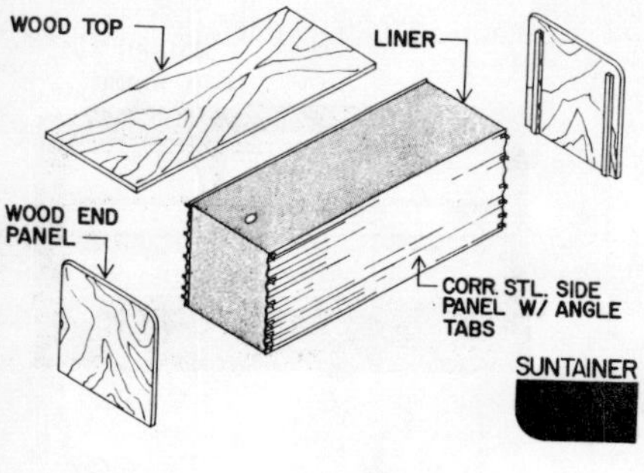


## 5.2 Masonry

Concrete blocks, stone, brick, tile, concrete are readily available and relatively easy to work with. We have included three items that may be of interest. For more information contact:

Intl. Masonry Institute
823 15th St. N.W. Suite 1001
Washington, D.C. 20005

National Concrete Masonry Association
P.O. 135
McLean, VA 22101

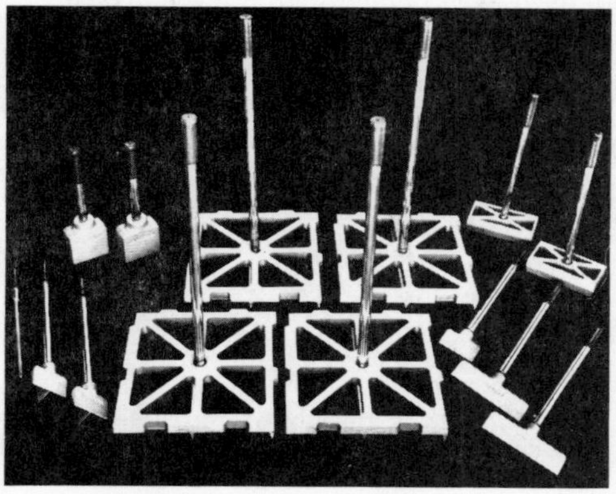

INDESCO (International Design Systems, LTD)
1929 Cross Highway
Fairfield, CT 06430
(203) 259-2900
Sales Department
Free upon inquiry
Direct Factory Sales Representatives and Distributors.
Tool Cost — $2,000-$5,000
$2.85 to $3.25/sq. ft.
Aluminum castings designed to allow finished concrete surface to appear as tile, stone, or brick, 14 types.
Entire flooring of passive solar homes. Also enhances driveways, sidewalks, patios, etc.
Support products such as colors, etc. (For enhancing the design appeal.)

Best's Blocks
34840 Alvarado-Niles Rd.
Union City, CA 94587
(415) 471-1111
Tomax ™ machine made concrete block walls up to 12' × 24'. Strength up to 5 times greater than hand laid-up walls.

Fuller O'Brien
South San Francisco, CA 94080
Epoxy fortified floor paint.

## 5.3 Phase Change Materials

Phase change materials offer even better energy storage than water. Most don't change phase below 80° so they are best used in sunspace or greenhouses.

Energy Materials, Inc.
2622 S. Zuni St.
Englewood, CO 80110
(303) 934-2444
Ken Hebb
N/C
Direct or dist.
Phase change storage rods. Stores 82 BTU/lb at 81°F phase change. Polyethylene tube. 3.5" × 6', 31 lbs of salts. Custom sizes also.

Architectural Research Corp.
13030 Wayne Rd.
Livonia, MI 48150
Phase change material in tiles.

Pace Corporation
555 Two Mile Road
Appleton, WI 54911
(414) 731-5281
The Thermol 81® Energy Storage Rod is an efficient and practical thermal storage concept. Heat is stored as latent heat of fusion in crystals. These crystals change phase at 81°F (27°C). While changing phase from a crystalline structure to a melted state, they absorb 82 BTU per pound of PCM. The PCM is congruent melting compound that will not stratify and is designed to retain efficiency for many years.

Heat Storage Rods
PSI-Thermol 81
1533 Fenpark Drive
Fenton, MO 63026
(314) 343-7666

Thermal Storage Rods
Texxor
P.O. Box 14337
Omaha, NE 68124

## 5.4 Other, for underground construction

B.F. Goodrich
Lexsuco Plant
F P DCP D/0415
500 South Main
Akron, OH 44318
(216) 379-2838
In Canada:
Lexsuco Canada Ltd.
Toronto
(416) 249-2227
Membrane roofing systems for such things as roofs, undeground houses, etc. Rolls 5'×120' or 5'×750'. Also related roof vents and sealants.

Burke Rubber
2250 South 10th
San Jose, CA
Hypalon sheet for roofing and water proofing.

Gates Engineering
100 S. West Street
Wilmington, Delaware, 19899
Sealants for underground construction.

Clearinghouse for Earth Covered Buildings
School of Architecture
University of Texas
Arlington, Tx 76019
817-273-2801

General information, research, information packets.

Earth Shelter Digest
479 Fort Rd
St Paul, Mn 55102

$15 yr US

Underground Space Center
11 Mines and Metallurgy Bldg.
221 Church St. SE
Minneapolis, Minn. 55455

Two Books: *Earth Sheltered Housing Design* (1978) U.S.C., Minneapolis, Minn. $10
This is the Book on underground so far. Excellent details and text, good references, and sources. A must if you'd like regular financing and code approval

*The $50 and Up Underground House Book* (1979) Mike Oehler, Mole Publishing Co. Rt. 1 Box 618, Bonners Ferry, Idaho 83805
A delightful and irreverent book on underground construction. Not all Code approved and better for it.
Nice stories, details, & photos.

# 6. Insulation

Insulation is critical for optimum passive solar heating and cooling. The insulation should be inert, non-toxic when burned, low in energy cost, recyclable, and have a high R-value. Alas, none offer all these for all purposes. For home insulatio cellulose is the first choice, followed by mineral wool and fiberglass, and finally the foams. For pipe insulation the foams are generally best. For people insulation wool and down are best for many uses but are being replaced with synthetic fibers.

## 6.1 Building Insulation

Cellulose is best for many building applications. It is cheap, stable, recycled, and effective. Make sure it is properly fire proofed. Then mineral wool or fiberglass. And for special purposes the various foams can be used. A 2"×6" wall with fiberglass batts and 1" of exterior foam sheathing is minimal, (R-25). Ceilings should be at least R-40.

Johns Manville
Service Center
1601 23rd St.
Denver, CO 80216
(303) 979-1000
Fiberglass insulation, batts-faced and unfaced R-11 to R-30.

Celotex
Box 22602
Tampa, Fla. 33622
(813) 871-4811
Advertising Dept., free lit.
Thermax foam sheathing, R 4.5 in ⅝".

BASF Wyandotte Corp.
Styropor Division
1609 Biddle Ave.
Wyandotte, MI 48192
Foam insulation.

Homasote
Box 7240
West Trenton, NJ 08628
Easy-ply roof decking and thermasote nail base roof insulation. Finished int. surface, fair R value.

Owens-Corning Fiberglass
Fiberglass Tower
Toledo, OH 43659
V.R. Meeks
Fiberglass insulation, all types up to R-38.

Dow-Styrofoam Insulation
1703 S. Saginaw Rd.
Midland, MI 48640
Styrofoam insulation for walls and ceilings.

Polyurethane Div.
Mobay Chemical Co.
Pittsburg, PA 15205
Foam insulation.

W.R. Grace
62 Whittemore Ave.
Cambridge, MA 02140
Zonolite panel foam and pour insulation. Pour material is vermiculite. Seals cracks and is inert.

National Cellulose Corp.
P.O. Box 45006
Houston, TX 77045
(713) 433-6701
Dan W. Kelly
N/C
Mail Order
Dealers Nationwide
Wallboard on one side nailed in place. All plumbing, electrical conduit, outlet boxes and duct work should be completed before application begins. Celbar to be sprayed according to manufacturer's recommendations at specified thickness. Celbar is a spray-on cellulose insulation for wall cavities and plumbing chases. Thermal insulation and sound transfer control are prime product benefits. Also available to control asbestos hazard.
Sprayed into place for thermal insulation and sound transfer control.
K-19 loose-fill insulation blown into attic areas for thermal insulation.
$ .14-$ .15 per sq. ft.
PSF $ .20-$ .22
PSF $ .20-$ .27 at 1000 mi.

THERMOCURVE
Box 205
Fishers, IN 46038
(317) 849-6350
Two inch thick, R-9+, polystyrene panels for use in cast concrete walls. Simply set in forms and pour over it. Special reinforcing available for added strength.

Korfil Block Insulation
Box 123
Chicopee, MA 01014
(413) 532-4400
Internal foam insulation for masonry block construction. R-4 in typical insulation.

Contech
ICS
3123 Weatherford
Independence, MO 64055
Pour insulation for concrete walls.

Scientific Applications
Box 615
Mt. Pleasant, Iowa 52641
(800) 553-5601
Carl Frank
N/C
Urea-Formaldehyde Foam insulation. Side wall cavity fill. UL listed.
Cold setting. Non-expansive. R-factor 4.9 per inch @ 25°.
25¢ PSF raw cost, 55¢ PSF installed.
Styrene cellulose.

Stanley Door Systems
1225 East Maple Rd.
Troy, MI 48084
(313) 528-1400
Barbara Levitz
N/C
Distributor only, no direct sales
Insulated doors, R-10.52. Pine frame with steel skin and foam cores.

Pease-Ever Straight
7110 Dixie Highway
Fairfield, OH 45023
Insulated doors, steel frame, triple bolt options.

Steel Craft
9017 Blue Ash Rd.
Cincinnati, OH 45242
Insulated steel doors.

Prest-on Clip Co.
Box 156
Libertyville, IL 60048
(312) 362-4615
Sheetrock clip allows sheetrock to be hung w/o special backing. This increases R-value and speed of construction.

## 6.2 Pipe Insulation

Pipe insulation is essential for energy conserving hot water system-solar or conventional.

RSA
P.O. Box CE
Jackson, CA 95642
(209) 293-7656
N/C
Dealer only
THE WRAPPITUP
Fiberglass with white vinyl thermal barrier. R-5/R-6/R-10 available in sies to fit water heater, 52-82-110 gallon capacity. UL listed materials FHC 25/50. The WRAPPITUP cuts home fuel costs. A do-it-yourself water heater insulation kit that cuts up to 20% of water heating costs. Fits all water heaters up to 52 gallons. Forms an effective barrier to keep stored water hot. Fast, easy to install. 15% federal income tax credit eligibility for users.
N/A
Suggested Retail Price: $19.95 to $24.95
Water Heater/Holding Tank Insulation Blanket
Also make pipe insulation.

Insultek Corp
82 Crestwood Rd.
Rockaway, NJ 17866
(201) 625-3828
N/C distributor or direct
Preinsulated copper pipe, isocyanurate urethane foam. Service range -50 + 375°F. Available with PVC jacket.
Also available with ceramic insulation R-70.9 in 3".

Teledyne Mono-Thane
1460 Industrial Parkway
Akron, OH 44310
Foam pipe insulation.

Wrap-on Co.
341 W. Superior St.
Chicago, IL 60610
Pipe insulation with vapor barrier, 3" or 6".

The Retro Co.
40 Oping Rd.
Pompton Plains, NJ 07444
Urethane pipe insulation.

Urethane Molding Inc.
RFD 3 Route 11
Laconia, NH 03246
(603) 524-7577
Staff
N/C
R-7 urethane foam pipe insulation. Supplied in 60" length. Outside and core diam. 3":1"; 4":1"; 5":2"; 6":3". Can be custom made to 36". Has PVC pipe exterior.

## 6.3 People Insulation

The place to begin in many houses is with the people. Some of my favorite suppliers are listed below.

Early Winters
110 Prefontaine Place South
Seattle, WA 98104

R.E.I.
Box 88125
Seattle, WA 98188

Eastern Mountain Sports
One Vose Farm Road
Peterborough, NY 03458

Damart
1811 Woodbury Ave.
Portsmouth, NH 03805

Thermoply Insulation
Box 10
3000 W. Beecher Road
Adrian, MI 49221

Olam-Pile garments
Box 460
Spring Hope, NC 27882

North Face
1234 Fifth St.
Berkeley, CA 94710

Frostline Kits
Frostline Circle
Denver, CO 80241

Sierra Design
247 Fourth St.
Oakland, CA 94607

L.L. Bean
Freeport, ME 04033

Marmot Mountain Works
331 South 13th St.
Grand Junction, CO 81501

Ski Hut
1615 University
Berkeley, CA 94703

Banana Equipment
Box 1076
Longmount, CO 80501

Borglite Pile
Borg Chemical
900 Commerce
Oakbrook, IL 60521

Caribou Mountaineering
Box 3696
Chico, CA 95927

Oddo Underwear
Sportco
69 North Main
Farmington, ME 04934

Black Ice
120 Woodland Ave.
Reno, NV 89523

Holubar
Box 7
Boulder, CO 80306

Jansport
Paine Field Industrial Park
Everett, WA 98204

# 7. Weatherizing

Weatherizing goes hand in hand with insulation. Each by itself is but a shadow of the two together. Caulking, weatherstripping, and storm windows are among the most cost effective investments you can make. Don't delay.

## 7.1 Caulk

The variety of caulk now available is almost staggering. You can get butyl, oil, acrylic, silicone, latex, and other base caulks. Make sure the one you choose is suitable for your application. Silicone seal is best for glass-glass. Butyl is generally best for wet areas or underground. Take a good look at the product literature. The foam caulks are very good for wider cracks around window and door frames.

Polycell One
Coplanar Corp
1631 San Pablo
Oakland, CA 94612
Foam caulk, in handy aerosol cans and larger sizes.

FOMOFILL
Fomo Products, Inc.
1090 N. Jacoby Rd., PO Box 4261
(216) 666-0021
Wilhelm Steding
N/C
Mail Order
In Canada:
Zeus Manufacturing
4 Buckingham St.
Toronto, Ontario M8Y 2W1
(416) 259-5474
One Component, Polyurethane Foam in Aerosol Dispenser.
$5 can retail (18 oz)
A non-toxic, flame retardant, non-shrinking foam for insulation. Sound-dampening-bonding.
Filling of gaps and cracks to stop air infiltration.

Dow Corning® 790 Building Sealant
Dow Corning® Corporation
Box 1767, Midland, MI 48640
Sweet's Buy-Line or (517) 496-4000
Solar Energy
None
Local Building Sealant Distributor
TT—S-001543
TT-S-00230
Canadian Spec. 19GP9A
Approx. $4.00 per 10.7 Fl. Oz.
One-part low-modulus silicon rubber building and glazing sealant. Outstanding performance in building joints that experience extreme movement up to ±100% of original joint width.
By hand or automatic dispensing equipment.
Glazing sealants for plastic glazing Dow Corning® 795.
Dow Corning® Silicone/Urethane Roofing System for retrofitting.

Silicone Products Division, G.E.
RTV Products Department
Waterford, NY 12188
(518) 237-3330
S.P. Doyle
Dist. Only
Silicone seal — the best for glass to glass. Also paintable silicone caulk and other silicone base products.

Macco Adhesives, SCM
Wickliffe, OH 44092
(800) 321-3647; Ohio (216) 943-6161
Butyl sealant and acrylic latex caulk

DETCO MARINE
3452 East Foothill Boulevard
Pasadena, CA 91107
(213) 681-2613
Detco/Grove polysulfide, polyurethane and epoxy sealants, adhesives and coatings, as well as specialized application tools.

See also SEARS, WARDS, local hardware stores.

RED DEVIL
Union, NJ 07083
Several varieties of caulk and a nice booklet on their uses.

## 7.2 Weatherstrip

The variety of weatherstrip is also incredible. The spring metal strip and interlock are most desirable but good results can be achieved with inexpensive foam.

Macklanburg Duncan
Box 25188
Oklahoma City, OK 73125
Weatherstrip, caulks, and thresholds.

Thermwell Products
2049 East 27th St.
Los Angeles, CA 90058
A wide range of weatherstrip, stormwindows, etc.

Pemko
Box 8216
Emeryville, CA 94662
A full range of weatherstripping, thresholds, etc. Include weatherstrip for louvre windows.

Certain Seal
Box 889
Waco, TX 76703
Various weatherstrip materials.

Schlegel Corp.
Box 23113
Rochester, NY 14692
(716) 244-1000
Fred Williams
In Canada:
P.O. Box 218
Oakville, Ontario L6J 5A2
(416) 845-6657
N/C
Excellent weatherstrip with pile, and a fin. Long life. Standard sizes .270" and .187" wide with .130" to .330" thick pile. Tests show good performance after 10 years exposure. Low water leakage. Also door seals and storm windows.

Gaska Tape, Inc.
1810 W. Lusher
Elkhart, IN 46514
(219) 294-5431
Milan Rakich
N/C
Mail Order, Dealer Only
Enclosed Brochure
Several types of foam tape for weatherproofing and sealing. Available with foil or vinyl skin also. Also pipe insulation.

Manco Tape, Inc.
P.O. Box 685
Cleveland, OH 44107
(216) 228-3100
Tom Corbo
N/C
Direct - retail outlets
(Catalog sheet enclosed)
Clear-Tite self-adhesive polyethylene tape. (Rope Caulk Substitute)
product # CC-1 1"×36 yds.
CC-2 2"×36 yds.
Boxed

Keystone Industries Inc.
P.O. Box 748
Elkader, IA
(319) 245-1471
Raymond L. Syvenson
N/C
Mail Order
Automatic retractible door seal. As door is shut seal locks into position.

## 7.3 Storm Windows

Quaker City Mfg. Co.
705 Chester Pike
Sharon Hill, PA 19079
(215) 727-5144
Barry M. Jones
N/C
Both Dealer and Mail Order
The WINDOW FIXER will save you money!
The WINDOW FIXER, Quaker Window Channels
Do-it-yourself product replaces window balances

Seasonall
Indiana, PA 15701
Stormwindows and doors, also repl. thermal pane windows. Free.

KSH Inc.
10091 Manchester Rd.
St. Louis, MO 63122
Self adhesive vinyl frame storm windows. "The Barrier" snaps apart for cleaning, glazing or replacing it. Inside only.

Elgar Products
P.O. 22348
Beachwood, OH 44122
Vinyl frame with turn buttons. Inside or outside.

Perkasie Ind.
50 East Spruce St.
Perkosie, PA 18944
Acrylic glazing. Stock-line windows. Come cut to fit.

Alumatrim
918 Stanley Ave.
Brooklyn, NY 11208
Precut alum. frame. Glazing held by removable plastic spline. Turn buttons secure frame. For outside or inside.

Flexoglass
Warp Bros.
Chicago IL 60651
Flexglass for storm windows. Cut and tack up.

Mortell Co.
550 Hobbie St.
Kankakee IL 60901
Storm windows kit.

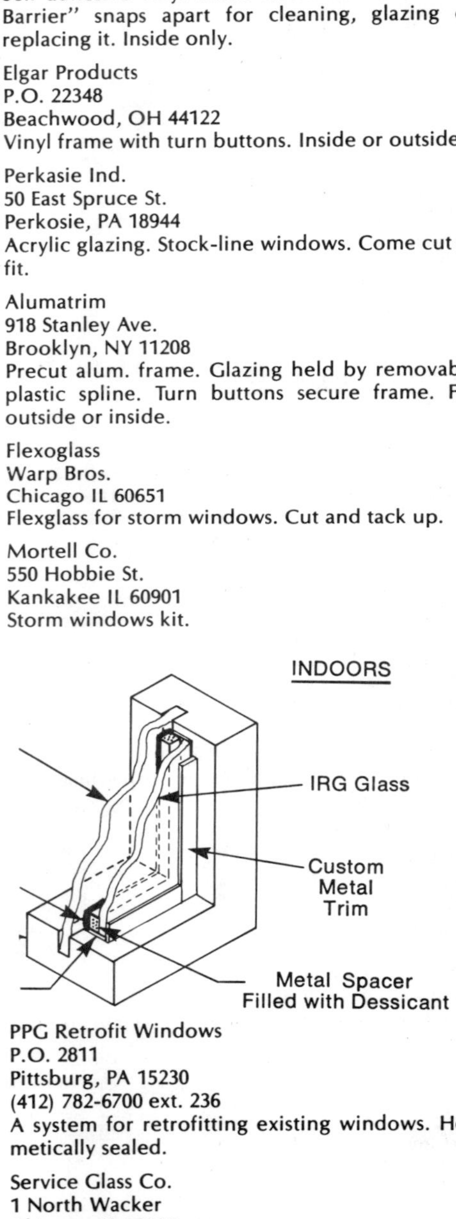


PPG Retrofit Windows
P.O. 2811
Pittsburg, PA 15230
(412) 782-6700 ext. 236
A system for retrofitting existing windows. Hermetically sealed.

Service Glass Co.
1 North Wacker
Chicago, IL 60606
(312) 726-5050
Add a pane system for double paneing single pane windows in place. Vulcanizing system. No need to remove windows. Particularly attractive for commercial buildings.

## 7.4 Vapor Barrier

A vapor barrier can help reduce infiltration heat loss. Use 6 mil. polyethylene or one of these products. Paint is best for retrofit.

Parsec Inc.
P.O. 38534
Dallas, TX 75238
(214) 324-2741
Aluminized plastic-pressure sensitive adhesiv e. Sheet or tape. For vapor barrier and infiltration stop.

TuTuff Vapor Barrier
Sto Cote Products
Drawer 310
Richmond IL 60071
(800) 435-2621
Plastic vapor barrier

US Gypsum
101 South Walker Dr.
Chicago IL 60606
Foil backed gypsum board

Glidden Paints
SCM
Cleveland, OH 44115
Insulaid vapor barrier paint.

Simplex Products Group
P.O. Box 10
Adrian, MI 49221
(517) 263-8881
Thermoply sheathing: $\frac{1}{8}$" multiple ply sheathing with foil face. Good for infiltration control. Stronger than builders paper.

## 7.5 Other

Two more products you might find of interest. The vent would be helpful in colder areas — or make your own vent cover.

Witten Automatic Vent Co., Inc.
P.O. Box 2244
Gastonia, NC 28052
(704) 866-8796
Howard W. Green
N/C
Mail Order
Size: $7\frac{3}{4}$"×$15\frac{3}{4}$" 65 sq. in. of net free area. One THERMATIC-VENT is required for each 65 sq. ft. or crawl space without ground cover. THERMATIC-VENT will open and close automatically. (Fully closed at approx. 40°F, fully open at approx. 70°F.) Designed to help elimiate moisture and save energy.
$19.95

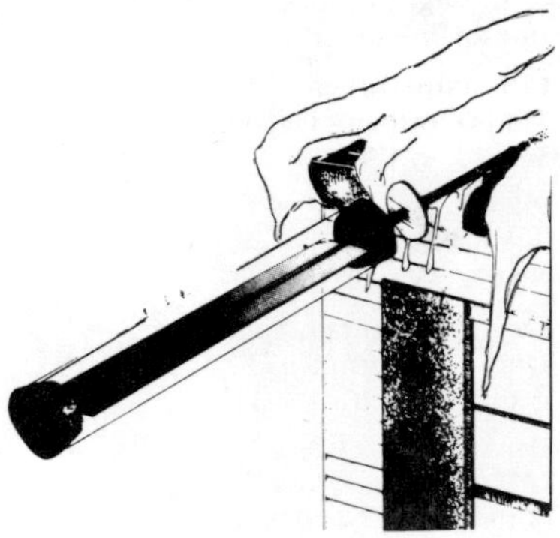

Tindev, Inc.
Laurel Drive
Lincoln, MA 01773
(617) 259-9641
Frederick M. Tingley
N/C
Mail Order, Dealer
SOROD Roof Ice Dam Lancer
$1\frac{3}{8}$"×4'   6 oz.
$40.00 per package of four, list.
An enclosed solar collecting fin attached to a four foot heat pipe. When positioned at the eave of a house, with the collector overhanging, it conducts the sun's energy into the frozen snow, melting a tunnel for draining off the dammed melt water.

# 8. PASSIVE SYSTEMS

The development of an appropriate passive design for your site is challenging yet not beyond the skills of almost everyone. A class or two on drafting and passive design at your local Junior College will help. See also Ken Kern's offer.

If you feel this is beyond your skills or you don't have time you might take a close look at the plans available now. These are in many cases very nicely done, aesthetic, and effective. Make sure it will work on your site.

Refer to the consultants and designers section if neither fits your needs. Expect to pay from 1-7% of construction cost for a good set of plans.

## 8.1 Passive Plans

These are the plans we managed to track down. Solstice and TVA have the best selections. P.S.I. and Positive Technology offer performance estimates in ten cities.

Tennessee Valley Authority — Plans
200 Haney Building
Chatanooga, TN 37401
(615) 755-2122
Contact: Jack Dodd
No Cost
Mail Order
Blueprints for passive solar homes (11 designs) $7 set.
$30.50 per sq. ft., construction cost.

Natural Architecture
8424 W. Agatite
Chicago, IL 60656
Underground house plans. 5 sets, $160.

Home Building Plan Service
2235 N.E. Sandy Blvd.
Portland, OR 97232
An energy efficient house plan. Would work well if oriented properly. $140.

Positive Technologies
P.O. Box 2356
Olympic Valley, CA 95730
N/C
Envelope house with greenhouse, 1500 sq. ft., 3 bd, 2 ba. Working drawings $285.

Dovetail Press
P.O. Box 1496
Boulder, CO 80306
303-449-2681
Stewart Elliott
Complete instruction manual for Solar Starr timber framed passive houses. Soft bound with 6 sets of blueprints and detailed manual. $1500.
Introduction and sampler, The Dovetail Primer. $10.00

Miller Solsearch
P.O. Box 1869
Charlottetown, PEI, C1A 5N7
902-894-3947
$2.00 info.
A low cost superinsulated home with integral greenhouse. Three sets of plans. $75.00.

Solstice Publications
P.O. Box 2043
Evergreen, CO 80439
Adobe, mountain, and urban passive solar house plans. $84-$145. Write for details.

Mother Earth News — #84010 Underground
P.O. Box 70
Hendersonville, NC 28739
Underground house plans. Can be good passive solar if oriented properly. $135.

Professional Builder
5 South Wabash St.
Chicago, IL 60603
PB-300 Plan set, 3 bd 2 ba. Five others being developed. $110.

Passive Solar Institute
P.O. Box 722
Davis, CA 95616
3bd, 2ba water wall house for the Sunbelt, $25 for three sets. Other plans being developed.

***The Passive Solar Institute***

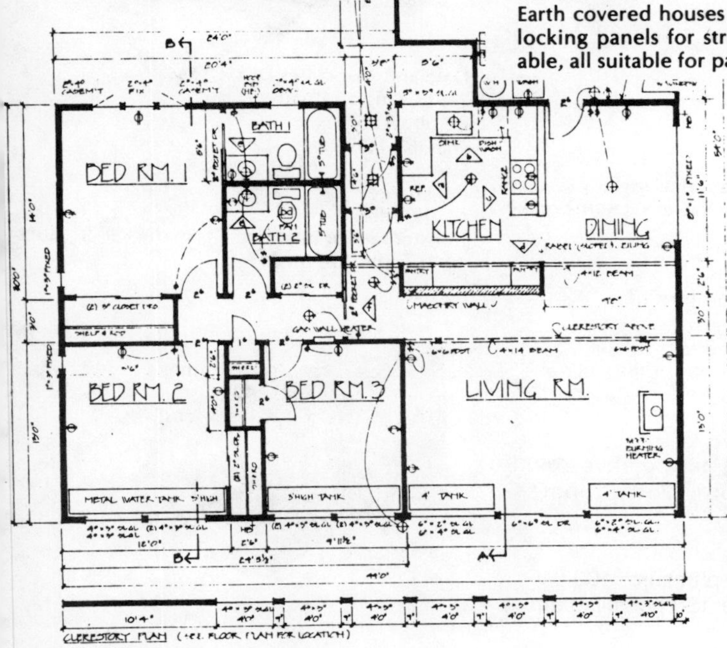


FLOOR PLAN

## 8.2 Passive Systems

Skytherm
2424 Wilshire Blvd.
Los Angeles, CA 90057
Roof pond houses are covered by Harold Hay's SKYTHERM patents. Plans should be available soon. The house for south and southeast U.S. SASE for info.

Passive Solar Variant Homes
Savell
575 Birch Court, Suite A
Colton, CA 92324
Exterior insulated tiltup concrete walls. Can work very well if oriented properly.

Green Mountain Homes, Inc.
Royalton, VT 05068
802-763-8384
Dolores Zick
$3.50
Mail Order or through regional dealers
Kit cost: $11/sq. ft., Site cost: $24/sq. ft.
Total cost of passive solar home? $35/sq. ft., plus standard trucking charges.
Direct gain hybrid passive solar buildings utilizing first floor patented Solar Slab Thermal Mass Heat Exchanger.
Single family and multifamily housing, light industrial.

The Energy Factory
1550 North Clark
Fresno, CA 93703
209-441-1833
Thomas W. Kristy
From District Offices
Passive Hybrid Space Heating and Cooling System Utilizing Built in Connector, Rock Storage and Computer Control. Collector is built by conventional subcontractors into a portion of attic space. Rock storage formed below floor. Most of System constructed with conventional building materials.
$2.50/sq. ft. of living area
$3.25/sq. ft. of living area
Conventional homes, light commercial.

American Solartron Corp.
R.R. #5, Box 170
Centralia, IL 62801
618-533-1842
Earth covered houses built using fiberglass interlocking panels for structure. Several plans available, all suitable for passive if oriented correctly.

## 8.3 Design Assistance

It's almost always a good idea to have someone check your plans. Most consultants are willing to do this including P.S.I. Most owner/builder schools will also help.

Ken Kern
P.O. Box 817
North Fork, CA 93643
Drafting design supervision in the offices of K.K. Owner does own plans. Send SASE for details.

Owner Builder Schools Include:

Shelter Institute
72 Front St.
Bath, ME 04530

Cornerstones
54 Cumberland St.
Brunswick, ME 04011

The Owner Builder Center
269 Cardinal Road
Mill Valley, CA 94941
415-383-6924

The Habitat Center
162 Christen Dr.
Pacheco, CA 94553

Sunworks
P.O. 455
Boulder, CO 80306

Heartwood
Owner-Builder School
Johnson Road
Washington, MA 01235

# 9. PASSIVE SOLAR WATER HEATER

The definition of passive solar water heaters has been expanded to include thermosiphoning collections as well as integral passive solar water heaters. Only the integral type is listed here.

## 9.1 Passive Solar Water Heater Plans

One of the beauties of most passive solar water heaters is their simplicity. Anyone can build one with a tolerable set of plans. The First Passive Solar Catalog, $7.50 ppd. from P.S.I., P.O. 722, Davis, CA 95616 has a good chapter on designing and building a PSWH. P.S.I. is currently working on a text book in PSWH, SASE r/e availability.

Passive Solar Institute
P.O. Box 722
Davis, CA 95616
Plans for a three tank PSWH. Includes instructing, diagrams and photos. SASE for information. Will cost around $500 to build in 1980. 90 gallons of storage, $10. ppd.

John C. Golder
P.O. Box 854
Santa Cruz, CA 95061
408-423-3064
John C. Golder
Information sheet — S.A.S.E., Technical report — S.A.S.E. + 50¢.
A. Mail Order
The Capsule Collector is a low cost, manually operated passive solar water preheater, designed as a simple D.I.Y. project. A variation of the "breadbox" design utilizing readily scavenged materials, it can be built for under $40 and will heat 30 gallons water up to 50°F gain on a clear sunny day. $45 to $90 per 30 or 40 gallon unit. $200 per unit. Capsule Collector plans including 1-18" × 24" plan sheet plus 6 pp. of instructions, materials list with sources, tool list, step-by-step procedures. Send $5 check or MO to John C. Golder, P.O. Box 854, Santa Cruz, CA 95061. Domestic water preheating. Greenhouse plans and Solar Oven Kits by Fall 1980.

Passive Hot Water Heaters
Two Tank Passive Solar Water Heater Plans
Integral Design
3825 Sebastopol Road
Santa Rosa, CA 95401
707-528-0616
John Burton
Free
Mail Order
Horizontal and Vertical Two Tank, 80 gallon, systems. Collector area 40 square feet. Uses standard gas or electric 40 gallon water heater tanks. $10-15 all materials. Do It Yourself: (total $400-600). Near Factory $1,200. 1000 miles away: $1,300. Complete installation by Integral Design. The design uses standard construction and building materials. The plans include complete and to the point drawings, plumbing schematics, materials list and installation notes. Water heating 100-150°F. Domestic, Commercial and Industrial. Weekend workshops to install a complete system offered. Workshop sponsor gets a completed installation for about $800.

Zomeworks
P.O. Box 712
Albuquerque, NM, 87103
Plans for a simple breadbox type PSWH, $5.00 ppd.

Ted Lucas
10371 Stone River Court
Fountain Valley, CA 92708
A one tank PSWH with reflector behind. Includes sources of material, $5. ppd.

Horace McCracken
329 W. Carlos
Alturas, CA 96101
Plans for a horizontal system with 1 or more tanks. Nicely prepared, $6. ppd.

### 9.2 Passive Solar Water Heaters

Two systems are now on the market. Standard Oil of California is to enter the market in late 1980-early 1981.

SAV SOLAR
P.O. Box 91277
Los Angeles, CA 90009
213-823-5432
Forty gallon system with reflector.

SUN WIZARD
1424 West 259th St.
Harbor City, CA 90710
213-539-8590
A one tank 40 gallon module

### 9.3 Passive Solar Water Heater Components

Most plumbing supply stores will have all you will need except for the tank. American Appliance has been the best source for new tanks.

American Appliance
2341 Michigan Ave.
Santa Monica, CA

Rheem (Quantity orders only — car load lots best)
7600 S. Kedzie
Chicago, IL 60652

Refrigeration Research
525 North Fifth St.
Brighton, MI 48116

Solar Future
P.O. 19177
Sacramento, CA 95819

A used tank may be your best bet. Look for an electric heater with ruined elements. Or find fire damaged or freight damaged.

Another component you may need is a vent/vacuum breaker.

Richdel Inc.
Solar Division
P.O. Drawer A
Carson City, NV 89701
A full range of solar components including valves, controls, and the R56-Vent Valve-Vacuum breaker.

2½" Dia. x 5⅜" long
1" hex wrench flats
½" male taper pipe thread

**Valve Materials:**

Thermoplastic body, glass-filled polysulfone

**Temperature:**

Max operating temperature 180°F under pressure, 250°F unpressurized

**Pressures:**

Max operating pressure, 100 psi
Proof Pressure 150 psi minimum

## 10. BACKUP SYSTEMS

Backup systems should be chosen with the same care that goes into the solar design. They can make or break a house in comfort, convenience, energy use, and cost.

### 10.1 Fans and Blowers

Although most passive systems will work without fans it is occasionally desirable to increase air flow (especially in retrofits). Check the CFM per watt on the fan. Very small fans will often prove sufficient.

Plexus 2 Symmetrical Controller
Weather Energy Systems, Inc.
P.O. Box 968
Pocasset, MA 02559
617-563-9337
James O. Brown, Jr., Sales Manager
Free.
Mail order or nearest dealer.
115 VAC; *Max. fan load* — 150W standard, 300W optional; *Auto functions* — fan start, continuous speed regula. and reverse 75°F to 95°F, activation of existing vent systems; Manual controls — on/off, SUMMER/WINTER, auto reverse temp. settings (45, 50, 55°F), auto rev. on/off. $349.
Automatic, variable speed rever. controller to be used with a fan to regulate air flow between a solar greenhouse or solarium and internal living areas. Greenhouses, Solaria, and Wood Stove settings. Plexus 1 — air motion control sys. for entire passive house. Plexus 3 — simplified, single direction air motion controller.

Dale and Associates Dist. Div.
1401 Cranston Road
Beloit, WI 53511
608-362-1495
Sue
No charge.
Mail Order
Booster fan for heat circ. fits in duct 5"-9", duct – 12-35 watts. $20 for 5"

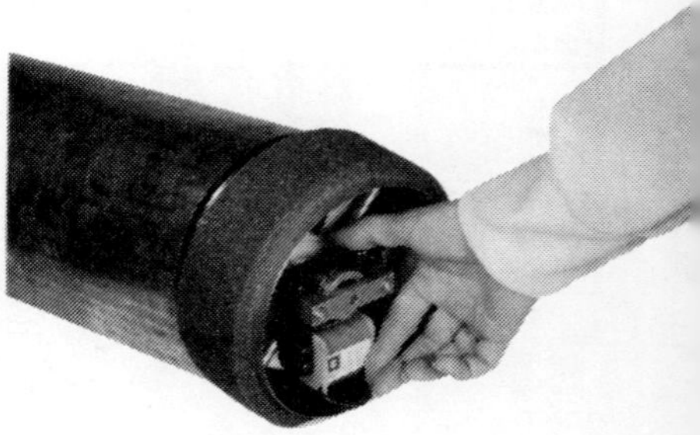

Standard Electric
West Berlin, NJ 08091
A wide variety of fans and pumps, including booster duct fans and circulating fans.

The Broan Co.
926 W. State Street
Hartford, WI 53027

Thermo-Attix
800 Industrial Road
San Carlos, CA 94070

W.W. Grainger, Inc.
5959 W. Howard
Chicago, IL 60648

The Muffin Fan
Rotran Inc.
Hasbrouch Lane
Woodstock, NY 12498

The Universal Circ. Fan
Martin Ind. Div.
Sheffield, AL 35660

## 10.2 Ceiling Fans

A ceiling fan will help both winter and summer. In winter it brings hot air down from the peak. In summer, it will increase your evaporation rate and keep you cool.

Hunter
2500 Frisco Ave.
Memphis, TN 38114
38" and 52" ceiling fans. Light kit available. Infinite speed control available. 36" uses 52 watts on low speed.

Brass Bed Factory
4120 Sepulveda
Century City, CA 90067

Casablanca Fan Co.
182 South Raymond
Pasadena, CA

Fly Fan Factory
2099 Westwood Blvd.
Westwood, CA 90025

George Town Fan Co.
P.O. Box 2128-T
Denver, CO 80201

Also Sears and Roebuck, Montgomery Wards, etc.

## 10.3 Heat Exchanger Ventilators

With tighter houses in cold climates some ventilation may be desirable. A heat exchanger ventilator is the way to go about it. Either buy or build one.

Melco Sales Inc.
3030 E. Victoria St.
Compton, CA 90221
800-421-1132
Mike Thomas
Free, Dealer Only.
Lossnay

| Model | Capacity | Power Source | Weight |
|---|---|---|---|
| VL1500 | 71 CFM | 117 V | 27 lbs |

| Power Consumption | Efficiency |
|---|---|
| 56 W | 70% |

Near factory: $278.00. Air to air static total heat exchanger. Residential to commercial buildings

Aldes
82 rue Feuillat
69008 Lyon, France
(7)874/51/62 Telex 340373
Freddy A. Baets
Air to air heat exchanger, 70% effic. 18"×57"×11". System for single family houses (residential).
Self balancing grilles, Circular fire dampers, Fans.

BAHCO Systems Inc.
P.O. Box 76121
Atlanta, GA 30328
B. Carlsson
Air to air heat exchanger for home ventilation in winter w/o heat loss. 27.5 cm × 60 × 70 cm. Inc. elec. heating element.

Div. of Ext. and Community Relations
Univ. of Saskatchewan
Saskatoon, Sask. S7N OWO
Air to air heat exchanger ventilation unit. To 95% effic. Plans for $2.

Dick Van Ee
RR #3
Saskatoon, Saskatchewan S7K 3J6
Prefab Sask. A to A heat exchanger. Stripped model $200. Complete $375.

Enercon
2073 Cornwall St.
Regina, Saskatchewan S4P 2K6
A to A heat exchanger ventilation. Up to 400 cfm, 85% effic., $995.

## 10.4 Vents and Ventilators

H.C. Products
P.O. Box 68
Prineville, IL 61559
309-385-4334
Ridgevent system.

Triangle Engineering
P.O. Box 7464
Houston, TX 77008

Penn Ventilator
11th and Alleghany Ave.
Philadelphia, PA 19140

Airstream Vents
3701-55 Sepviva St. at Colston Ave.
Philadelphia, PA 19137

Hart & Cooley — Registers & Grills
Alliedf Thermal Cap
Holland, MI 49423

See also Sears and Roebuck, Montgomery Wards, etc. McCraes directory has extensive listing.

## 10.5 Backup Water Heaters

The three most attractive alternations for backup water heating are: flash water heaters, wood water heaters, and active solar water heaters. All are worth considering.

### 10.5.A. Flash Water Heaters

There are two basic types of flash water heater, those with a thermostat (necessary for backup use) and those with constant temperature rise (50-90°F over ambient). Make sure you get the right kind for your purpose.

Thorn Flash WH
The Kiley Co.
25 Valley Drain
Greenwich, CT 06830

Paloma
John Condon Co.
1103 N. 36th St.
Seattle, WA 98103

Silverstar RGS
41019 Solieva
Modena, Italy

Lorenzetti S.A.I.B.E.
Av. Pres. Wilson 1230
Sao Paulo, SP
Brazil

Sitam Modena
Viale Indipendenza 5
51010 Modena Est.
Italy

Junkers
Pressure Cleaning Systems
612 N. 16th Avenue
Yakima, WA 98902

Ascot
Lauren Nichols
8285 S.W. Cashmore Lane
Portland, OR

Chronomite Labs.
21011 South Figuerora St.
Carson, CA 90745

Silvio Serin
Via Col De. Rosso
35100 Padona,
Italy

### 10.5.B. Wood Water Heating

If you use wood for much of your heating you may be able to get hot water as well.

Woodburning Water Heaters
Walden Foundation
P.O. Box 5
El Rito, NM 87530
Domestic hot water heater — wood.

Aqua Heater
P.O. Box 815
Clark, CO 80428
Domestic waterheater, wood.

Tirolia
169 Dunning Road
Middletown, NY 10950
914-343-5900
Water jacket kitchen wood or coal ranges

Thermo Grate
P.O. Box 43021
St. Paul, MN 55164
Water heating grate for fireplace.

Blazing Showers
Elaine Walsh
P.O. Box 327
Pt. Arena, CA 95468
882-2592 on Mondays and Thursdays
*Handmade Hot Water Systems* $4.95
Mail Order
Wood water heating components for stack or fire box.

Holly Hydro Heater
Frizelle Enos Co.
265 Petaluma Ave.
P.O. Box 339
Sebastopol, CA 95472
(707) 823-6557
Jim Huff
$1.00 for owners manual brochure free.
Mail Order.
Firebox water heat exchanger. Several sizes. For Domestic Water Heating Hydronic Heating. $89-$110 per. Near Factory: $100-150 per unit. 1000 Miles Away: Add $3-5.

T.B. Stevens
10 Wayfin Circle
Newport News, VA 23606
804-595-4152
Tom Stevens
$5.00 plans.
Mail Order.
Self contained, automatic, no controls or electric power required. $75.00. 1000 miles away: $125.00
Automatic heat recovery system for gas fired water heater. Gas fired water heater or furnace (with modification)

### 10.5.C. Active Solar Hot Water

It may sound like a strange backup but in many cases it will work. Or use active solar hot water and a passive solar space heating system.

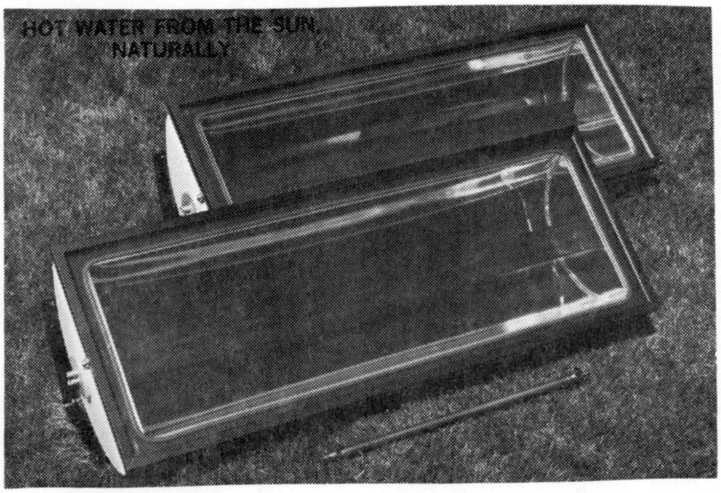


Sunpump/Suncycle
Entropy Limited
5735 Arapahoe
Boulder, CO 80303
303-443-5103
Jerry Moon.
Brochure: 50¢
Mail Order or Dealers, Distributors
Standard system contains 2 square meters of Sunpump concentrating, nontracking collectors and Suncycle heat exchanger for interface with 40-60 gallon hot water tank. Additional collectors can be added for larger storage capacity. System can provide up to 70% hot water requirement for typical family. Contact Mfg. for local dealer name.
Gravity Powered and water Powered Sunpump/Suncycle Systems use only sun and water to operate. Package includes 2 collectors, heat exchanger, all special parts and fittings, insulation (including hot water tank blanket) and instructions. Used with existing hot water tank regardless of backup fuel source.

Island Solar Supply
2915 Prior St.
Victoria, B.C. V8T 3Y4
Active domestic systems and collectors, finished or kit.

Revere Solar
P.O. Box 151
Rome, NY 13440
Active solar systems.

A.O. Smith
Consumer Products
P.O. Box 28
Kanakee, IL 60901
Energy conserving active solar DHW

Earth Services
P.O. Box 99, Route 30
Pawlet, VT 05761
Active solar water systems.

Solex
P.O. Box 245
Scarsdale, NY 10583
914-723-6356
Active solar water heaters. Application manual $10.95 ppd.

Myson Inc.
P.O. Box 5025
Falmouth, VA 22401
(703) 371-4331
John Bernard
Circulator Pump. Liquid circulation.

## 10.6 BACKUP HEATING

The possibilities are almost endless. Some of the more attractive candidates are: wood fired boilers (hydronic systems are most comfortable and flexible; multifuel furnaces; airtight wood stoves, and others.

### 10.6.A. Wood Fired Boilers

The stoves in 10.5.B may also prove workable. A thermosiphon to storage tank is often feasible. Pumps may be required for some systems.

Hydroheat
P.O. Box 382
Ridgway, PA 15853
(814) 776-1323
Tom Herbster
Free
A. mail order; B. dealer only
Hydrostone. $1000.00
Airtight wood stove with water heat exchanger built in.
Hydronic, forced air, solar heating systems.

MW100 (MEMCO) Wood Boiler
Melvin Mfg. Corp.
Rte. 133 & 156, Box 38D, RFD 1
Jay, ME 04239
Morton A. Melvin - Kevin L. Melvin.
Mail Order
MW100 Wood Boiler
$1575.00 per boiler. Installed near factory: $2300; 1000 miles away $2400.
Wood boiler (Hot Water Heat)
Wood Burning, Hot Water Heat.

Solatherm Fireplace Waterheater
Solatherm Corp.
1255 Timber Lake Drive
Lynchburg, VA 24502
(804) 237-3249
W.W. Hays III
Single copy: Free
Combination fireplace grate/heat exchanger for heating recirculating water. Can be used forced flow or thermal syphon.
List price for RM36A (3 ft. fireplace, hi output) = $341.00
Near Factory: Wide range, depending on system and labor source; 1000 miles away: $500-$2500.
Primary/Backup space heating and hot water.

MASCOT MW 100

Mascot MW100 Wood Fired Boiler
Hiestand Distributors
Rt. #1, Box 96
Marietta, PA 17547
(717) 426-1921
Galen Hiestand
Free
Dealer Only, Mail Order on UPS orders when dealer not near. (Only on some items)
85,000 BTU, 12 Gallon Water Capacity, 24 volt Automatic Draft Control, Quarter inch boiler plate construction, A.S.M.E. Certified, use as solar backup.
The Mascot MW 100 is a wood fired hot water boiler that can be adapted to solar backup use. Heat output is 85,000 BTU, making it ideal for this application. The firebox is firebrick lined, and the boiler works on fuel-oil theory, using UL listed Honeywell controls.
$1195 FOB Marietta, PA
Near factory: $200-$400 plus chimney. 1000 miles away: Approximately the same.
Passive backup.

La Font
1319 Town St.
Prentice, WI 54556
(715) 428-2881
Water heat exchanger grill for fireplace use. Transfers heat to existing boiler.

Tasso Universal
ITS, Inc.
P.O. Box 900
Exeter, NH 03833
(603) 778-7373
Ms. Virginia Andrew, VP-Product Manager.
No Charge.
Dealer only, Plumbing & Heating Supply Wholesalers
BTU Range; 68-196,000. Fuels; wood, coal, oil, gas, and pelletized fuel.
Imported from Denmark. Sectional cast iron boiler utilizing a gassification combustion principle. Approximately 73% efficient on solid fuel. Hydronic Space Heating and Process Heat Applications.
$1,676-$2,864. Near factory: $1,975-$3,370.
Mixing values, baseboard distibution systems, circulators.

**10.6.B. Multifuel Furnaces**

Multifuel furnaces allow you to use wood when you are home. Gas, oil, or coal when you are gone or lazy.

Roycraft
Box 23C Star Route
Merrifield, MN 56465
(218) 765-3111
Automatic switch multifuel furnace, wood/oil, wood gas. 120 KBTU.

Multi-Fuel Energy Converter
Multi-Fuel Energy Systems
2185 N. Sherman Drive
Indianapolis, IN 46218
(317) 356-6358
Richard E. Lang
Free. Mail order or Dealer
Forced-air furnace which efficiently burns *any* clean, safe solid fuel — wood, coal, compacted trash, and crop residues. Suitable for use as supplementary furnace or free standing space heater. 60,000 BTUs/hour, heats 1500-2000 square feet
Suggested retail $899.00, FOB Indianaplis, IN
Back-up furnace.

Thermokon
Dept PSC
Galesville, WI 54630

Riteway Multifuel
Dept PSC
P.O. Box 153
Harrisonburg, VA 22801

Forestiere Central Heat Boiler
Franco-Belge
Dept. PSC
70 Pine Street
New York, NY 10005

HS Tarm Multifuel Boiler
AES Inc.
P.O. Box 208
Underhill, VT 05487

Wood/oil Furnace
Arotek Corp. Dept. PSC
1703 East Main
Torrington, CT 06790

Wood/oil Furnace
Charmaster Products
Dept. PSC
Grand Rapids, MN 55744

Wood/Gas Furnace
Dept PSC
Longwood Furnace Corp.
Gallatin, MO 64640

Multifuel Furnace
Northern Heating and Sheet Metal
Dept. PSC
Grand Rapids, MN 55744

Multifuel Furnace
S/A Energy Division
Dept PSC
Cross Roads Park Drive
Liverpool, NY 13088

Yukon W/O, W/G
Wilson Industries
Dept. PSC
2296 Wycliff Road
St. Paul, MN 55114

Multifuel Furnace
Bellway Mfg.
Dept. PSC
Grafton, VT 05146

DuoMatic
Dept PSC
Waterford, Ontario
Canada N0E 1Y0

Combo Furnace Co.
Div. of Northern Heating
1707 West Fourth St.
Grand Rapids, MN 55744

The Defiance Co.
Chassell, MI 49916

Franco-Belge Foundries of America, Inc.
70 Pine St.
New York, NY 10005

Johnson Energy Systems, Inc.
7350 North 76th St.
Milwaukee, WI 53223

Kickapoo Stove Works Ltd.
Box 127
LaFarge, WI 54639

Monarch Ranges and Heater
Div. of Malleable Iron Range Co.
Beaver Dam, WI 53916

National Stove Works Inc.
Box 640
Cobleskill, NY 12043

Oneida Heater Co., Inc.
Oneida, NY 13421

Powermatic Inc.
2906 Baltimore Blvd.
Finksburg, MD 21048

Dual Fuel
Dept PSC
Box 514
Simcoe, Ontario N3Y 4L5

**10.6.C. Airtight Wood Stoves**

An airtight wood stove can approach 60-70% efficiency. The best traditionally were the European stoves but the U.S. now offers many solid stoves made here. The cheapest airtights are from Taiwan — some good, some bad. Even a cheaper one may be okay if you use furnace cement carefully.

Kristia Associates
P.O. Box 1118
Portland, ME
Jotul stoves and stove accessories, including cleaning equipment, $2.

SHENANDOAH
P.O. Box 839
Harrisonburg, VA 22801
Air tight stoves and fireplace inserts.

Upland Wood Stove
P.O. Box 87
Green, NY 13778
(607) 656-4156
N/C
Air tight wood stoves.

Vermont Iron Stove Works
At The Bobbin Mill
Warren, VT 05674
Air tight wood stoves, $1.

Hinckley Foundry
13 Water Street
New Market, NH 03857
Air tight box stove, 50¢.

Freeflow
P.O. 300
Beloeil, Quebec J3G 5S9
(514) 931-6035
or
Freeflow
South Stratford, VT 05070
(802) 765-4022
Air tight, passive circulation stove.

Waterford Iron Founder
Bilberry
Waterford, Ireland
Nice series, includes waterback kitchen stove.

Duncan's Chimney Service
P.O. Box 454
Ross, CA 94957
(415) 453-4644, 388-0560
The Ulefos airtight stove. Triple tier stove. Proven in use since 1657.

Arizona Forest Supply
P.O. Box 188
Flagstaff, AZ 86002
Airtight wood stove. Catalog: Free.

Heathdelle Associates, Inc.
Box 1039
Meredith, NJ 03253
Nashua Doubleheat Woodstove, $1.00

Bow & Arrow Imports
14 Arrow Street
Cambridge, MA 02138
PETIT GODIN Wood/coal stove. Color brochure: 35¢

Energy Harvesters Corp.
P.O. Box 19
Fitzwilliam, NJ 03447
Airtight wood stoves, Free

Schrader Stoves
P.O. Box 2898
Eugene, OR 97477
Air tight wood stoves, Free.

Sierra Marketing, Inc.
P.O. Box 346
Harrisonburg, VA 22801
Air tight stove, hearthstoves, fireplace inserts. Brochures sent on request.

Garrison Stove Works
P.O. Box 42067
Portland, OR 97242
Garrison One™ airtight stove, also Garrison two and three models. Write for brochure, 50¢ f/2 page brochure, $1.00 f/installation and operation manual.

The Cawley Stove Co., Inc.
27 North Washington St.
Boyertown, PA 19512
Cawley/ Le May airtight stoves (cast iron) Catalog: $1.00.

Lynndale International, Inc.
P.O. Box 1154
Harrison, AR 72601
Airtight wood furnaces. Catalog: Free.

NHWS Inc.
38 Haywood St.
Greenfield, MA 01301
(413) 772-0328
Homewarmer Airtight Stove and Home Warmer Fireplace (free standing fireplace), $1.

The Hearthmate
C&D Distributor Inc.
P.O. Box 766
Old Saybrook, CT 06475
Airtight wood stoves and fireplace inserts. Catalog: free.

G&S Mill
75 Otis Street
75 Northbord, MA
(617) 393-9266
Sales Eng. Dept. Mail order
1,250,000 BTU/Hr. Waste wood/cord wood, furnace.
Burns any waste wood, solid regardless of moisture content.
The Waste Wood Producing Industry.

HIGH EFFICIENCY INDUSTRIAL WOOD FURNACE RATED AT 1.25 MILLION BTU/HOUR

Super Patriot
270 Farmington Ave.
Farmington, CT 06032
Downdraft Stove, free.

The Majestic Co.
P.O. Box 800
Huntington, IN 46750
Airtight wood stoves. Catalog: free.

Tempwood
Mohawk Industries
P.O. Box 71-0
Adams, MA 01220
Airtight wood stove. Catalog: free.

Fire Magic
Whittier Steel
P.O. Box 221
Shelbyville, KY 40065
Air tight wood stove. Catalog: free.

HASHA
Box 112
Falls, PA 18615
717-587-5565

$1 info. $20 plans
Remote heating room with high mass provides clean and safe wood heat. 18 tons of sand provide considerable heat storage.

ARC Industries, Inc.
P.O. Box 60
Hartford, CT 06101
Fyroor Fireplace Heating Unit. Catalog: free.

The Lorena Cookstove
ATP—Volunteers in Asia
Box 4543
Stanford, CA 94305
A homebuilt mud stove developed in Guatemala. Manual describes design and construction. $3.00 ppd.

### 10.6.D Active Solar Systems

An active solar system may work very well with a passive system. It can provide distribution to the north of a deep house and added storage.

Solar Forced Air Heating System Plans
Solstice Designs, Inc.
P.O. Box 2043
Evergreen, CO 80439
(303) 674-1597
Lynda Lillywhite. Brochure: free
Mail order
$3/sq. ft. materials (labor only)
Solar Forced Air Heating System Plans

Solaron Corp.
440 Galleria Tower
720 Colorado Blvd.
Denver, CO 80222
Air active solar. Catalog: free.

Solarcrete
Columbia Sessex Bldg.
Florence, KY 41042
(800) 354-9818
A solar collector system using pipes imbedded in south facing concrete walls.

Solar Control Corp
5721 Arapahoe Road
Boulder, CO 80303
(303) 449-9180
System controls and air movers for active solar systems.

LENNOX
P.O. 400450
Dallas, TX 75240
Active solar system and heat pump.

American Solarize
139 South St.
Murray Hill, NJ 07974
Active air solar collectors. Honeycomb increases efficiency.

The Guide
Solar Age
Box 4934
Manchester, NJ 03108
403 products from 226 firms in solar design. Best source of active solar information. Cost $120, includes 6 up dates.

### 10.6.E. Miscellaneous

Intertherm
3800 Park Ave.
St. Louis, MO 63110
Electric water baseboard system.

Heatilator
Des Moines, IA 50306
Fireplace systems.

Camelot
801 Greenbay Road
Lake Bluff, IL 60044
Electric radiant heater — oil filled radiator. $99.95.

Berry Metal Fabricating
Molasses Hill Road
Lebanon, NJ 08833
(201) 735-8330
Jack Chase
Hearth Shield floor and wall shields for wood stoves. Also fireplace covers and adapters for airtight stoves. Metal with non-asbestos insulation.

Improved Consumer Products
Box 1264
Attleboro Falls, MA 02763
(617) 695-7000
Stainless Steel Vacu-Stack™ wind operated turbovent for chimneys. Eliminates back draft and reduces creosote buildup.

Stovo - Stove Patch
Rutland Products
Curtis Ave.
Rutland, VT 05701

Black Magic Chimney Sweeps
Box 977
Stowe, VT 05672

J.H. Stroneberg
Brockport, PA 15823
Flue radiator, gas only. $139.75.

PROJET
Country Stove & Shelter, Ltd.
270 Farmington Ave.
Farmington, CT 06032
Insulated pipe, free

## 10.7 Backup Cooling

An evaporative cooler is generally the last choice for cooling — but it should be second right after whole house fans.

### 10.7.A.Whole House Fans

If night temperatures are cool and you have sufficient mass and insulation a whole house fan may do the whole job! Can be intertied with air conditioning in warmer climates. Still worthwhile. Air conditioning goes off as fan goes on.

KooloMatic
1831 Terminal Road
Niles, MI 49120
(616) 683-2600
Whole house fans.

White Rodgers
9797 Reavis Road
St. Louis, MO 63123
Whole house fans, blowers, heat pumps, hydronics, etc.

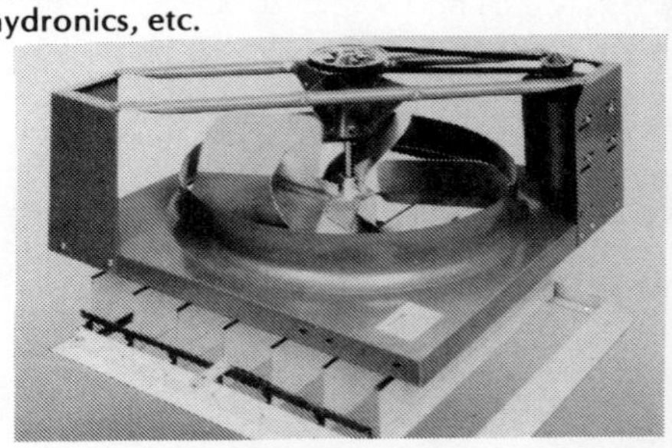

NUTONE DIVISION, SCOVILL
Madison and Red Bank Roads
Cinicinnati, OH 45227
(513) 527-5347
Mrs. Marian Finney, $1.00
Dealer only, or thru distributors nationally.
One or two-speed models. HVI rated with CFM from 4350 to 7000 three models.
$300. Near factory depends on labor cost in area for electricians.
Attic roof and wall ventilating fans. Paddle fans.

**0.7.B. Evaporative Cooler**
If a whole house fan won't do the job get an ›vaporative cooler.

\rctic Circle Evaporative Cooler
ᴧcGraw-Edison Cooler, Co. IMPD Div.
'.O. Box 20188
'hoenix, AZ 85036
602) 949-9717
ᴧarc Lee. Free, dealer only.
\n economical space conditioning system utilizing he evaporation of water for heat absorption and perating at a cost approximately one-fourth that of conventional air conditioning. Fresh air-cooled, :leaned and filtered — continuously flows into the :onditioned space, pushing out odors, heat, moke and other pollutants.
375-$600.
ᴧear factory $600-$1500. 1000 miles away: same.
Residential — Southwest. Light commercial — U.S.

You might wish to make an indirect evaporative :ooler to reduce humidity in the house. To do it /ou will need a water/air heat exchanger.

Trane
3600 Pammel Creek Road
.a Crosse, WI 54601
608) 787-3187
Water to air heat exchanger.

## 0.8 Controls and Etc.

Honeywell Inc.
0400 Yellow Circle Drive
ᴧinnetonka, MN 55343
612) 931-4184
Neil Larson, Solar Market Manager
Write & request free brochure
UEL SAVER THERMOSTAT. Saves 9 to 30% on eating costs and 7 to 25% on cooling costs. Uses xisting thermostat wires. Rechargeable battery-perated clock, multiple setback/setup capability, xclusive program advance button for schedule hanges.
Heating, cooling, or heating-cooling systems, .7412 Differential Temp. Con. C773 Temperature ensors

COMPUSTAT
SG Industries, Inc.
225 Tunnel Road
erkasie, PA 18944
215) 257-3621
Carl Gearhart
ree, Dealer Only, HVAC Contractors
.utomatic setback thermostat with preset mercury ensors, and digital microprocessor for programming setback periods. Additional controls for omplete control of HVAC system.
79.99 - $99.95. Heating only, heating/cooling.
79.99 - $99.95 plus installation. 1000 miles away: ame as factory.
ack-up heating, or heating/cooling system ontrol.
.ccustat set-point thermostat with preset mercury ensors.

Demand Manager
nertrol Corporation
418 E. Cholla Lane
cottsdale, AZ 85253
602) 947-4942
ames W. Ross
.pplication manual: $5.50 Mail Order
ᴧicroprocessor based peak load controller. For se in homes and small businesses in areas with eak pricing.
7.50. Near factory: $1500, 1000 miles away: $1500.
ime of day and time of week electronic programmer.

# 11. Daylighting and Solar Exposure

Natural lighting is the light we evolved with and it's still the best. The following instruments will help calculate daylighting, solar exposure, and shading.

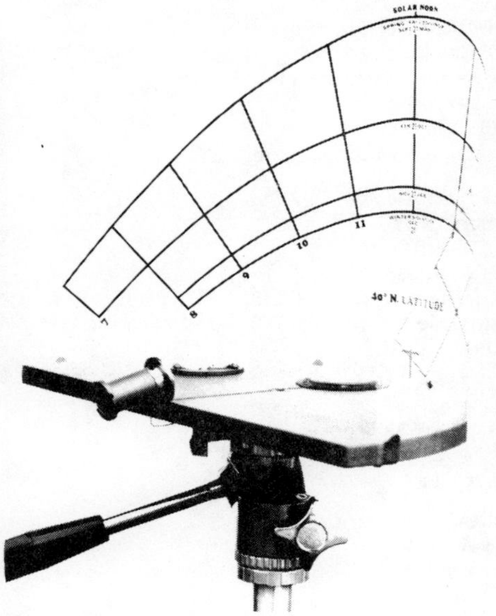

Solar Site Selector
Lewis and Associates
105 Rockwood Drive
Grass Valley, CA 95945
(916) 272-2077
Sheri Lewis
Both mail order/dealer
Baseplate includes bubble level, compensating compass, 180° distortion optic, grids in 2° even latitudes. Threaded insert fits photographic tripods.

The Solar Site Selector calculates solar dayhour from sunpaths and hour segments superimposed on the panorama being studied. Shading patterns are also instantly determined. Correct siting and orientation can be instantly determined.

The Solar Pathfinder™
Solar Pathways, Inc.
3710 Highway 82
Glenwood Springs, CO
(303) 945-6503
Robbi Dagenhart. Free, mail order or telephone orders
The Solar Pathfinder™ is a manual, recording site-survey device. Printed tracing sheets provide latitude-specific half-hour percentages of daily total radiation available at the site. Worksheets and data supplements included.

The Solar Pathfinder™ is a patented sitesurvey device which utilizes the reflected image off a transparent dome to display sunrise and sunset times, sunrise and sunset directions, interim shading patterns and half-hour energy values. One quick tracing permanently records all data.
$124.00

The Sun Machine
331 Harvard
Cambridge, MA 02139

IN-AR Faculte de l'amenagent
Univ de Montreal 5620
Darlington, Montreal
Quebec

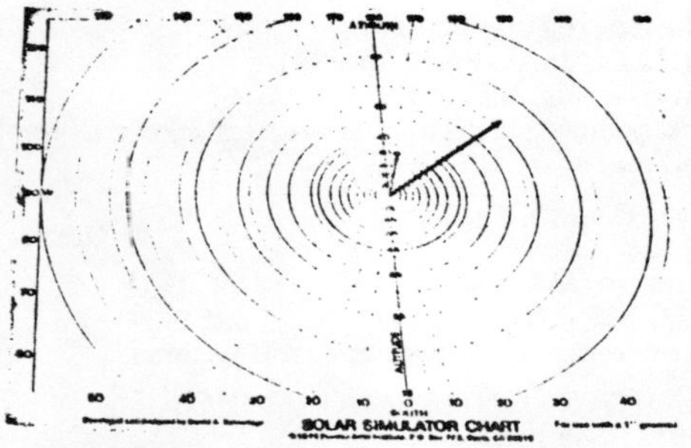


Solar Simulator
PSI
P.O. Box 722
Davis, CA 95616
$10 kit, $39.95 built, $699.95 deluxe for shading, solar penetration, and daylighting studies.

# 12. SOLAR ELECTRIC

Developing an independent energy system may be desirable for a number of reasons. It is considerably more challenging than building a solar house but it can be done. The best candidate will depend on your site. Generally, water power is most cost effective and reliable. Photovotaics are most expensive but relatively simple. Wind is the most demanding but can be inexpensive and very cost effective. And a wood power or fuel powered generator is somewhere between.

## 12.1 Water Power

A water power system can't be beat if you have flowing water nearby. You can build you own mill or buy one.

Independent Power Developers, Inc.
Rte. 3, Box 174H
Sandpoint, ID 83864
Bill Delp
$2.00

Judd Dist. Co.
8600 Lake City Way NW
Seattle, WA 98115

Pumps and Power
1380 Napier
PO 2048
Vancouver, B.C.

Marine Ind. Ltd.
1405 Peel St.
Mont. 1110
Canada

Barber Turbines
PO 370
Meaford, Ont.
Canada

James Leffel and Co.
Springfield, OH 45501

Small Hydro
Box 54
Silverton, B.C. V0G 2B0

## 12.2 Photovoltaics

If you have money and sun, go photovoltaic. Price is dropping rapidly. May be best in conjunction with wind or backup.

Solenergy Corporation
23 North Avenue
Wakefield, MA 01880
(617) 246-1855
Photovoltaic cells and equipment.

Chesepake Solar Systems
10750 Columbia Pike, Suite 601
Silver Spring, MD 20901
Photovoltaic panels for marine use, trickel chargers and more.

Solar Power Corp.
20 Cabot Road
Woburn, MA 01801
(617) 935-4600
Photovoltaic cells, regultors, and systems.

PHOTOWATT INTERNATIONAL INC.
21012 Lassen St.
Chatsworth, CA 91311
(213) 882-4100
Photovoltaics.

Sollos
2231 Carmelina Ave.
Los Angeles, Ca 95064

Solec International
12533 Chadron Avenue
Hawthorne, CA 90250
(213) 970-0065
Robert Crawford
Photovoltaic solar cells, panels and systems.

## 12.3 Wind

In a windy site even a small mill may suffice. Designing a windmill is not easy as forces are very strong and complex. If in doubt, buy one instead.

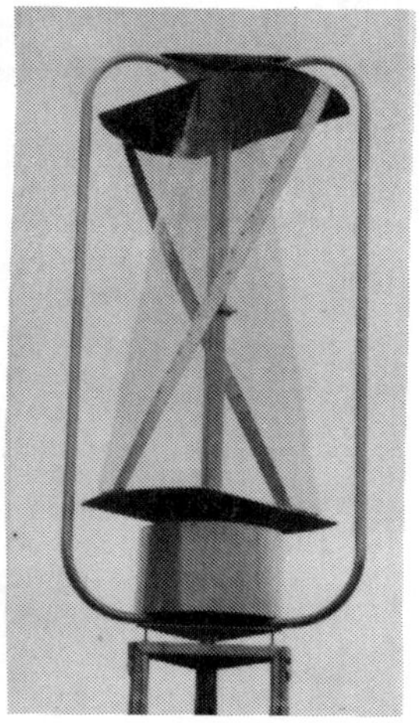

Helius Rotor Kit
Thermax Canada Limited
Box 275
Hawkesbury, Ont.
Canada K6A 2R9
(613) 632-0134
E. Higginson
Free, Mail order
Maximum output: 150 watts; rated 15 mph: 25 watts, $379
Kit: easy assembly, shipping weight, 68 lbs.

AeroPower Systems Inc.
2398 4th Street
Berkeley, CA 94710
(415) 848-2710
Wind generators, 1 and 1.5 kw. Cost installed to $7,000. Proven and reliable.

North Wind Power Co.
Box 315
Warren, VT 05674
(802) 496-2955, Telex 940933
HR2-$7000, 2 kw system for severe environments. Design service & complete systems.

IES
6043 Sterrettania Road
Fairview, PA 16415
(814) 833-3567
Sky Hawk wind generators, 2000 W to 4000 W. Cost for complete system $8-12,000. Delivery 120 days. Other products — wind instruments, composting toilets, etc., $2.50 info.

DAF-INDAL LTD.
3570 Hawkstone Road
Mississauga, Ontario
L5C 2V8
(416) 275-5300
Vertical axis wind machines.

Crowfield Assoc.
Crapand RR1
PEI, COA IJO
(902) 658-2233
Wind Machines, Heat Churn 12KW @ 130 rpm (adjustable to order) $2500. Wind machine WH-1 $12,000.

WTG Energy Systems Inc.
251 Elm Street
Buffalo, NY 14203
(716) 856-1620
Info. $3.00

Whirlwind Power Co.
2458 W. 29th Ave.
Denver, CO 80211
(303) 477-6436
2 Kw wind machines and equipment. No charge.

Natural Power Inc.
Francetown Turnpike
New Boston, NJ 03070
(603) 487-5512
Wind machines, wind instruments, system components. Free.

Wind/Power/Systems, Inc.
P.O. Box 17323
San Diego, Ca 92117
Wind Machines, Storm Master Model 10. $3.00

Enertech
Box 420
Norwich, VT 05055
(802) 649-1145
Wind machines. Free flyer.

Dakota Wind and Sun
Box 1781
Aberdeen, SD
(605) 229-0815
High quality Jacobs based machines.

Winco
7850 Metro Parkway
Minneapolis, MN 55420
(612) 853-8400
Small 12v system. Inexpensive and reliable.

Massawippi Wind
RR 3
Ayers Cliff, Quebec
J0B 1C0

Maier Still
5985 Atlantic Drive, Unit 6
Mesissaugua, Ont.
L4W 1F4

Zephyr Wind Dynamo
P.O. 241
Brunswick, ME 04011
(207) 725-6534
Direct drive alternators for use in wind machines.

Millville Windmills
P.O. Box 32
Millville, CA 96062
(916) 547-4302

Sencenbaugh Wind Electric
P.O. Box 11174
Palo Alto, CA 94306
Wind Machines, 500W, $3

KEDCO
9016 Aviation Blvd.
Inglewood, CA 90301
(213) 776-6636
Wind powered electric generators. Information package: $2.00, Owners manual: $7.50.

Oakridge Windpower
Rte. 1
Underwood, MN 56586
Wind Machines, $1.00

## 12.4 Generators

You may need a generator in any case. My favorite is the Lincoln Weldan Power. Most can be adapted to run on alcohol, natural gas, methane, or wood gas. You might also build your own. Use the heat from the engine to warm water and heat air for a total energy system.

UNIGEN Inc.
194 West Stone St.
Almont, MI 48003
(313) 798-3150
A wide variety of alternators to 5 kw with gas engines, can use natural gas for home hookup.

Lincoln Electric
Cleveland, OH 44117
Toronto, M4G 2B9
Makers of Weldanpower 150 a 4.5 kw generator and 150 amp ac welder. Also exc. welding reference book and welding equipment.

Ag-Tronic-Inc.
364 Airport Road
Kearny, Neb. 68847
(308) 237-2181
Generators from 1250 to 4500 watts. Built for continuous service.

Homelight
P.O. 7047
Charlotte, NC 28217
Gas generators for a variety of uses. Including standby for home in own case and 5 gal. gas supply.

## 12.5 Other

Two other solar products are worth considering. These are solar pumps from:

Sofretes/Mengin
Zone Industrielle d'Amilly B.P. 163
45203 Montargis
France
A variety of sun pumps. In use all over the world. Over 12 years in the field. Use vapor to drive pump.

SEI-15 Sun Pump
Solar Electric International
4837 Del Ray Avenue
Bethesda, MD 20014
(301) 656-9614
Stephen V. Allison or Theodore K. Osgood
No charge Mail Order
The Sun Pump will deliver about 40 gallons per minute under clear skies (insulation of 100mW/$cm^2$) at a 15′ head.
Single units $5,250 FOB Bedford, MA.
No installation costs.
The Sun Pump consists of a: 250 watt pv array (8′×4′); a low voltage, submersible D.C. motor, directly coupled to a single stage vertical turbine pump, and a Maximum Power Controller (power conditioning unit).
Water pumping for irrigation, livestock or domestic use in remote areas.
I-V Curve Tracers, Solar-rechargeable lanterns.

## 13. Instruments

It is always good to keep track of what your house, solar system, or site is doing. These instruments may prove helpful. Full instrumentation desirable but installation and monitoring can equal the cost of the project. Even fairly crude measurements are better than none.

DuPont Energy Management Corp.
3301 Conflans — Suite 102
Irving, TX 75061
(214) 258-0054
Energy monitor gives readout of energy use and cost. Knowing how much you use can help you save.

Fitch Energy Monitor
P.O. Box 111
Chapel Hill, NJ 27514
House data, displays energy use. Recorded 12% reduction in use after installation in one test. Monitored by UC Boulder, Inst. of Beh. Science.

Condor Co.
Box 6
Hiram, OH 44234
Chimgaurd surface thermometer, $9.95.

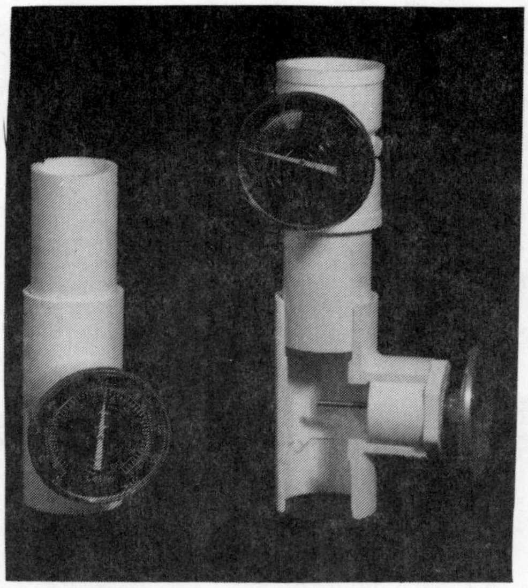

Letro Thermometer, Inc.
8311 Airport Rd.
Redding, CA 96001
(916) 365-5424
Jeri L. Campbell
Free.
Distributor, OEM.
SL-1D thermometer or Swimming Pool Solar. SL-2D Thermometer for Hot Water Systems. ST-2 Thermometer for storage tanks.

AQUATEMP
Rodco Products Co., Inc.
P.O. Box 944, 2565 16th Avenue
Columbus, NB 68601
(402) 563-2009
Mary Kolm
No charge. Mail Order. Dealer Only.
Power: 117 VAC, 9 watts. Prove length is 75′ and is expandable with kit. Size is 6½″ wide × 2¾″ high × 4½″ deep. Accuracy: ±2°F. 2-6 units: $77.00. Near factory $119.95 retail. LED display of two temperatures from -30° to +150°F, F or C display capabilities, and time of day, remembers high and low temperature and times they occurred. Temperature alarm and time alarm.
Pools, spas, hot tubs, solar. Computemp II, Ag-Temp.

Blue White
14931 Chestnut St.
Westminster, CA 92683
(714) 893-8529
Robin or Wanda.
Free, Dealer only.
Flowmeters for solar systems.

THE EPPLEY LABORATORY, INC.
12 Sheffield Ave.
Newport, RI 02840
(401) 847-1020
George L. Kirk, President
No charge, order direct
Pyranometers and other met. instruments.

Cole-Parmer
7425 North Oak Park Ave.
Chicago, IL 60648
A broad range of instruments, equipment and supplies for research and development work.

Belfort Instrument Company
1600 S. Clinton
Baltimore, MD 21224
(301) 342-2626
A full range of meteorological instruments, including totalizing anemometer 5-349C $360, 5-385 Pyranograph for solar radiation. $535.

Climet
P.O. 151
Redlands, CA 92373
(714) 793-2788
Fred Nikithser
Research grade meteorological sensors and systems including wind and pyranometers.

Weathertronics
P.O. 1286
West Sacramento, CA 95691
A full range of weather instruments and monitoring systems including remote thermographs, pyranometers, and wind recorders. *"Good company to deal with." (ed.)*

Aeolian Kinetics
Box 100
Providence, RI 02901
Wind resource analysis instruments.

Ecotope
2332 East Madison
Seattle, WA 98112
Eco-Dau user programmable data acquisition unit. Calculator keyboard designed for monitoring passive buildings.

WeatherMeasure Corp.
P.O. Box 41257
Sacramento, CA 95841
(916) 481-7565
Stan Burmich
Mail Order
R413-Star Pyranometer, Size 13.5 cm diameter, 9 cm height, Sensitivity: Apx 8mV gm-cal/$cm^2$/min. Thermopile: 72 junction CrNi-Constantan
$895.00
A sturdy pyranometer developed to measure solar and scattered radiation and reflected radiation from the earth's surface in the wave range from 0-3-3u. Frequent use in agrometeorology, and forest science. Continuous reading as well as spot measurements.
Weather instruments and systems.

## 14. APPLIANCES

The choice of appliances can considerably affect your home energy use. If you are planning your own power system you must be particularly careful and may have to remodel your appliances to make them more energy efficient.

### 14.1 Refrigerators

The refrigerator is usually the second or third largest user of energy in the home, after space heating and cooling, and is often greater than water heating. This needn't be so as 15 to 20 kwh month will be sufficient for a well insulated, well designed refrigerator. This should be built in with R-40 insulation and a remote compressor/coil.

If you are buying a refrigerator look for a 1950's model or get one of these on the list here. By no means should refrigerator energy use exceed 50 kwh/mo. If you don't have electricity get a gas or kerosene refrigerator.

| Refrigerators: | Model | cu. ft. | kwh/mo. |
|---|---|---|---|
| Philco-Ford<br>Union Meeting Rd<br>Blue Hill, PA 19422 | RS14B3 | 13.7 | 42 |
| Amana Ref. Co.<br>Amana, IA 52203 | ESR-12A | 12.2 | 51 |
| G.E. | TA-14SW | 13.9 | 47 |
| Whirlpool | 7671210 | 12.9 | 45 |
| Frigidaire | D-12 | 11.6 | 49 |
| G.P. | Rc13M1HGB | 13.0 | 49 |
| IGNIS | C05140 | 4.7 | 29 |
| Freezers: | | | |
| Frigidaire | F-101-C | 10.1 | 59 |
| Frigidaire (AS) | 203H8 | 20.3 | 98 |
| Amana | ESU-13B | 13.1 | 60 |
| CS | CU21AE | 21.1 | 99 |
| Checkpoint | 7282030 | 20 | 90 |

Other refrigerator/freezers:
Dometic Refrigerator (Gas or Kerosene)
12 cu. ft. 5 gal./mo.
Dometic Freezer (Gas or Kerosene)
5.5 cu. ft.
from Lehman's Hardware
Box 41, Kidron, OH 44636
or:
Dometic
14441 Bonelli St.
City of Industry, CA 91746
(213) 968-9431

Norcold (12V/110V/gas 27 cu. ft.)

from: Camping World, Beech Bend Rd.
Bowling Green, KY 42101
LNG Refrigerator:
from:
Gas Systems, Inc.
5361 Production Dr.,
Huntington Beach, CA
from:
Koolatron (Peltier effect)
58 Harvester Ave.
Batavia, NY 14020
or
230 Bayview Drive
Barrie, Ontario, L4N 4y8

12vdc Refrigerators, several sites
Adlera Barbour
42 SE Lawton ST.
New Rochelle, NY 10801
(pulley drive refrig.)

Look for many advances in the future including:

Solar Refrigerator from
Zeo Power Co.
Natick, MA

Power Factor Controller, from:

Electronic Relays, Inc.
1438 Brook Drive
Downers Grove, IL 60515
(312) 620-4646
Joseph E. Pascente, President

Free brochure by mail order.
Handles motors up to ½ horsepower, 115VAC
Motor control system to decrease energy usage.
Can reduce use considerably for refrig.-freezer.
Application: pumps and fans
Approximate cost: $29.95 each
Other products: Solid-State relays

### 14.2 Miscellaneous

1 Qt. — Hot Water (212°F) tap
In-Sink-Erator (I.S.R.)
4700 21st St.
Racine, WI 53406

Sudsaver-Washing Machine
The Maytag Company
17065 E. Green Drive
City of Industry, CA 91744

Efficient Dishwasher (but you really don't need it!)

Kitchen Aid
Hobart Mfg. Co.
1501 W. 8th St.
Los Angeles, CA 90017

Thermadoor Waste King
5114 District Blvd.
Los Angeles, CA 90040
Time delay dishwasher, use off peak power.

### 14.3 Toilets

The major villain in water wasting and major contributer to America's sewage problems. We've invested billions in inefficient, expensive, and unreliable water treatment. In many ways the privy was better. Now the composting toilet is much better, does require maintenance tho. Ideally a maintenance district could be set up. For more information, see: *Goodbye to the Flush Toilet*, Carol Stoner, Rodale Press.

Composting Toilets:
A.B. Gustasvsbergs Fabriker
Flemingatan 62 B, Box 12159
10224 Stockholm 12, SWEDEN

Clivus Multrum USA Inc.
14A Eliot Street
Cambridge, MA 02138
Include the full size Clirus and the smaller bio-led composting toilet.

Enviroscope, Inc.
Box 752
Corona Del Mar, CA 92625
Carry the Carousel composting toilet. Compact and proven. Uses elec. heater and fan.

Mini use toilets
Microphor
P.O. Box 490
Willits, CA 95410
707-459-5563
2 qt. flush toilet, uses compressed air to increase flush action.

Humphrey
243 12th St.
Del Mar, CA 92014

Danforth
500 Riverside Ind. Pkwy.
Portland, ME 04103

Mansfield Vacu Flush
201 Wayne St.
Big Praire, Ohio 44611

A better standard:
American Standard
P.O. Box 2003
New Brusnwick, N.J. 08703
Make the Lexington one piece water saver toilet
Uses 3.5 gallons.

### 14.4 Water Purifiers

Sad to say the water in much of America is unfit for drinking. Seems we'll never learn it's easier and better to stop pollution at the source than clean it up later. Water is particularly bad at the end of major river systems where chlorination reacts with pesticide and chemical residue to form carcinogens and mutagens. Like Europe it's better not to drink the water—unless you clean it. Activated carbon looks best. Distillation is good for many impurities but is energy costly. Ideally you'd distill then carbon filter. Everpure charcoal filter seems to be the best.

Write: Frank Bell WH550, E.P.A. Washington, D.C. 20460 for latest test results and addresses.

Also write for activated carbon filter:
General Ecology Inc.
P.O. Box 320
Paoli, PA 19301

American Water Purification
1900 Olivera Rd.
Concord, CA 94520

LT Marketing Inc.
4724 Browns Blvd. Rd.
Louisville, KY 40207

For distillers:

Home Still
Box A
Bismarck, ND 58501
701-258-7740

Pure Water Inc.
P.O. Box 83226
Lincoln, NE 68501
402-467-2577

New World Distiller
BPS
P.O. Box 199
Mountain Home, NC 28758

### 14.5 Water Conservation

Flow restricters and low use fixtures are easier to find these days. Some brands to look for are:

Nova — ion flow shower restrictor
Ondine — water saver shower head
American Standard — flow restrictor, aquamizer
Spearman — flow restrictor, auto flow

The best water conservation is simply behavioral. Examples: Shower: Conventional 27 gallon, Conservation: 5 gallon (wet, soap up, wash); Wash hands: water running 2 gallons, wet soap up, rinse 1 quart, etc.

## 14.6 Lights

The Alladin mantle kerosene lamp is the standard, available from many sources, including:

Cumberland General Store, $3 catalog
Rt. 3
Crossville, TN 38555

For low voltage lights look for:

REC Specialties Inc.
15155 Stagg St.
Van Nuys, CA 91405
Thinlites™
12 and 24 VDC Lighting fixtures and inverter ballasts and through 75 watts.

# 15. THE GOOD LIFE

A collection of items for Living the Good Life, my picks and recommendations for the 80's.

## 15.1 Bikes

Bicycling
33 East Minor
Emmaus, PA 18049

Sturmey Archer
1014 Carolina Drive
West Chicago, IL 60185
312-231-5150
Internal drum brakes for bicycles, front and 3 speed rear. Well built—work in rain or shine.

Bell Helmets
15301 Shoemaker Avenue
Norwalk, CA 90650

Avocet Seats
P.O.B. 7615
Menlo Park, CA 94025

Blackburn Racks
1080 Florence Way
Campbell, CA 95008

Bicycle Lighting Systems $1 info.
718 N. Vermont St.
Arlington, VA 22203

Bellweather Panniers
1161 Mission St.
San Francisco, CA 94103

Sun Tour USA—Derailleurs and Components
60 Chapin Road
Pine Brook, NJ 07058

Bickerton Cycle Ltd. The Ultimate Portable Bike
1314 N.W. Glisan St.
Portland, OR 97209

Bike Centennial
P.O. 8308
Missoula, MT 59807
Tours and maps

## 15.2 Running

Runners World
Box 366
Mountain View, CA 94042

J.E. Systems
1995 Broadway
New York, NY 10023
217-787-1152
Ultra sound dog chaser

Nike Shoes
3285 S.W. Nimbus, Suite 115
Beaverton, OR 97005
LDV-shoe, one of the best.

## 15.3 Cars and Trucks for the 80's

Best of all, no car at all. Ride a bike! If you must, in order of my judgement of value, economy, comfort and durability.

**New Cars**

| Model | List Price | EPA Mileage City | Motor Trend Magazine | Other |
|---|---|---|---|---|
| 1. Toyota Tercel | $3,698 | 31 | 46 | |
| 2. Honda Civic | $3,699 | 36 | 46 | |
| 3. Mazda GLC | $3,695 | 30 | 43 | |
| 4. Plymouth Champ (or Dodge Colt) | $4,450 | 33 | 46 | |
| 5. VW Rabbit, diesel | $5,000 | 43 | | |
| 6. Datsun 210 | $3,749 | 31 | 49 | |
| 7. Renault Le Car | $4,269 | 30 | 41 | |
| 8. Toyota Corolla | $4,098 | 27 | | |
| 9. Honda Accord | $6,365 | 30 | | |
| 10. VW Rabbit, diesel turbo | $8,000 | | | 50-65 |
| **New Trucks** | | | | |
| 1. VW Diesel ½ ton | — | 43 | | |
| 2. Mazda ½ ton | $4,695 | 26 | | |
| 3. Toyota ¾ ton | $5,124 | 18 | | |
| 4. Volkswagon — Vanagon | $9,140 | 17 | | |
| 5. Subaru 4 WD | $6,000 | 17 | | |
| 6. Suzuki Brute 4 WD | $5,000 | | | 35 |
| **Used** | | | | |
| 1. Datsun 510 (68-72) | $900-1,500 | | | 25-30 |
| 2. Toyota Corolla (any) | $900-5,000 | | | 25-40 |
| 3. Morris Minor (58-67) | $750-6,000 | | | 30-40 |
| 4. Datsun 521 pickup | $900-2,000 | | | 25-30 |

FREEWAY CAR
H-M Vehciles
9341 Penn. Ave. South
Bloomington, MN 55421
100 MPG guaranteed

Cyberlite
770 Welch Road, Suite 154
Palo Alto, CA 94304
415-854-1242
A brake light which flashes—intensity changes as brake effort changes. Superb for motorcycles and car. Reduced taxi rear enders 60%.

ICE-WING
ANSAR
Upplandsgatan 19B
S 113 60 Stockholm
(08) 300 802
Mr. Anders Ansar
Free, Mail order; Also kits and/or plans.
(Ice) Skate Sail shaped like a standing aircraft wing with the sailor inside, is an efficient and easy sport to learn. Sail area: 50 sq. ft. Weight: 22 lbs. Top speed around 50+ mph or similar to DN Ice Boats. US $250, Plans US $10. Near Factory Kit: US $295, Complete: US $790. The sail folds together for storage or car top transport. When building only simple tools are required.

Pterodactyl , Ltd.
847 Airport Road
Monterrey, CA 93940
Jack McCornack
$1.00, Mail order
Microlight aircraft, Span: 33 feet, Weight: 155 pounds, Speed: 25055 mph. Fuel Consumption: 5 qts per hour. This is a single place aircraft, fully collapsible for storage or transport. It is designed to fly from pastures and other unimproved fields. No license is required. Gas or alcohol powered. $3250.

Pterodactyl Fledgling

## 15.5 Gathering Wood

Chain saws, get the best. Stihl, Husquavarna, and Jonsereds. Use safety equipment and be kind to the forest. Trees are a renewable resource only if we treat them right. Major suppliers:

Baileys
P.O.B. 550
Laytonville, CA 95454
707-984-6133
Excellent prices, selection. Very good safety equipment. N/C.

Zip Penn
P.O.B. 179
Erie, PA 16512
Chainsaws, chain, and wood stoves. N/C.

Tilton Equipment
4612 North Chatsworth
St. Paul, MN 55112
Distributor: Jonsereds, chain saws, brushcutter, cut-off saw.

## 15.6 Miscellaneous

Golden Nagas
3103 Geary
San Francisco, CA 94118
415-752-7693
Folding foam beds in the Japanese style. A space saver. Attractive and comfortable as well.

SICO Inc.
P.O. Box 1169
Minneapolis, MN 55440
A wide range of fold-to-wall tables, beds, etc. Look well built.

Westerbeke Diesel
33 Aveon Industrial Park
Avon, MA 02322
7-100 HP, highly regarded. Used by Amish for refrigerator motors.

Janca's Jojoba Oil and Seed Co.
20 E. Southern
Mesa, AZ 85202
602-833-4940
B.J. Bianco

The Safe Energy Fund
Continental Savings and Loan
2109 Market St.
San Francisco, CA 94114
Solar-T-bills. All funds used to make solar loans.

The Solar Lobby
1001 Connecticut Ave. NW, Suite 510
Washington, D.C. 20036
202-466-6350
Membership $15.

Wooden Boat
Box 4943
Manchester, NH 03102
$15 US/$16 Canada s/O.S.

Folk Legacy
Sharon, CT 06069
The best collection of folk I've found. I've worn out some of my Gordon Bok records from using them so much. Also get the Golden Ring series.

AEU
8143 Big Bend
St. Louis, MO 63119
Natural sound records, thunderstorm, sea, etc.

# 16. SIMULATION

Computers have their uses in evaluating design. Here are some of the programs now available.

PEGFIX/PEGFLOAT
Princeton Energy Group
729 Alexander Road
Princeton, NJ 08540
609-452-8235
Lawrence Lindsey or William Glennie
Self-addressed stamped envelope, Mail order
Hand calculator Programs for hour-by-hour simulation of Direct Gain Passive Solar spaces: running time 5-9 min. per 24-hour simulation. Accuracy within 3°, verified against data published by Balcomb, Palmiter.
$75 complete, both programs.
Pre-recorded cards, detailed PEGFIX/GEFLOAT HANDBOOK (user manual), printed Worksheets. CARD-PROGRAMMABLE HAND CALCULATOR (HP-67, HP-97, HP 41OC, or TI-59) REQUIRED, and not included in program purchase.
Other programs for hand calculators and microcomputers under development: passive solar heating, daylighting, natural cooling and ventilation.

Stereatronics
15 Ellic Court
Weston, MA 02193
Norman Saunders
N/C
Calculation program for NBS and annual solar fraction and passive building prediction. Other products include: Solar Staircase™ heating system, glazing system, house plans, proportional controls, data logger.

D.A.T.A.
P.O. Box 503
Davis, CA 95616
LASL 6 for passive system evaluation. $50. in basic. Also program for envelope houses.

University of Texas
Attn: Francisco Arumi
2604 Parkview Drive
Austin, TX 78757
DEROB: Output: daily, monthly and peak day load profiles. Service: computer. Materials: program source code, user manual; program operates on UNIVAC 1108 computer. Cost: nominal price.

Ross F. Meriwether & Associates
1600 NE Loop 410
San Antonio, TX 78209
512-842-5302
ESAS: Output: hour by hour annual energy consumption, economic analysis. Service: output data. Materials: printouts, user manual. Cost: $350 to $400 (one set of runs).

Colorado State University
Mechanical Engineering Dept.
Attn: Dr. C. Byron Winn
Fort Collins, CO 80523
303-491-6983
FREHEAT: Output: detailed daily passive energy flow analysis and solar fraction. Service: computer. Materials: magnetic tape transmittal program operates on CDC computer.

GK Associates
157 Stanton Ave.
Auburndale, MA 02166
GLASIM: Output: hourly, daily or monthly temperature loads. Service: computer. Materials: magnetic tape transmittal; program operates on IBM 370/158 computer.

Total Environmental Action, Inc.
24 Church Hill
Harrisville, NJ
603-827-3374

PASCALC: Output: monthly and annual energy consumption of passive buildings. Service: hand held calculator. Materials: prerecorded magnetic program cards; program operates on TI-59 hand held calculator. Cost: $50.

National Solar Heating and Cooling Information Center
P.O. Box 1607
Rockville, MD 20850
202-223-8105
RSVP: Ouput: heating, hot water and cooling loads, economic analysis. Service: computer. Materials: computer tape (containing copy of program and weather data), user manual and programmer's manual; program operates on CDC and UNIVAC computers. Cost: $175.

Guard, Incorporated
Attn: Robert Henninger
7440 N. Natchez Ave.
Niles, IL 60648
312-647-9000
SCOUT: Output: heating and cooling loads. Service: computer. Materials: see contact.

Guard, Incorporated
Attn: Robert Henninger
7440 N. Natchez Ave.
Niles, IL 60648
312-647-9000
SEE: Output: heating and cooling loads. Service: output data. Materials: printouts, worksheets, user manual. Cost: $180. (base price) × $474.60 (per system) × 1 to 4 (number of zones) × $0.80 (fee).

Solar Environmental Engineering
2524 E. Vine Dr.
Ft. Collins, CO 80524
SEEC IV: Output: annual heating load, economic analysis. Service: hand held calculator. Materials: prerecorded magnetic program; program operates on HP-67, HP-97 or TI-59 hand held calculator. Cost: $125.

Solarcon, Inc.
Attn: Roderick W. Groff
607 Church St.
Ann Arbor, MI 48103
ST-33 and ST-34: Output: Hourly heating loads with various vents and walls. Service: hand held calculator. Materials: prerecorded magnetic program cards; program operates on TI-59 hand held calculator with printer. Cost: $135/each.

Berkeley Solar Group
Attn: Bruce Wilcox
1215 Francisco St.
Berkeley, CA 94703
415-843-7600
SUN: Output: daily, monthly and annual heating load. Service: output data. Materials: program source code; programs operate on CDC 6400 computer. Cost: $50 plus $25 hourly.

National Center for Appropriate Technology
Attn: Larry Palmiter or Terry Wheeling
P.O. Box 3838
Butte, MT 59701
406-723-5475
SUNCAT: Output: heating loads in buildings, solar fraction. Service: computer. Materials: program source code, user manual; program operates on Data General Eclipse. Cost: nominal charge.

Sun-Pulse II Joint Venture
138 Mt. Auburn St.
Cambridge, MA 02138
617-491-0961
SUN-PULSE II: Output: daily and monthly heating loads. Service: hand held calculator. Materials: prerecorded magnetic program cards: program operates on TI-59 hand held calculator, user manual, weather tables (based on TMY-SOLMET). Cost: $100.

Syracuse University
Department Mechanical Engineering
Attn: Dr. Manas Ucar
Syracuse, NY 13210
315-423-3038
SYRSOL: Output: daily, monthly, and annual heating loads and energy consumptions, economic analysis. Service: computer.

Total Environmental Action Inc.
Church Hill
Harrisville, NH 03450
603-827-3374
TEANET: Output: hourly heating load in building with Trombe wall. Service: hand held calculator. Materials: prerecorded magnetic program cards; program operates on TI-59 hand held calculator with PC-100A printer. Costs: $95.

GK Associates
157 Stanton Ave.
Auburndale, MA
THERMAL SYSTEMS DYNAMICS: Output: systems performance. Service: computer. Materials: source deck (cards), tape or disc; program operates on IBM 360/75, IBM 370/158 computer.

The Trane Company
Attn: Neil R. Patterson
3600 Pammel Creek Road
La Crosse, WI 54601
608-782-8000
TRACE: Output: heating, hot water and cooling loads, monthly fuel and electrical consumption, economic analysis. Service: computer. Materials: program will be available on a time-sharing basis as of the fall of 1979; program operates on an IBM 370/168 system. Cost: $500 to $3,000. (program is not presently available on a purchase or lease arrangement).

University of Wisconsin
Solar Energy Lab.
Attn: John C. Mitchell
1500 Johnson Drive
Madison, WI 53706
608-263-1589
TRNSYS: Output: hourly heating, hot water and cooling loads and system and component performance. Service: hand held calculator, computer. Materials: prerecorded magnetic program cards, source deck (cards), user manual, program operates on CDC, IBM, UNIVAC computers or uses hand held calculator, time-sharing is available. Cost: $200; $40 additional for cards.

University of California
Attn: Arthur H. Rosenfeld
Lawrence Berkeley Laboratory
Berkeley, CA 94720
415-842-2740, X5711 or 486-4000
TWO-ZONE: Output: monthly heating and cooling loads for the residence as well as breakdown by contribution (windows, walls, infiltrations, etc.) Service: computer.

University of Washington
Dept. Mechanical Engineering, FU-10
Attn: Dr. A.F. Emery
Seattle, WA 58195
206-683-5338
UWENSOL: Output: heating and cooling loads. Service: computer. Materials: 7 track tape, source deck (cards), user manual: program operates on CDC 6000 computer. Cost: $200 (tape), $250 (cards), $20 (user manual).

California Energy Commission
Residential Programs and Standards Office
CAL/CON
1111 Howe Avenue
Sacramento, CA 95825
916-920-6321

Gibson Yackey Trinidade/Assoc.
Robert Z. Gibson, M.E.
3118 Fulton Avenue, Suite 1
Sacramento, CA 95821
916-483-4369
AC PROG 1 & ENERGY 1

Boeing Computer Service Co.
Richard A. Chesbrough
530 Beraut Drive, Suite B
Sacramento, CA 95814
916-447-2113
CAL/CON 1

Energy Management Service
Robert M. Helm
0435 S.W. Iowa
Portland, Oregon 97201
503-244-3613
EP 3.3

## 17. More Product Information, Sources of information

Your library should have some. If it doesn't ask them to order.

### 17.1 Guides

Standard & Poor's Register
345 Hudson St.
New York, NY 10014

Thomas Register
One Penn Plaza
New York, NY 10001

Sweets
McGraw Hill Info. Systems
1221 Ave. of the Americas
New York, NY 10020

McRaes Blue Book
Hinsdale, IL 60521
Includes corporate, product, and trade name indexes. *"I like this one best."* ed.

ASHRAE
345 East 47th St.
New York, NY 10017

Handbook of Fundamentals (1977) $42
Applications Handbook (1978) $42
Systems Handbook (1976) $42
Equipment Handbook (revised June 1979) $42

Wood Stove Directory
P.O. Box 4474
Manchester, NH 03108

Solar Energy AV Materials Catalog
free — NSHCIC
800-523-2929
From building a solar greenhouse to Village Homes solar houses. Over 100 listings.

### 17.2 Distributors and Warehouses (see also intro)

Pelikan and Lindsay Inc.
2594 Leghorn St.
Mountain View, CA 94043
415-961-7600
Bob Pelikan
Includes active solar, passive components, and hydronic systems.

Real Gas and Electric
P.O.B. 5
Santa Rosa, CA 95402
707-526-3400
Wind and solar equipment.

Camping World
Beech Bend Road
Bowling Green, KY 42191 (free)
12VDC system parts

Cumberland General Stove, $3.
Route 3
Crossville, TN 38555
Stoves, castiron, buggys, etc. Super!

Lemmon Hardware and Appliances, $2.
Kidron, OH 44636
215-857-5441
Exec. access for hard to find items including coal and wood kitchen ranges, gas refrigerators, and electric bed sheet (better than electric blanket).

Airborne Sales, $1
P.O. 2727 Dept. PSI
Culver City, CA 90230
Surplus source for motors, generator, etc. Several PM motors in stock — cheap!

M & E Marine Supply Co., $2.
P.O. Box 601
Camden, NJ 08101
Marine Supply Catalog, sail runners & sail track for insulated drapes. Exc.

### 17.3 Book Sellers

Solar Data
13 Evergreen Road
Hampton, NJ 03842
N/C
One of the best sources for solar and energy books, mail order

Mothers Bookshelf
P.O. Box 70
Hendersonville, NC 28739
Over 600 books and plans. $1.

Owner Builder Publications
P.O. Box 817
North Fork, CA 93643
The best series of books. The Owner Built Home, Stone Masonry, etc. SASE.

Garden Way Publishing
Charlotte, VT 05445
A fine series of homesteading, solar, wind and water power books. N/C.

The Can Do Bookstore
83 Bloor Street West
Toronto, Ontario M5S 1M1
Over 7,000 titles.

EARS
Box 545
La Veta, CO 81055
A fine selection of books. N/C.

Solstice Publications
Box 2043
Evergreen, CO 80439
A good series of solar and AT books.

Whole Earth Truck Store
c/o Zen Center
300 Page St.
San Francisco, CA 94102
A wide range of A.T.

Harrowsmith Books
Camden East
Ontario, K0K 1J0
Gardening, Country Living, Solar, etc.

Magisal Books
1588 Barrington St.
Halifax, N.S. B3J 1Z6
A good AT source.

## 17.4 Source Books

Books worth finding but hard to find.

Publications currently available from NCAT
P.O. Box 3838
Butte, MT 59701

*The Values of Appropriate Technology and Visions for a Saner World*, $.50

*Furnace Efficiency: Space & Water Heating Energy Saving Measures*, $.65

*Passive Solar Heat for New Home Construction*, $.65

*The Push Method for Fast, Simple Caulking*, $.50

*Heating With Wood: Wood Stove & Fireplace Heating Guidelines*, $.75

*Solar Bibliography (2nd Edition)*, $.25

*Wind Bibliography*, $.25

*Organizing Community Gardens Bibliography*, $.25

*Economic Development Bibliography*, $.25

*Building and Energy Bibliography*, $.25

*Alternative Waste Systems Bibliography*, $.25

*NCAT Briefs*, Vol. 1, No. 1, August, 1978. Something Old, Something New, Something Borrowed, Something Due; Women & Appropriate Technology, $.75 per copy
*NCAT Briefs*, back issue Jan. '78 Agriculture, Free
*NCAT Briefs*, back issue May '78 Solar, Free
*NCAT News*, back issues, Free

Minimum Cost Housing Group
School of Architecture
McGill University
3438 University Street
Montreal, Quebec H3A 2A7
Payment must accompany all orders.

*The Ecol Operation — Ecology × Building × Common Sense*. Describes the experience of developing and building a habitable low-cost house which incorporates asbestos super-tiles of sewer pipes, sulphur lock-blocks, modular logs, roof-top solar stills and a re-cycling toilet. Also includes excerpts from "The Problem Is". Revised edition 128 pp, 1975. Also available in French. $5.00

*Stop the Five Gallon Flush! — A Survey of Alternative Waste Disposal Systems*. A classification and description of sanitation methods and a catalog of 72 systems. New chapter on composting. 88 pp. Revised edition December 1979. $4.00.

*Use It Again, Sam*. An investigation into the design of consumer containers for re-use as building components. 60 pp., May 1975. $3.00.

*Rooftop Wastelands*. A description of a two year project in community rooftop gardening, includes sections on organics, solar cold frames and container gardening. 32 pp, November 1976. $2.00

*Water Conservation and the Mist Experience*. Describes atomization and mist spraying for showering and washing. Nozzle and shower testing. Plans for a do-it-yourself mist shower that uses only one litre of water. A catalogue of domestic water-saving devices. An extensive abstracted bibliography. 60 pp., July 1978. $4.00.

Living Systems
Rt. 1, Box 170
Winters, CA 95694

*A Strategy for Energy Conservation*. (1974) Jonathan Hammond, Marshall Hunt, Richard Cramer, and Loren Neubauer. (52 pages) $5.30. This work describes energy use in Davis, CA, including monitoring and research work conducted in apartment buildings. It includes a program for reducing energy use in existing buildings; an energy conservative building code is described; solar house design for the Davis climate is described; planning policies to reduce energy use are suggested.

*The Davis Energy Conservation Report: Practical Use of the Sun*. (1977) Jonathan Hammond, David Bainbridge, Marshall Hunt, Bill Kopper, Bruce T. Maeda, and Jim Plumb. 128 pp. $10.00. This is a summary of the work conducted for the City of Davis during 1975-76. It includes a description of work on passive heating and cooling; two model low cost passive solar houses with 100% natural cooling and 90% natural heating; land use and transportation planning for every conservation; public education efforts; and an evaluation of the code implementation process.

"The Davis Residential Energy Conservation Building Code: Resolution and Ordinance as Passed" (1975) Jonathan Hammond and Marshall Hunt. 20 pp. $2.50. This is the energy conserving building code as passed by the City of Davis.

"Bibliography: Energy Conservation" (1976) David Bainbridge. A list of the best articles and books on solar energy and energy conservation. Free with self addressed stamped envelope; $1.00 otherwise.

Earthmind
4844 Hirsh Rd.
Mariposa, CA 95338

*Wind and Windspinners — A Nuts-'n'-Bolts Approach to Wind-Electric Systems* by Michael Hackleman, 1974 with 1975 update newsletter, 140 pp. A do-it-yourself manual on building working wind-electric units, simply and inexpensively. Donation $8.

*The Homebuilt, Wind-Generated Electricity Handbook* by Michael Hackleman, 1975, 200 pp. Wind Power Digest says, "...superb. A logical and much-needed extension of the first book...strongly recommend adding it to your home bookshelf." Enjoyable reading, and *usable*. Donation $8.

*Electric Vehicles — Design and Build Your Own* by Michael Hackleman, 1977, 202 pp. First in a planned series of three EV books, this is a manual to keep on your workbench while building. It takes a simple approach to the electronics, and has extensive safety and street-legal info. Donation $8.

Ecotope Group
2332 E. Madison
Seattle, WA 98112

*A Solar Water Heater Workshop Manual*. Ken Eklund, et al., revised March 1978, 40 pp. $5.00 A step-by-step instruction manual on the construction of a flat plate, tube-on-sheet, parallel pipe, thermo-siphon solar hot water system. Includes complete list of tools and materials needed for construction and provides guidelines for producing your own "hands-on" workshop.

*A Solar Greenhouse Guide for the Northwest*. Tim Magee, May 1978, 66 pp. + working drawings. $5.00 An instruction guide to passive attached solar greenhouses for food and home heating in the Pacific Northwest. Includes tools and materials, guide for heat loss analysis of a home, how to do a solar site survey, and more!

*Report on the Design and First Year Operation of a 50,000 Gallon Anaerobic Digester at the State Honor Farm Dairy, Monroe, WA*. Elizabeth Coppinger, et al., July 1978, 88 pp + engineering drawings. $5.00. The engineering specifications, biological system, net energy production and economics of a full-scale methane digester are evaluated in the context of the first year's operation.

*A Preliminary Evaluation of Potential Solar/Conservation Remodel Candidates*. David Baylon, et al., November 1977, 66 pp. $5.00. A report prepared for the Office of Conservation, Seattle City Light on the potential for solar remodels as contributors to conservation. Includes projected costs, materials and expected payback for five specific remodel designs.

*A Preliminary Assessment of the Unique Features of a Parabolic Aquaculture/Greenhouse*, Davis Straub and Evan Brown, with *The Effect of Solar Energy on Aquaculture Support: A Preliminary Assessment*, Elizabeth Coppinger, September 1977, 22 pp. $2.00. Report on first year operating experience and research at the Pragtree Farm greenhouse.

*A Feasibility Study and Business Organization Plan for Solar Greenhouse Production, Distribution and Support Services*. Ecotope Group, January 1978, 104 pp.

*Operation of a 50,000 Gallon Anaerobic Digester, Monroe State Dairy Farm: Quarterly Technical Status Reports*. $1.50 per quarter. Ecotope Group reports on research at the Monroe Digester.

# The Passive Solar Institute
## P.O. Box 722
## Davis, CA 95706

# Progress Report: 1979

Publishing:

### The First Passive Solar Catalog, January

The Catalog includes not only access to passive products but also the design and theory involved in passive solar heating and cooling, passive solar water heating, daylightling, solar access planning, and much more. It was very well received by the critics.

### Village Homes' Solar House Designs, April

Rodale Press published this guide to 43 of the solar homes in this innovative solar, energy conservation subdivision near Sacramento, California. Energy use is less than 1/2 that of surrounding developments. The development is a practical demonstration of passive solar design and was privately funded and built.

### Village Homes Slide Set

A descriptive set of 20 slides was prepared to accompany the Village Homes book. The slides were prepared to illustrate many of the passive systems in Village Homes and their transferability.

### 1980 Year of the Sun Calendar

Prepared to promote all solar uses in 1980.

### Technical Report #1: Water Wall Passive Solar Homes

This report was presented in January to the 3rd National Passive Solar Conference. It was revised and printed in *Solar Age* magazine in August and is now available from PSI.

### Technical Report #2: Insulated Shutters and Drapes.

The report was carried to the draft stage by the end of 1979. The section on how to build shutters was completed in mid-year and accepted by Rodale press. It appeared in the January 1980 issue of *Organic Gardening* magazine.

### Residential Street Design for Energy Conservation

The key role streets play in energy use in residential developments is high-lighted in this paper. It was reprinted from an earlier report completed by the President of the Passive Solar Institute.

### Patwin: An Energy Self-sufficient New Town

PSI was the lead group in the energy studies for a self-sufficient new town near Davis, California. The results of these studies will be published in 1980.

## Water Wall Solar House Plans

Design:

### Passive Solar House Plan for the Sun Belt

A mass market design for a three bedroom, two bath water wall house. Plans include design, specifications, and performance estimates for ten cities.

PERFORMANCE OF THE SOLAR CLERESTORY WATER WALL HOUSE

| Area | Solar Heating | Cooling |
|---|---|---|
| Los Angeles, CA | 100% | 100% |
| Sacramento, CA | 94% | 100% |
| Medford, OR | 72% | 100% |
| Phoenix, AZ | 100% | 50% |
| Denver, CO | 82% | 100% |

## A Passive Solar Doctor's Office

### Passive Solar Manual

The development of a convenient and complete passive solar design manual has been of critical importance for the last three years. In 1979, we worked very closely with the solar group at California Polytechnic University's Architecture Department on the development of this type of handbook. It will be published in 1980.

### Solar Simulator

A simple technique was developed to undertake complicated shading studies, sun exposure studies, and daylighting studies. The chart and plans for construction were tested and finally published.

### Passive Solar Doctors Office: Grants Pass

PSI provided the solar design consulting for a small commercial building in Grants Pass, Oregon. The building was completed in late 1979.

### Miscellaneous:

PSI also undertook many other projects in 1979. These included a Seminar on Energy Efficient Neighborhood Design for the Sacramento Municipal Utility District, solar design consultations, and solar tax credit consultations.

# The Passive Solar Institute
## P.O. Box 722
## Davis, CA 95706

# Proposed Projects: 1980

Self-supporting: Funded by PSI

For details on any of these projects, please write PSI. We will provide a complete proposal to any legitimate sponsor. We are and will remain a private profit organization however and do not qualify for any tax credit or tax deduction programs.

### The Second Passive Solar Catalog

The Second Catalog will feature retrofit passive solar applications and design for cold climates. It will also include expanded product listings and more complete consultation information.

### Passive Solar House Plan for the Cold North

A mass market design for the cold north. This house will feature super-insulation and super-weatherization. Preliminary calculations indicate 90% heating in Winnipeg.

### A Child's Guide to Passive Solar Design

A much needed introduction to passive solar design for children of all ages. Includes many experiments to help demonstrate passive solar heating and cooling and hot water heating.

### Passive Solar Water Heater Plan Set

Plans for a three tank PSWH for temperate and hot climates.

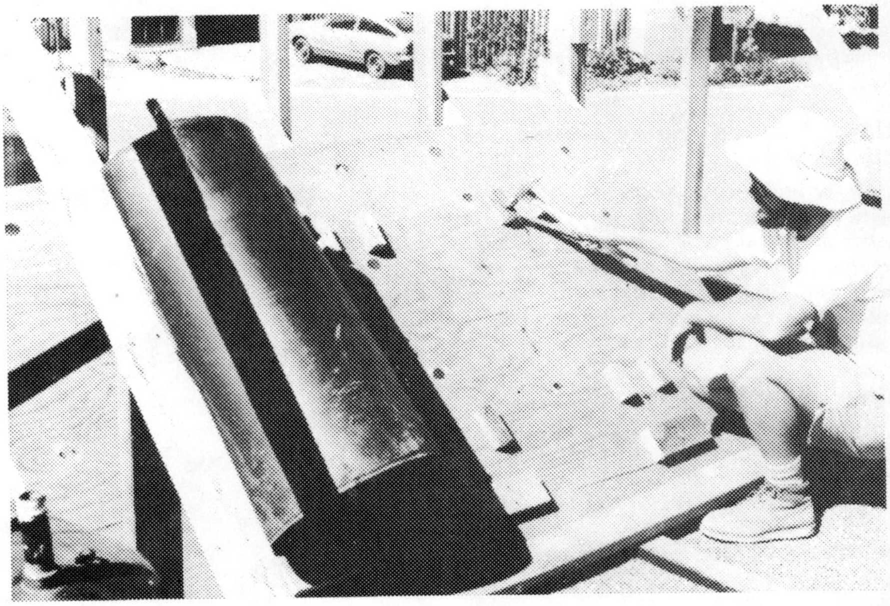

**A Passive Solar Water Heater**

## In Search of Support: Funding Required

Technical Reports: Passive solar water heaters, air-to-air heat exchangers, energy self-sufficiency for the home, super-insulation, super-weatherization. **$5,000 each**

An Energy Self-sufficient New Town, continuation of work started in 1979. To include technical backup and references. **$7,500**

Research and Evaluation of Passive Solar Water Heaters in Three Climates, continuation of work started in 1978. Test will include evaluation of new designs in hot, temperate, and cold climates. **$14,000**

Subdivision Retrofit study, an evaluation of a large subdivision built in the energy intensive era. Study will evaluate cost and strategies for major retrofit and potential savings. **$18,000**

Energy Efficient Subdivision Design Manual, a short, crisply written guide for developers on passive solar design, solar access, and energy saving subdivision design. **$6,000**

Low-cost Vertical Axis Wind Machine, design and evaluation of two configurations of vertical axis machines that promise to be inexpensive and durable. **$6,000**

Design and evaluation of a prototype watermill of a new design that promises to be inexpensive and adaptable to a variety of applications. The low head mill should be of interest in third world countries and for rural applications in the U.S. and Canada. **$6,000**

Low-cost Passive Solar Design for the Third World, this report will provide details on application of passive solar-heating and cooling and hot water system. Designs will be developed for construction with available materials and construction techniques. **$8,500**

Operation of a Passive Solar Resource and Research Center. The work of the Passive Solar Institute would be facilitated if a central well equipped center for operations were available The project would involve the acquisition of an old commercial or residential building, its retrofit with energy conservation and passive solar design, and its operation as a resource center for the community. The retrofit would also be used for a case study and nucleus for a retrofit manual. **$150,000**

Design Studies for an Autonomous House with working models of key elements. The practicality and performance of passive solar houses is now well established. This project would begin work on the next step, the autonomous house, fully self-sufficient in power and food. Power system to be based on super energy efficient appliances, 12 vdc supply from wind and photo-voltaics. Water and waste to be autonomous as well, with cistern roof collection, composting toilet, and low water use facilities. **$12,000**

# Chapter 9: Consultants

The following listing of passive solar consultants is alphabetical by state and province. We have done this to help you find a lcoal source when possible although it is not uncommon to work with an architect in another state or across the country. Each lising includes the type of work the consultant has done. The key for this is the bottom of this page.

The listing was complete as we could make it, houwever, we are certain there are many more consultants out there and we would like to include them in next years catalog. If you are, or know a passive consultant or builder who is not listed, please send a self-addressed stamped envelope to PSI, Attention: Forms.

The listing of a consultant here in no way constitutes a warranty or recommendatoin. Like all professional consultants, they vary widely in terms of their ability, knowledge, and cost. Before you hire anyone for design or consultation, ask to talk with previous clients and look at his previous designs. Did they come in at cost/ Was the builder satisfied with the plans? Did it work? Does it look like you would like it to look on your project? etc. Remember these words from Vitruvius:

*"(in the Greek city of Ephesus) When an architect accepts the charge of a public work, he has to promise what the cost of it will be. His estimate is handed to the magistrate, and his property is pledged as security until the work is doen. When it is finished, if the outlay agrees with his statement, he is complimented by decrees and marks of honor. If no more than a fourth has to be added to his estimate, it is furnished by the treasury and no penalty is inflicted. But when more than one fourth has to be spent in addition to the work, the money required to finish it is taken from his property.*

*Would to God that this were also a law for the Roman people, not merely for public, but also for private buildings . . . architects themselves could be forced, by fear of the penalty, to be more careful in calculating and starting the limit of expense."*

We do a wide range of consulting at PSI and will work by mail on the following topics:

- Passive solar heating and cooling
- Passive solar water heater design
- Passive solar subdivision design
- Energy conservation and energy audits
- Information referrals and much more.

Our charge rates are typically $25 hr. for residential projects, $35 hr. for commercial projects, and $45 hr. for government projects. One of our most popular services is a plan check and quick review of solar design. This is usuallly very helpful and can be done for less than $50 in most cases. A more detailed review iwth computer calculations will cost considerably more but can be very worth while. These reviews may save many thousands of dollars in some cases.

## Consultant Listing Key:

**5. Type**
- A. Consulting
- B. Design
- C. Engineering
- D. Research & Development
- E. Energy Audit
- F. Energy Conservation
- G. Daylighting
- H. Other Passive

**6. System Used**
- A. Green House/Soalrium
- B. Water Wall
- C. Trombe Wall
- D. Underground
- E. Simple Passive
- F. Roof Pond
- G. Hybrid
- H. Other Passive

**7. Emphasis**
- A. Residential
- B. Commercial
- C. Industrial

**8. Passive Water Heating Sytem Used**
- A. Bread Box
- B. Thermosiphon
- C. Wood Assisted

**9. Other Specialties**
- A. Movable Insulation
- B. Shading
- C. Ventilation
- D. Weatherization
- E. Wood Heating
- F. Photo Voltaics
- G. Wind
- H. Hydropower
- I. Active Soalr
- J. Back Up System

**10. Articles & Books**

**11. Years in Business**

**12. % Work**

**13. Licenses & Registration**

**14. Hourly Fee (opt.)**

In categories 6 - 9, the letterd **d** stands for designed, **u** means underway, and **c** indicates completed.

# United States

## ALASKA

ALASKA RENEWABLE ENERGY ASSOCIATES
308 G Street, Suite 311
Anchorage, AK 99501

1-907-279-5904

5. AB,DEFGH 6. A-d, C-d, D-d, E-d, H-d 7. A-c 8. A-d, C-d 9. A-d, D-u, E-c, G-c 11. 8 months 12. 50% private, 50% government

Super Insulated Homes
P.O. Box 73876
Fairbanks, AK 99707
1-907-452-5323

5. ABC,EFGH 6. E 8d, 2u, 3c 7. A-u, -c; B-d 10. *How to Build A Super Insulated House*, Ed McGrath, *Bob Roggasch's Workshop Book*, Bob Roggasch 11. 1 year 12. 100% private 13. Assoc. License including Engineering and Arch.

## CALIFORNIA

Dennis A. Andrejko, AIA SEAgroup
418 Broad Street
Nevada City, CA 95959
1-916-265-9458

5. BC,F,H 6. A 3d, lu 2c; C 2d, 2c; D 2d, lu, lc; E 10d, 2u, 8c, G 5d, 2u, 3c 7. A 45d, 5u, 16c; B 3d, lu. 8. A 10d, 5u, 2c; B 20d 10u, 10c. 10. *Natural Solar Architecture*, illustrator, 1978.*Passive Solar Architecture*, Author, 1980."Energy and Office Architecture," Thesis, MIT, 1977."Lightweight Building Materials for the Storage of Solar Energy," MIT, 1977. 11.3 years 12. 95% private, 5% government
13. California Architect, AIA member

Pete Antoniak. PE
SOL AIR ENGINEERING
1056 Crestview Drive
Millbrae, CA 94030
1-415-583-1939 (W) 597-4003 (H)

5. ABCDEFGH 6. A 1d, E 1d, F 1d, G 1d 7. A 1d, B 1d 8. B 1d 9. A 1d, C 1d, E 1d, F 1d, F 1d, 1c; G lu, I 3c, J 1c 10. Photovoltaics for Pipeline Cathotic Protection 11. 2 years 12. 100% private 13. PE Calif. #C 23301

T. BALLARD BUILDING & DESIGN
P.O. Box 60
Duncans Mills, CA 95430
1-707-865-1497

5. ABCDEFGH 6. A-c, B-d, C-d, D-d, E-c, H-c 7. A-c, B-c 8. A-c, B-c, C-c 9. A-c, B-c, C-c, D-c, E-c,F-c,G-u, H-d, I-c, J-c 11. 3 years 12. 100% private 13. Calif. Contractor #352314 MFA CSET

David Bainbridge
P.O. Box 722
Davis, CA 95616
1-916-544-3106 or 415-526-1549

5. AB,DEFGH 6. A 5d, 2u, B 2d, 1c; E 1d, 1u 7. A 8d, 3u, 1c, B 2d, lc 8. A 3d, 1u, 1c; C ld 10. *Village Homes Solar House Designs*, Rodale Press, 1979. *The 1st Passive Solar Catalog*, PSI, 1979. How to Build Water Walls, *Solar Age*, order from PSI *Planning for Energy Conservation*, City of Davis, CA 11. 7 years 12. 90% private, 10% government 13. Solar Specialist I, II 14. $15-50

Eric N. Best, Ph.D
Solar Consultants
501 West Mariposa
El Segundo, CA 90245
1-213-322-5876

5. ABCD 7. A-d, B-d 11. 2½ years 12. 50% private, 50% government

Joseph C. Broyles
2100 Mt. Diablo Scenic Blvd.
Danville, CA 94526
1-415-837-1279

5. AB,F.H (generally, consulting on building and retrofit options) 11. 11 months part-time 12. Private

John W. Burton
Integral Design
3825 Sebastapol Rd.
Santa Rosa, CA 95401
1-707-528-0616

5. AB,D 6. A c2, Ed1, Hd1c1 7. A d1, c8, B d c1 8. a d2 c9 9. Fd1 10. 3rd Passive Conference "Passive Solar Retrofit, Sacramento" Calif. Energy Commission "Solar Domestic Hot Water" and "Solar Information Packet" Solar Energy Advocates News, April 77 - March 78 11. one year 12. 100% private 13. Contractor 14. $10 hr up

Victor Conforti
755 Broadway
Sonoma, CA 95476
1-707-996-7923

5. ABC,EFG H (Industrial space cooling, wine storage) 6. A 3c, C 1c, D 1c, G 1c, H 1u 7. A 2d, 2u, 4c; B 1u; C 1u 8. A 1u; B 1c 9. C 5c, E 3c 10. 3rd Passive Conference "Mild Climate Passive" 11. 7 years 12. 100% private 13. Architect (Calif.) 14. $35

Richard Conrat
Sunworks
1501 Felta Rd.
Healdsburg, CA 95448
1-707-433-3693

5. AB,F 6. A 2u, B 4u, G 6d, 2c 7. A6 8. C 6d, 6u 9. C 6d, E 6d, 3c, I, 75d, 75c 10. "Solar Woodstove Connection"*Solar Age*, Sept. 79. 4th Passive Conference, "Solar-Woodstove Domestic Water Heating Systems" 11. 3 years 12. 90% private, 10% government 13. Contractor 14. $20 hr.

Polly Cooper
2065 McCollum
San Luis Obispo, CA 93401
1-805-543-4228

5. AB,FG 6. A-d, -u, -c; B-d, -u,-c; E-d, -u, -c 7. A-d, -u, -c; B-d 8. B-c 10. 2nd Passive Solar Conference Proceedings, pp. 133 &223 3rd Passive Solar Conference Proceedings, p. 616. "Retrofitting a 1910 California Bungalow." 11. 4 years 12. 100% private 14. $30

Richard Cushing, Principal
ARC SUN DESIGN ASSOCIATES
5907 Garlough Drive
San Jose, CA 95123
1-408-225-9550

5. AB,EFGH 6. A 14d, 4u, 3c; B 1d, 1u; C 2d, 1c; D 1d; G 3d, 1c; H 2d, 2u 7. A 15d, 4u, 3c; B 1d, 1u 8. A 2d, B 3d; C 2d 9. A 10d, 4u, 3c; B 16d, 5u, 3c; C 15d, 4u, 3c; D 16d, 5u, 3c; E 5d, 1u; I 8d, 6c 11. 4 years 12. Private

DAYSTAR SOLAR ASSOC.
2489 Gravenstein Highway
Sebastopol, CA 95472
1-707-829-1202

5. ABCDEFGH 6. A -c, B -u, E-c 7. A-c, C-u 8 A-c 9. A-c, B-u, C-c, D-c, E-c, F-c, I-c, J-c 10. 2nd National Passive Conference Paper 11. 1 year 12. Private 100% 13. General Contractor license, Solar license, & Resale license

Davis Alternative Technology Associates (D.A.T.A.)
P.O. Box 503
Davis, CA 95616
1-916-756-9300

5. ABCDEFGH 6. A 85d, 10u, 10c; B 78d, 5u, 25c; C 15d, 9u; D 5d, 2u; E 92d, F 10d; G 4d, 1c; H 8d, 4u, 2c 7. A (involved in over 500 designs) 15u, 100+c 8. 20d, 2u, 3c 10. "Suncatcher Monitoring Project Final Report" for September 1977 to December 31, 1978 *Breadbox Design for Fred Young Farm Labor Center in Indio*(CA) Solar "Breadbox" Simulation Using D.A.T.A.'s TBBX Code Simulation and Performance of Double Envelope Houses in Cold Climates 11. 3 years 12. 40% private 60% Gov. 13. Registered architect & Engineer 14. $40 (less 5 hr.), $35 (5-10 hrs.), $25 (more than 10 hrs.)

Daystar Solar Associates
2489 Gravenstein Hwy.
Sebastopol, CA 95472
1-707-829-1202

5. AB,EF,H 6. A 2d, 1u, 2c; B 1u; E 1u 7. A 2d, 1u 8. A 2c, C 1u, 1c 9. A 1d, 2c; I 1d, 4c 11. 5 months 12. 100% private 13. State Contractors Lic. #367448 (Tim O'Conner)

Deloitte Hasdins & Sells
101 Park Center Plaza
San Jose, CA 95113
1-408-998-4000

5. A,E 7. A-c, C-c 11. 4 years 12. 100% private

Cynthia Easton
1723 J Street
Sacramento, CA 95814
1-916-443-5911

5. B 6. B-c 7. A-c, B-d 8. -c 9. B-c, C-c 10. Davis Energy Conservation Report 11. 5 years 12. 75% private, 25% government 13. California Architect August, #C10344

Ecological Life Systems Institute
P.O. Box 15991
San Diego CA 92115
1-714-286-4301

5. AB,D,F,H, 6. A-c 7. A-c 8. -c 11. 3 years 12. 100% private 14. Ability to pay

Charles Eley Associates
342 Green Street
San Francisco, CA
1-415-398-6535

5. AB, EFGH 6. A-c, B-c, C-c, D-d, E-c, F-d, G-u, H-u 7. A-c 8. A-d, B-d 9. A-c, B-c, C-c, D-c, E-u, G-d, I-c, J-c 10. "Encouraging Passive Solar Applications Through Regulation," *Daily Pacific Builder*, 7/27/79. *Energy Design Guidelines*, *Zingel Canyon Development*, Kennewick, WA 5/8/79. CCAIA Position Paper on the California Non-Residential Energy Standards, 11/79. "Energy Efficient Buildings without Regulation?" *Daily Pacific Builder*, 7/27/79 11. 3 years 12. 75% private, 25% government 13. Architect (Calif) 13. $18-$35

Energy Specialties
9312 Greenback Lane
Orangevale, CA 95662
1-916-988-1208

5. A,F 6. A-d, -u, -c 7. A-d, -c; B-d, -c 8. B-d, -u, -c; C-d, -u,-c 9 F-d, -u, -c; G-d, -u, -c; H-d, I-d, -u, -c 11. 3+ years 12. 90% private, 10% government 13. General Contractor 14. $20, $35 w crane truck

Jim Finley
Solar Designs
2864 Ray Lawyer Drive
Suite 209
Placerville, CA 95667
1-916-626-3345 or 1-916-626-0364

5. ABC,F 6. A-d, -u, -c; B-d, C-d; E-d, -u, -c 7. A-d, -u, -c, B-d 9. A-d, -u; B-d, -u, -c; C-d, -u,-c; E-d, -u, -c; I-d, -u,-c, J, -d,-u,-c 2 years 12. 100% private 13. Calif. Contractors, Calif. Brokers 14. Fees determined on a project basis

Good Earth Glass & Sun
729 Heinz Ave. #10
Berkeley, Ca 94710
1-415-843-2019

5. ABGH 6. Ad 7. A 2d 11. 2 years 12 100% private

The Habitat Center
162 Christen Drive
Pacheco, CA 94553
1-415-825-8434

5. ABCDEFGH 6. A 16d, 7u, 15c; C 1d, E 1d 7. A 18d, 7u, 15c 9. B 18d, 7u, 15c; C 18d, 7u, 15; I 1d 10. *Solar For Your Present Home* In cooperation with Berkeley Solar Group Chapter 7 of *The Sol Home Book* by Anderson/Riordan A Step-by-step manual on th construction of passive greenhouses should be available by the summer of 1980 11. 3 years 12. 85% private, 15% government 13. Non-profit corporation 14. $25-$35/hour

Kenneth Haggard
San Luis Solar Group
2065 McCollum
San Luis Obispo, CA 93401
1-805-543-4228

5. AB,D 6. A-d, -u, -c; B-d, -u, -c; E-d, -u, -c; F-d, -u, -c 7. A-d, -u -c; B-d 8. B-d, -u, -c 9. A-d, -u, -c; B-d, -u, -c; C-d, -u, -c; D-d, -u, -c 10. *Solar Energy*, Vol 19 #4, 1977 Ch. 17-18, *The Passive Solar Book* 1st, 2nd, and 3rd Passive Conferences *Research Evaluation of a System of Natural Air Conditioning* 11. 7 years 12. 50% private, 50% government 13. Arch. license 14. $30

Interactive Resources, Inc.
117 Park Place
Point Richmond, Ca 94801
1-415-236-7435

5. ABCDEFGH 6. A-c, C-c, D-u, E-c, G-c, H-c 7. A-c, B-c 11. 6 years 12. 60% private, 40% government 13. Architecture/Engineering 14. $20-45

David C. Kantz
Bailey and Kantz
Geologic-Structural and Solar Engineering
P.O. Box 305
Garden Valley, CA 95633
1-916-333-1444 or 622-7725

5. ABC,F,H 6. A 9d, 5u; C 7d, 3u; D 2d, E3d, G 2d, 2u 7. A 9d, 5u,; B 2d, 2u 8. B 1d, C 1d 9. A 1d, B 1d, 1u; E 1d, I 1d 11. 1 year 12. 100% private 13. Registered Civil Engineer 14. Depends on scope of project

Lawrence Kasparowitz
P.O. Box 160
Monte Rio, CA 95462
1-707-865-2175

5. B,F 6. A 1c, B 1u, E 2d 7. A 3d, 1u, 1c 9. B 1u 10. Upcoming book w/Ken Kern "The Owner Designed Home" 11. 4 years 12. 100% private 13. Arch. (June 1980), Cont. (Nov. 1979)

Tom Kenison, the R.E.U.S.E.D. Project
1640 Chase Lane
El Cajon, CA 92020
1-714-444-3337

5. A,F 6. H (food dryer) -c, (water distiller) 1u 8. A -u, -c 9. B -c, C -u, -c; D -u K. Restoration of used hot water tanks, for residential passive solar applications 10. Three reports of local solar training and educational agencies for the California Solar Action Network's Solar Survey 1979 11. 6 month 12. Private 13. Member of California Solar Energy Assn. and California Solar Action Network.

Stanley Keniston, AIA
666 State Street
San Diego, Ca 92101
1-714-231-1312

5. ABCDEFGH 6. A 4d, B 1d, C 1d, E 2d, G 2d 7. A 40d 1c, B 2d, C 11d 8. C 1d 9. A 2d, B 35d, C 35d, E 3d, I 2d 10. "The Third Overlay" 11. 3 years 12. 90% private, 10% government 13. Architect, CA

Fred Klammt
127 La Granada
Moss Beach, CA
1-415-728-7046

5. ABC,EFGH 6. A 1d, B 1c, E 1c, H 1d, 1c 7. A 2d, 1u, 1c; B 1d 8. A 1u, B 1d 9. A 1d, B 1d, C 1d, 1c. E 1d, 1c. F 1c G 1u 11. 2 years 12. 100% private 13. California Certified energy auditor, Sellers license #SR BB 20-643400 14. Varies: $10-$50/hour

Living Systems
Route 1, Box 170
Winters, CA 95694
10916-795-2111

5. AB,D,FGH 6. A 1c, B 1d, 16c, C 2d, 1c, F 1c, H 1d 7. A 2d, 22c; B 1d 1u, C 1d 8. A 15c 9. A 15c, B 1c 10. *Site Planning for Solar Access* 1979 *The Davis Energy Conservation Report* 1977 *A Strategy for Energy Conservation* 1974 11. 5 years 12. 50% private, 50% govenrment 13. Landscape Architecture, Architect

Paul H. Lutton
406 Main St.
Vacaville, CA 95688
1-707-446-8341

5. ABCDEFGH 6. A 3d, D 1d, E 3d, 1c; F 1u, 1c 7. A -c, B 1u, 1c 9. A 6d, 3u, 3c; B 6d, 3u, 3c 11. 4 years 12. 100% private

Michael E. Luttrell
132 Soscol Avenue
Napa, CA 94558
1-707-224-4222

5. ABCD,FGH 6. A 6d, 2u, 1c; B 1d, E 10d, 2u, 1c; G 40d, 4u, 12c 7. A 40d, 6u, 15c; C 1d, 1c 8. C 15d, 6u, 2c 9. A 6d; B 6d, 3u C 6d, 2u; D 4d, 2c; E 12d, 4u, 1c; F 10d; I 40d, 4u, 12c; J 40d, 4u, 10c 11. 3 years 12. 100% private 13. General Contractor B-1, SC-44, C20, C4 Appl'd for 14. $30/hour

Gilbert M. Masters
1450 Mills Court
Menlo Park, CA 94025
1-415-497-3658

5. ABC 6. A 4d, E 8d, 4u, 2c 7. A 10d, 5u, 2c 9. E 1d, 1u, 1c, F 1d, G 1d, I 10d, 5u, 2c 11. 6 mos. independent, 3 years in another firm 12. 75% private, 25% government 13. PhD., P.E.

Horace McCraken
329 W. Carlos
Alturas, CA 96101
1-916-233-3175 or 1-714-755-1637

5. A 6. E 1c, G 3u 7. A 3u, 1c 8. A 8c, B 1c 9. G 1d, I 1u, 10c 10. *How to Build a Passive Solar Water Heater* 11. 21 years 12. 100% private 14. $35

Meckler Energy Group
Suite 1023
15760 Ventura Blvd.
Encino, CA 91436
1-213-995-7672

5. ABC, EF,H 6. C -d, E -d, G-d 7. B -d, -u, -c; C -d 9. C -d, -u, -c; I -d, -u, -c 10. *Energy Conservation for Buildings and Industry*, being published 1980, McGraw-Hill, *The Retrofitting for Energy Conservation Handbook*, being published 1980 Marcel Dekker, Inc. 11. 2 years 12. 70% private, 30% government 13. All engineering fields, construction

Murray Milne, Prof.
UCLA School of Architecture
Los Angeles, CA 90024
1-213-825-7370

5. ABC,G 6. A -d, D -d, -u; E -d, -u, -c, G -d H -d, -u 7. A -d, -u, -c; B -d, C -d 9. A -d, B -d, -u, -c; C -d, -u, -c; E -d, I -d, J -d 10. "Solar City 2025: The Impact of Passive Buildings," with Joe Lakin, *Proceedings of the Fourth National Passive Solar Conference*, International Soalr Energy Society, Kansas City, MO, Oct. 3-5, 1979 "Climate Based Building Design," with Baruch Givoni, Ch. 6, in *Energy Conservation Through Building Design*, ed. Donald Watson, McGraw Hill, 1979 "Solar-5: An Interactive Computer-Aided Passive Solar Design System," with Shin Yoshikawa, *Proc. 3rd Annual Passive Solar Conference, AS/ISES*, Jan. 1979 "Coastline Condominiums," *Architectural Record: Record Homes and Multi-Family Projects of 1978*, mid May 1978, p. 124 (Publication of Malibu project selected as one of six best multi-family projects of the year) 12. 80% private, 20% government

Mogavero and Unruh
811 J Street
Sacramento, CA 95814
1-916-443-1033

5. AB,D 6. A -d, -u, -c; B -d, -c; C -d; E -d, -c 7. A -d, -u, -c; B -d 8. A -d, -c 11. 2 years 12. 100% private 13. California Architect

Glen H. Mortensen, Inc.
1036 W. Robinhood Drive, #201
Stockton, Ca 95207
1-209-478-2670

5. ABC,EFG 6. A -d, -u, -c; B -c; C -d; E -d, -c; G-d, -c 7. A -d, -c; B -d 9. A -c, B -c, C -c, D -c, E -u, I -c, J -d, -c 11. 30 years 12. 100% private 13. Arch. Calif. C 2037 14. $35-40

Dick Munday
P.O. Box 1307
Tahoe City, Ca 95730
1-916-587-6802

5. ABCD,FGH 6. A 2d, 1u; B 1d, C 1d, 1u, 1c; D 1d, E 2c 7. A 4d, 2u, 2c 9. A 2d, 1u, 2c; B 1d, 1u, 1c; C 4d, 2u, 3c; E 4d, 2u, 3c; I 4d, 2u, 2c 10. "Construction and Performance of an Inclined Mass Trombe Wall w/Moveable Insulation," *3rd National Passive Conference Proceedings* 11. 5 years 12. 80% private, 20% government 13. Calif. Architect C-7916 14. $50

Net Energy Inc.
854 Ninth St.
Arcata, CA 95521
1-707-822-5926

5. AB,DEF,H 6. A 10d, 10c; E -d, -c; H -d, -c 7. A -d, -c 8. A 10d, 10c; B 2d, 2c; C -u 9. A -d, -c; D -d, -c; I -d, -c, J -d, -u 11. 2½ years 12. 20% private, 80% government

Dr. Loren W. Neubauer
3422 Monte Vista
Davis, CA 95616
1-916-573-2271 or 752-0529

5. ABC 6. E 1c 7. A 1c 8. A 1c 10. 30 technical articles published Book partly on Solar Energy, *Farm Building Design* 11. 30 years 12. 100% university 14. Variable

David B Norton
Moonshine Road, #71
Camptonville, CA 95922
1-916-288-3462

5. AB,FG 6. A -d, -u, -c; B -d, -u, -c; C -d; E -d, -u, -c 7. A -d, -u, -c 8. A -d, B -d, C -c 9. A -d, -c; B -d, -u, -c; C -d, -u, -c; D -d, -u, -c; E -d, -u, -c 10. *Grass Valley Union* "Solar Greenhouse Project for North San Juan School" 11. 3½ years 12. 100% private 13. General Building Contractor "B" #368990

Pacific Sun Incorporated
439 Tasso St.
Palo Alto, CA 94301
1-415-328-4588

5. ABCDE 7. B, C 10. *Other Homes and Garbage* 11. 4 years 12. 40% private, 60% government 13. Calif. PE/Mechanical

Paintridge Design and Development, Inc.
803 Russell Blvd, # 6
Davis, CA 95616
1-916-756-2148

5. A,B,H 6. A -d, -u, C -d, -u, D -d, E -d, -u, -c, G -d 7. A -d, -u, -c 11. 3 years 12. 100% private 13. Contractors License

Neil Pinney, AIA, Architect
Earth/Life Systems Design
1637 Seal Way, Seal Beach, CA 90740
1-213-431-0077

5. ABCD,FGH 6. A -c, B -c, D -c, E -c, H -c 7. A -u, -c, B -d 8. B -c 9. A -u, B -u, C -u, D -u, E -u, G -d, I -u, J-u 10 Energy and Architecture: *The American Dowser*; Feb. 1980 A New Nocturnal Air Cooling System, *Proceedings 2nd Passive Solar Conference* The Sun's Influence on Architectonic Form; *Energies*; Sept. 1975 11. 7 years 15. 50% private, 50% government 13. Calif. Arch. Cert. C 3064

E. Gary Schloh, AIA
213 Bean Avenue
Los Gatos, CA 95030
1-408-354-4551

5. AB 6. A -u, B -d, C -u, D -d, E -d, G -d 7. A -d, -u, B -d 8. A -u, B -u 9. A -d, B -d, C -d 11. 8 years 12. 95% private 13. Architect C9420

SEAgroup
418 Broad Street
Nevada City, CA 95959
1-916-265-9458

5. BC, FGH 6. A 3d, 1u, 2c; C 2d, 2u; D 3d, 1u, 2c; E 36d, 14u, 22c; G 5d, 2u, 3c 7. A 210d, 20u, 30c; B 3c, 2u, 1c 8. A 10d, 5u, 2c; B 20d, 10u, 10c 10. *Natural Solar Architecture*, authors, 1978 *Passive Solar Architecture*, authors, 1980 11. 7 years 12. 95% private, 5% government 13. Architects in: California, Colorado, New Mexico, NCARB (National) Certificate holders

Jacobson/Silverstein, Archs.
3026 Shattuck Ave.
Berkeley, CA 94705
1-415-848-8861

5. B 6. A 3d, 2c; E 6d, 2c 7. A 6d, 3c, B 2d 9. B 2d, 1c; C 2c 10. Pattern Language 11. 10 years 12. 100% private 13. California Architects

Sloan and Associates, Inc.
11505 Douglas Road
Rancho Cordova, Ca 95670
1-916-351-0150

5. ABF 6. E -c 7. A -c 11. 10 years 12. 100% private 13. A & B Contractor's Licenses

Gerald B. Steel
P.O. Box 2359
Rancho Santa Fe, CA 92067
1-714-789-0014, 753-9022, 744-1946

5. ABCDEF,H 6. A 17d, 6u, 5c; B 12d, 3u, 2c; C 1d, 1u; D 1d, 1u; E 6d, 1u, 2c; C 6d, 2u, 1c 7. A 34d, 10u, 5c; B 8d, 3u, 4c; C 1d, 1c 8. B 3d, 3c; C 1d, 1c 9. A 3d, 1u; B 21d, 6u, 5c; C 30d, 13u, 8c; D 30d, 13u, 8c; E 16d, 2u, 6c; G 3d, 1u, 2c; H 1d; I 2d, 1c; J 16d, 4u, 4c 10. "Solar Aquafarms, Inc., Greenhouse Lab," *3rd National Passive Solar Conference* "A Magnificient Entrance, Energy Wing: 1979 *ISES International Conference* "Solar ACTS, Air-Inflated Cable Tensile Structure," *Solar Greenhouse Digest* "4" Thick Masonry Walls," *4th National Passive Solar Conference* 11. 7 years 12. 100% private 13. Registered Civil Engineer C22445 14. $65

Jonathan Stoumen, Architect
P.O. Box 587
Miranda, CA 95553

5. ABCD,FGH 6. A 5d, 4u, 1c; C 1d, 1u; D 3d, 2u, 1c; E 5d, 3u, 2c 7. B 2d, 2c; C 1d, 1c 8. B 10d, 4c; C 10d, 4c 9. B 8d, 4u, 4c; C 8d, 4u, 4c; E 8d, 4u, 4c; G 1d, 1c; I 1d, 1c 10. Design Awards-H.U.D.-D.O.E., 1st Passive Solar Home Awards 11. 5 years 12. 100% private 13. Registered Architect, California

Sun Light & Power Co.
115 Park Place
Pt. Richmond, Ca
1-415-232-0277

5. AB,F,H 6. A 1d, 1u, 2c 7. A 35-40d 9. I 41d, 6u, 35c 11. 3 years 12. 100% private

Superior California Solar Consultants (SC²)
926 "J" St., Suite 1013
Sacramento, CA 95814
1-916-448-5840

5. A 10. Solar Tax Credits-Reference/Workbook 11. 2 years 12. 100% private 14. $60, negotiable with firm clients

Christian Sweningson
Solar Construction Co.
326 Alta Avenue
Santa Cruz, CA 95060
1-408-427-1437

5. AB,EF 6. A -c, B -d, G -c 7. A -c 9. A -d, B -c, C -c, D -c, I -c 11. 3 years 13. General Contractor

Sun Energy Builders
5838 Robertson Avenue
Carmichael, CA 95608
1-916-486-1801

5. ABH 6. A 6d, 6u; B 6d, 1u, 1c; D 1d, 1c; E 10d, F 1d 7. A -d, 7u, 2c 8. A -d, 80u, 2c; B 2d 10. *Builder*, August 1979 11. 3 years 12. 100% private 13. Contractor B-1 C-36

Susan and Ted Van Syckel
2414 Elendil Lane
Davis, CA 95616
1-916-758-4534

5. AB 6. A 2d, 2u, 2c 7. A 2d, 2u, 2c 11. 1 year 12. 100% private 13. B-1

Virginia Thigpen
2312 Elendil Lane
Davis, CA 95616

5. AB,F,H 6. A 1c, B 10c, E 1c 7. A 2d, 10c 8. A 7c 9. A -c, B -c, C-c, E-c 11. 5 years 12. 100% private 13. Calif. General Contractor #335726 14. $25.00

Paul Timberman
P.O. Box 130
Occidental, CA
1-707-874-3325

5. AB 6. A -c, C -c,D -c, E -c, G -c 7. A -d, -c 8. B -c, C -c 11. 5 years 12. 100% private 13. Calif. Arch. C 10263

Thomas Tollefsen
139 E. Napa St.
Sonoma, CA 95476
1-707-938-5596

5. ABC,F,H 6. E 1c 7. A 1c 12. 100% private 13. Architect

Allan Van Vleet
2600 Dwight Way #204
Berkeley, CA 94704
1-415-849-3816, 845-3629

5. AB,DEFGH 6. A 6c, 1u, 2c; B 2d, 1u; C 2d, 1u; D 1d, 1u; E 12d, 6u, 6c; G 2d, 1u 7. A 15d, 4u, 7c; B 1d 8. A 10d, 8c; B 2d, 1c 9. A 3d, 1u, B -d, -u; C 3d, 2u; D 12d, 3u, 3c E; 1d, 1c; I 2d; J 2d 10. *Steps to Energy Self-Reliance*, A.E.C. *Alternative Currents*, A.E.C. publications Column in Co-op News "Heating Water in Breadbox?", A.S.E. mag. 11. 2 years 12. 30% private, 70% non-profit Corp. 13. Calif. Certified Energy Auditor for Public Bldgs. 14. $30

David Wright
SEAgroup
418 Broad Street
Nevada City, CA 95959
1-916-265-9458

5. BC,FGH 6. A 3d, 1u, 2c; C 2d, 2u; D 3d, 1u, 2c; E 36d, 14u, 22c; G 5d, 2u, 3c 7. A 210d, 20u, 30c; B 3d, 2u, 1c 8. A 10d, 5u, 2c; B 20d, 10u, 10c 10. *Natural Solar Architecture*, Author, 1978 *Pasive Solar Architecture*, Author, 1980 11. 7 years 12. 95% private, 5% government 13. Architect in: California, Colorado, New Mexico, NCARB Certificate holder

Peter R. Zweig
Farallones Rural Center
15290 Coleman Valley Road
Occidental, CA 95465

5. AB,F 6. A 4d, 1u, 3c; B 10d, 9u, 1c; C 1d, 1c 8. A 3d, 1u, 2c; B 3d, 1u, 2c; C 3d, 2u, 1c 9. A 2d, 1u, 1c; D 4d, 4c; F 1d, 1c; G 1d, 1c; I 3d, 3c 11. 5 years 12. 50% private, 50% government 14. $15-30

Pacific Coast Greenhouses
8360 Industrial Ave.
Cotati, Calif. 94928
707-795-2164

5. ABCD 6. A -c 7. A -c 52 years in business 100% private

Jeff Reiss
Solar Future
PO 19177
Sacramento, CA 95819

5. ABDFH 6. A 1d 1c 7. A 1d 1c 8. A 1d 1c Retrofit article, *3rd Passive Conference* 2 years (greenhouse and breadbox source

Bruce Brewer
311 Main St.
Winters, CA 95694
1-916-795-2604

5. ABDFH 6. B 1d 1c, E 1d 1c, H 1d 1c 9. ACE 2 years 100% private Contractors license

Dick Bourne
Solar Concept Development
2655 Portage Bay
Davis, CA 95616
1-916-753-1100

5. ABCDEFH 6. AB 7. A -d -u -c

John Golder
PO 854
Santa Cruz, CA 95061
1-408-423-3064

5. ABDEFH 6. A 3d 2u, B 4d 2u, C 2d, E 4d 1u, G 1d 1u, H 2d, 2u 7. A 3d 1u 8. A 3d 1u 10c B 3d 1u 2c, C 2d 2u 9. A 4d 1u 3c, B 5d 3c, C 6d 2u 2c D 3c E 4d 1u 1c 18d 3c

"A Low Cost Passive Solar Water Heater" *3rd Passive Conf.* 4 yrs 100% private $15-25

Denny Long
RT 1 Box 158
Woodland, CA 95695

5. ABCF 6. B 21c 7. A 21c 9. A 12c 10. "Heat Trapping Insulated Shutters" in *Organic Gardening*

Marshall Hunt
2313 Elendil
Davis, CA 95616

5. ABDEFG 6. A 3c, B 2c 7. A 4c, B 1u 8. A 4c

# COLORADO

Atelier Associates, Charles R. Bahlman & H.T. Brunleve III
845 15th Street
Boulder, CO 80302
1-202-447-8900

5. AB,GH 6. A -c, C -d, E -d, 2c; H -d, 1c 7. -d, 3c; B -d, 1c 8. B 1u 9. B -d, 2c; C -d, 2u, 1c; D -u, I -d, 2u 11. ½ year 12. 80% private, 20% government 13. Architects: Colorado and California

Downing-Leach Architects
2305 Broadway
Boulder, CO 80302
1-303-443-7533

5. ABCDEF,H 6. A 10c, C 1c, E -c, G 25u, -c; H -c 7. A 150c, B 3c, C 3c 9. A 1c, B 150c, C 150c, D 150c, E 150c, I 125c, J -c 10. *AIA Journal*, Sept. 1978 *Professional Builder*, Oct. 1977 *Housing Magazine*, March 1979 *3rd Natural Passive Solar Conference*, Jan. 1979, p. 805 11. 11 years 12. Private 13. Colorado #'s 651, 786,882,1423 14. $30-50

Dovetail Press., Ltd. and Associates
P.O. Box 1496
Boulder, CO 80306
1-303-449-2681

5. ABCDEFGH 6. A 20c, E 20c, H 5d, 15c 7. A 30c 8. A 4d, 2u, 2c 10. *The Complete House Manual: Solar Starr House* Timber-frame working blueprints: 12 designs *The Timber Frame Raising Book*, by Stewart Elliott *The Timber Framing Book*, by Stewart Elliott 11. 15 years 12. 100% private 13. AIA, Colorado

David Finholm
P.O. Box 2839
Aspen, CO 81611
1-303-925-5713

5. AB,F,H 6. A 2c, B 2c, C 1c, D 1c E 3d, 1u, G 1d 7. A 6c, B 1d, 1d, 1c 8. A 1d, 4c; B 1d, 2c 9. A 9c, B 9c, C 9c, D 9c, E 1c, F 1c, G 1c, I 1c 10. Process Architecture/Solar Issue Paper in 2nd Passive Solar Conference Sunset "Solar Heating Book" Sunset Magazine Award Winner in A.I.A. design awards for energy efficiency 11. 4 years 122. Private 13. Colorado, have applied for NCARB

Malcolm Lillywhite
P.O. Box 2043
Evergreen, CO 80439
1-303-674-1597

5. ABC,H 6. A 60d, 7u, 53c; B 16d, 16c; C 6d, 4u, 2c, D 3d, 1u, 2c; E 5d, 5c; G 12d, 12c 7. A 14d, 4u, 10c; B 2d, 2c 8. A 12d, 12c, B 60d, 60c 9. E 70d, 70c; F 1d, 1c; G 2d, 2u; H 6d, 6u; I 10d, 10c; J -d 10. No published books, however, I authored several documents on renewable energy techniques for the U.S. State Department on developing nations. All other publication are in blueprint format 11. 10 years 12. 70% private, 30% government 13. Contact us for specific states 14. $45.00

Solar Environmental Engineering Co. (SEEC)
2524 E. Vine Drive
Fort Collins, CO 80524
1-303-221-5166

5. ABCDEF 6. A 10d, 5u, 5c; B 5d, 2u; C 10d, 5u, 5c; E 20d, 10u, 20c; G 20d, 5u, 20c; H 2d, 2u, 2c 7. A 15d, 10u, 30c; B 2d, 5u, 10c; C 5d 8. A 2d, B 5d, 2u, 1c; C 2d, 1u, 1c 9. A 5d, 2u; B 10d, 10u, 5c; C 5d, 2u; D 5d, 5u, 10c; E 5d, 5u, 2c; I 20d, 5u, 50c; J 20d, 5u, 60c 10. Passive Solar Design for handheld Programmable Calculators (HHPC) Sun Angles and Shading Analysis for HHPC FCHART Active Solar Design for HHPC Relative Areas Active Solar Design for HHPC 11. 7 years 12. 30% private, 70% government

Solar Technology Corporation (SOLTEC)
2160 Clay Street
Denver, CO 80211
1-303-455-3309

5. D,F,H 6. A 70c 7. A 60c, B 2u 9. A -d, -u, -c; I -d, -c 11. 7 years 12. 100% private

Peggy Wrenn
Colorado Office of Energy Conservation
1600 Downing Street
Denver, CO 80218
1-303-839-2507

5. A,F,H 6. A 1c, C 1u 7. A 1c 10. *Colorado Solar and Renewable Energy Directory* *Colorado Solar Handbook* Roaring Forks Solar Conference Proceedings, 1978 talk Solar is Hot: Don't Get Burned (brochure) 1979 11. 3 years 12. State and Local

# DISTRICT OF COLUMBIA

Design Alternatives, Inc.
1313-18th St., NW
Washington, D.C. 20036
1-202-223-0830

5. AB,DEF,H 6. A 3c, C 3c, D 1c, E 5c, G 5c 7. 3c 8. A 1c, B 1c, C 1c 9. A 5c, B 3c, C 5c, D 20c, E 6c, J 6c 10. Low-Cost, Energy-Efficient Shelter Save Energy: Save Money! Feasibility of an Energy-Related Loan Program for Low-Income Homeowners 11. 3 years 12. 90% government 13. Use registered architects and engineers as consultants

David R. Wolcott
International Business Services
1010 Vermont Avenue, NW
Washington, D.C. 20005
1-202-789-5360

5. AH (Market development) 10. *A Passive Solar Communication Assessment* "A Communication Strategy to Commercialize Passive Solar" 11. 2 years 12. 10% private, 90% government

# ILLINOIS

Hawkweed Group
4643 N. Clark Street
Chicago, Il 60640

5. ABC, FG 6. A -c, C -c, E -c 7. A 7d, 12u, 19c, B 8d, 2u, 7c 11. 20 years

# KANSAS

Eric Kjelshus
Box 172, Route 1
Garner, KS 66030
1-913-884-7949

5. ABC,EF,H 6. A -c, C -c, D -u, E -c, H -c 7. A -c, B -u 8. A -c, B -d, C -d 9. A -c, B -c, C -c, D -c, E -c, G -u, I -c, J -c 11. 6 years 12. 100% private 14. $5-$30

Eldon B. Swensson, AIA, Architect
Box 1320
Wichita, KS 67201
1-316-267-1351

5. AB,EF 6. D -d, -u; E 2d, 2u, 4c 7. A 2d, 2u, 4c 11. 30 years 12. 10% private, 90% government 13. Kansas, Tennessee

# MARYLAND

Engercon, Earl Adkins Jr.
5604 Frankford Avenue
Baltimore, MD 21206
1-301-448-3366

5. AB,DEF,H 6. A 5d, 2u, 3c; C 4d, 2u, 2c; H 1d 7. A 6d, 2u, 4c; B 1d 9. A 6d, 2u, 4c; B 6d, 2u, 4c; C 6d, 2u, 4c; D 12d, 4u, 8c; I 2d, 1u 11. 2.5 years 12. 100% private 13. Contractors 14. Upon request, varies with job and conditions, etc.

# MASSACHUSETS

Ann Gordon
P.O. Box 929
Brewster, MA 02631
1-617-897-3204

5. AB,D,F 6. H Design appropriate landscapes for rooftops 7. A 6d, 1u; B 2d 9. Appropriate, self-maintaining, low-energy rooftop gardens for energy-conserving passively-heated earth-covered buildings 11. 1 year 12. 100% private 13. None 14. Negotiated

Professor Thomas P. Bligh
Massachusetts Institute of Technology, 3-457
Cambridge, MA 02139
1-617-253-2018

5. AB,D,FGH 6. D -d, -u, -c; E -d, -u, -c; G -d, -u, -c 7. B -d, -u, -c; C -d, -u, -c 9. A -d, B -d, -u, -c; I -d, -u, -c 10 Energy Conservation by Building Underground Solar Heating and Cooling for an Underground Building at the University of Minnesota Energy Comparisons and Where to Insulate Earth Sheltered Buildings and Basements Earth Sheltered Housing Design 11. 12 years 12. 10% private, 90% government

William Shurcliff
19 Appleton Street
Cambridge, MA 02138

5. A; 10 *Solar Buildings of North America, Insulated Shutters and Drapes*, and *New Inventions in Low Cost Solar Heating*, Brick House Publishing.

Norman B. Saunders
15 Ellis Psi.
Weston, Mass 02193

5. ABCDEFGH 6. A 4c, C 1c, D 1u, E 5c, G 1u, 3c, H 2c 7. A 6u, 7c, B 2d, C 1u 9. A 2c, C 2c, F 1u, I 3c 10. Solar Heating Basics $7 SHB Supplement $4 All Solar House $22 11. 26 years in business 99% private 1% government Physical Engineer in Mass. $35 hr

Weather Energy Systems
PO 968
Pocasset, MA 02559
1-617-563-9337

5. ABCDEFGH 6. A -d -u -c, B -d, C -d, G -d -u, H -d 7. A -d -u -c, B -d 8. A -d 9. C -d -u -c 2 yrs 100% private

# MINNESOTA

Architectural Alliance Energy
400 Clifton Avenue South
Minneapolis, MN 55403
1-612-871-5703
Attn: Peter Pfister, AIA

5. ABCDEFGH 6. A 5d, 1u, 3c; D 1d, 1c, E 2d, 1u, 1; H 1d, 1c (Retrofit) 7. B 2d, 2u 9. A 3d, 1u, 2c; B 5d, 1u, 4c; C 3d, 2u, 1c; D 2d, 2c I 3d, 3c 11. 9 years 12. 90% private, 10% government 13. NCARB-New York, Minn., Wisc., Texas, Florida, Penns., Indiana 14. Varies

Underground Space Center
221 Church St., S.E.
University of Minnesota
Minneapolis, MN 55455
1-612-376-5341

5. A,D,F 10. *Earth Sheltered Housing Design* *Potential Use of Underground Space* *Preliminary Design for Underground Space* *Cost Code Study of Underground Buildings* 11. 3 years

# MISSISSIPPI

Pablo Okhuysen
225 W. Lampkin Road
Starkville, MS 39759
1-601-323-7246

5. AB,D 6. A -c, C -c, E -c 7. A 30c, B 1c, C 2c 9. D -c 10. "Energy Conserving Houses," *The Mississippi Architect, 4* (1974) "Croft House" in Shurcliff's *Solar Buildings of North America, 120 Outstanding Examples,* (1977) 11. 8 years 12. 100% private 14. $30.00

# MISSOURI

Herb Wade
Route 1, Box 85
Hartsburg, MO 65039
1-314-657-9735

5. AB,F,H 6. A -d, -u, -c; C -d, -u, -c; D -d, -u, E -d, -c; G -d, -c; H -d, -u, -c 7. A -d, -u, -c; B -d, -u 8. A -d, -c; B -d, -c; C -d, -u 9. A -d, -c; C -d, -c; D -d, -u, -c; E -d, -u, -c; G -d, I -d, -c; J -d, -u, -c 10. *The Solar Greenhouse Book* (Tech. Consultant) *Solar Architecture* (Chapter author) Numerous articles in ISES publications 11. 17 years solar, 11 consultant in solar 12. 100% private

Timothy E. Montgomery
46 North Allen
Bonne Terre, Missouri 63628
314-358-2424

5. ABF 6. D -c

# MONTANNA

(Design Build II) DBII
Box 854
Miles City, MT 59301
1-406-232-3230

5. AB,F,H 6. A -d, -u, D -d, E -d, -u, G -d, -u, -c 7. A -d, -u, -c; B -d 9. A -c, B -c, I -d 11. 3 years 12. 99% private, 1% government

Paul Fellers
800 South Third
Bozeman, Mt 59715
1-406-587-2788

5. AB 6. A, B 2c, E Architect, MT, CA

# NEW HAMPSHIRE

Don Booth
Community Builders
Canterbury, NH 03224
1-603-783-4743

5. AB, D,F,H 6. A 3d 3c; B 1d, 1c; C 1c; D 1d, 1c; E 3d, 1u, 1c; F 1c, G 1d, 1c; H 6d, 1u, 2c 8. A -d, -c; B -d, -u; C -d, -c 9. A 2d, 1u, 3c; E 8d, 1u, 8c 10. "Envelope House," Conference Report 11. 20 years 12. 98% private, 2% government

Total Environmental Action, Inc.
Church Hill
Harrisville, NH 03450
1-603-827-3374

5. ABCDEF,H 6. A -d, B -c, C -c, E -c, G -c 7. A -c, B -u 8. A -d, B -c 9. A -c, B -c, C -c, D -c, I -c 10. *Solar Energy: Fundamentals in Building Design* *The Solar Home Book* *The Fuel Savers* *Thermal Delight in Architecture* 11. 6 years 12. 50% private, 50% government 13. 2 reegistered architects on staff

# NEW JERSEY

Princeton Energy Group/Harrison Fraker, Architect
729 Alexander Road
Princeton, New Jersey 08540
1-609-452-8235/609-452-8231

5. AB,DEFGH 6. A 17d, -u, -c; B 15d, -u, -c; C 5d, -u, -c; D 1d; E 20d, -u, -c; F 1d; G 20d, -u, -c; H 20d, -u, -c 7. A 30d, -u, -c; B 8d, -u, -c; C 5d, -u, -c 8. A -d, B -d, -u 9. 20d, -u, -c; B 8d, -u, -c; C 8d, -u, -c; D 20d, -u, -c; E 11d, -u, -c; F 1d; G 4d, -u, -c; H 2d, -u, I 13d, -u, -c; J 20d, -u, -c; K (solar-assisted heat pump) d, u, c, L (water-source heat pump) 1d 10. PEGFIX/PEGFLOAT (hand calculator program for passive solar design, including *Handbook* Several article references, *Research & Design*, Vol. II, No. 3, Fall 1979 Several articles, *AIA Journal*, *Solar News* (MASEA Newsletter) Numerous papers at Solar conferences 11. 6 years 12. 80% private, 20% government 13. New Jersey, New York, Pennsylvania, North Carolina, and NCARB 14. Principals $30-50/hour, Staff $15-30 hour, plus travel and other reimbursable expenses

Frank Thornburgh Company
44 Beverly Drive
Belle Mead, NJ 08502
1-201-874-5115

5. AB,D,FGH 6. A -c, D -c, E -c 7. A -c, B -c, C -c 8. C -c 9. B -c, C -c, D -c, E -c, G -c, H -c, I -c, J -c 11. 7 years 12. Private and commercial 100% 14. $10 and up

# NEW MEXICO

Steve Baer, Zomeworks Corp.
P.O. Box 712
Albuquerque, NM 87103
1-505-242-5354

5. A,D,G 6. E -u, -c, H -u, -c 7. A -u, -c 8. A -d, -u, -c; B -d, -u, -c 9. A -d, -u, -c; B -d, -u, -c; C -u, -c; F -u, -c 10. *Sunspots* NMSEA Bulletin Alternative Sources of Energy CoEvolution Quarterly 11. 10 years 12. 100% private 13. NM Contractor's License 14. $50.00

Bruce Davis, Architect
R.R. 3, Box 90 B.D.
Santa Fe, NM 87501
1-505-983-9178

5. B 6. A -c, C -c, E -c, G -c 7. 36d, -u, -c 8. A -c, B -c, C -c 9. E -d, F -d, G -d 11. 4 years 12. 100% private 13. New Mexico Architect

Fred Hopman, President
Ecolink, Inc.
P.O. Box 171
Arroyo Seco, NM 87514
1-505-758-4051
and
Taos Solar Energy Assoc,
P.O. Box 2334
Taos, NM 87571
1-505-758-4051

5. ABH 6. A 8d, 6c; B 1d, 1u; C 6d, 4c; H 1d, 1c (Convective Loop)7. A 16d, 12c 8. A 2d, 1c; B 1d, 1c 9. A -d; K Patent: Self-insulating Water wall, architectural module, -c and mechanical adobe construction, -c 10. "The Self-insulating Water Wall-A Passive Solar Module for Heating and Cooling," Proceedings, *3rd National Passive Solar Conference*, "Innovative Controls Using Natural Energies," Proceedings, *2nd National ISES Passive Conference* "Night Sky Cooling, Two Passive Strategies," *Proceedings 3rd National* New Mexico Solar Energy Association, "Southwest Bulletin," various articles 11. 8 years 12. 40% private, 60% government 13. 501C3 14. $12.00

William Lumpkins
118½ West San Francisco
Santa Fe, New Mexico 87501
1-505-893-2618

5. ABCDEFGH 6. A-d -u -c, B. -d -u -c, C -d -u -c, 7. a -d, -u, -c; B -d, -u, -c 8. B -d, -c; C -u 11. 35 years 12. 100% private

E&M Lumsdaine Solar Consultants, Inc.
4825 Panorama Drive
Las Cruces, NM 88001
1-505-524-4771

5. AB,EF,H 6. A 2d, C 3d, E 5d, G 3d, 2c 7. A 9d, 2c; B 1d, 1u, 1c 9. C -d, D -c, E -d, -c; F -d, G -d, I -d, -c; J -d, -c 10. M. Lumsdaine *Passive Solar House Design for New Mexico Climates* (Part I: Design Manual), Rpt. NMEI 22-1, New Mexico Energy Institute, Box 3EI, New Mexico State Univ., Las Cruces, NM 88003. SERI Operational Results Conference Proceedings, "Lumsdaine House," SERI/TP-245-430, 1979 11. 1 year 12. 50% private, 50% government

Ron McClure, Architect Modular Structures Ltd.
P.O. Box 128
Tijeras, NM 87059
1-505-243-0878

5. ABCD,FGH 6. A -d, -u; C -d, -u, D -d, -c; E -d, -c 7. A -d, -c, B -d, -c 8. C -d, -u, -c 9. A -d, -u, -c 10. New Mexico Mag. Feb. 77, "Underground House," "Earth Sheltered Housing Design," by the Univ. of Minnesota Albuquerque Journal, Oct. 1977, Feature article New Mexico Bldg, Mag. Jan. 1978, Feature article 11. 12 years 12. 100% 13. New Mexico Architect, New Mexico Contractor

W. Scott Morris
P.O. Box 4815
Santa Fe, NM 87502
1-505-455-2645

5. ABCD 7. A -c 8. D (thermosiphon air collectors for domestic hot water) 10. Window-Mounted Solar Collector, 1/77 (NMSEA-POB 2004, Santa Fe, NM 87501) Natural Covection Collectors, *Solar Age*, 9/78 Natural Convection: No Moving Parts, *Solar Age*, 1/79 "Solar Hot Air Balloons." *3rd Passive Conference* 11. 5 years 12. 50% private, 50% government 14. $25-$35 plus expenses

N.M.S.E.A., B Stickney, A.T. Wilson
P.O. 2004
Santa Fe, N.M. 87501

5. B,F 6. A -d, -c 7. A -d, -c; B -d, -c 8. A -d, -c; B -d, -u, -c; C -d, -c 9. A -d, -c; B -d, -c; C -d, -c; D -d, -c; E -d, -c; F -d, -u; I -d, -c; J -d, -c 10. *Thermal Storage Wall Design Manual* A.T. Wilson "The Thermal Performance of an Attached Solar Greenhouse" B. Stickney 11. 5 years 12. 60% private, 40% government

# NEW YORK

Ed Mac Dougall
9 Academy Lane
Bellport, NY 11713
1-516-286-8091

5. A,D 6. A -c, H (solar hen house) -c 7. C -d, -u, -c 8. B -d 10. "Solar Hen House," ISES Pub. "Use of Solar Greenhouse as Water Preheater," ISES Pub. "Solar Greenhouse," Marlboro, VT 11. 5 years 12. 100% private 13. P.E. in Calif. 14. $10 plus expenses

Fred S. Dubin
42 West 39th St.
New York, NY 10018
1-21-840-6700

. ABCDEFGH 6. A, B, C, D, E, F, G 7. B 9. A, B, C, D, E, F, G, H, I, J 10. *Energy Conservation Standards*, McGraw-Hill *The Energy Efficient Home* *Energy Conservation for Existing Homes*, ECM 1/2 for FEA, 1974 *Total Energy*, TR2 Ford Foundation, New York 11. 30 years 12. 40% private, 60% government 13. 24 states PE

# NORTH CAROLINA

Richard Kennedy
Sunspace
Box 318, Rt. 5
Burnsville, NC 28714
1-704-675-5286

5. ABC,H 6. A 3d, 3c; D 1c, E 3d, 2c 7. A 6d 9. A 2d, 2c; B 1d, 1c; C 1d, 1c 10. 2nd National Passive Conference 3rd National Passive Conference Earth Shelter Digest Many slide lectures 11. 4 years 12. 100% private 14. $ 14.00

William Langdon
50 College Street
Asheville, NC
1-704-252-0296

5. AB,D,F,H 6. A -c, B -d, C -u, E -c, G -u 7. A -c 8. A -d, B -d 9. A -c, B -c 10. *Movable Insulation*, author *The First Passive Solar Home Awards*, editor & contributor *Low-Cost Energy Efficient Shelter*, contributor 11. 5 years 12. 10% private 13. N.C. Licensed Architect

Bruce Johnson
411 N. Boylan Ave
Raleigh, NC 27603
1-919-834-9178

5. ABEFGH 6. A-c, B-c, C-c, D-c, E-c, G-c 7. A-c, B-d 9. ABCDE Nov. *Solar Age*, 5 years, 100% private NCARB

# OREGON

Gregory Acker, Architect
1645 Liberty S.E.
Salem, Oregon 97302
1-503-364-9246

5. AB,F 6. A 1u, B 4c, C 1u, 1c; E 1c 7. A 2u, 5c 8. A 2c, B 1u 11. 4 years 12. 100% private 13. Oregon and California

Bob Burke
2812 N.W. Thurman
Portland, OR 97210
1-503-224-0210

5. ABC 6. A -d, -u, -c; C -d, -u, -c; D -d, -u, -c 9. I -c 11. 2 years 12. 95% private 13. in partnership with reg. arch. 14. $10-$20/hr.

Equinox Design, Inc.
P.O. Box 30085
Eugene, OR 97403
1-503-345-6161

5. AB,G 6. A -d, -u, -c, B -d, -u, -c, C -d, -u, -c 7. A 10d, 4u, 5c, B 2d, 2u 8. B-d, -u, -c 9. A -d, B -d, C -d 10. McGuinnes, Stein, Reynolds-*Mechanical & Electrical Equipment for Buildings*, 6th ed. 11. 2 years 12. 80% private, 20% government 13. Arch.

Arden Handshy
Box 1200
Jacksonville, OR 97530
1-503-899-1250

5. AB,F 11. 19 years (construction), 2 years (passive) 13. Oregon Builder

Michael S. Stevens
67325 Fryrear Road
Bend, OR 97701
1-503-388-2493

5. AB,F 6. A -c, E -d 7. A 1d 8. A 1d, 1c; B 1d, 1c; C 1d, 1c 9. A -c, D -c, E -c 10. Heating with Wood, *Seriatim*, Vol. 3, No. 2 11. 1 year 12. Private 13. Working with contractors

Sunshelter Natural Energy systems
Stephen Kujawa
90472 Sheffler Road
Elmira, OR 97437
1-503-935-7664

5. AB,EF 6. A 3d, 1u, 2c; B 1d, D, E, G 3d, 2u, 1c; H 1d (Commercial Greehouse) 7. A 3d, 3u, 2c; B 1d 8. A 2d, 1u, B 2d, 2u, C 3d, 2u 9. A -d, B -d, -c; C, D, E, I 10. 4th Passive Conference 11. 6 years contruction, 3½ years solar 12. 100% private 13. Bonded, insured, licensed 14. $15.00-$25.00

## PENNSYLVANIA

Shelter Design Group
Stony Run, PA 19557
215-756-6112

5. ABEF 6. A 15d, 2u, 6c, B 1d, 1c, C 1u, 1d, D 9d, 3u, 3c, E 13d, 3u, 9c, G 2d, 1u, 1c 7. A 41d, 9u, 18c 8. A 6d, 1u, 3c, B 2d, C 10d, 3u, 4c 9. A 13d, 3u, 2c, B 35d, 9u, 12d, C 41d, 9u, 19d, D 6d, 4u, 2d, E 39d, 9u, 16c, I 6d, 4c, J 16d, 5u, 11c Articles in *New Shelter*, houses in Alex Wade *30 Energy Efficient Houses*, Rodale Press 4 years in business 100% private PA, FLA 30hr

Burt Hill Kosar Rittelmann Associates
400 Morgan Center
Butler, PA 16001
1-412-285-4761

5. ABCDEFGH 6. A 2d, 1u, C 1c, D2u, G2d. 7. A 1d, 3u, B 5d, 1u, 1c; C 1d. 8. C -c. 9. E 1u, 1c; F 2d; I 5d, 1u, 16c. 10. *Minimum Energy Dwelling Workbook. Solar Energy Systems Design Handbook for Copper Development Association. Introduction to Solar Energy in Pennsylvania. Energy Conservation Design Guidelines for New Office Buildings.* 11. 20 years. 12. 50% private, 50% government. 13. Yes. 14. $30.00 preliminary.

Joe Carter
227 South 5th Avenue
Emmaus, PA 18049
214-967-5171 or 965-6889

5. AB,DEF 6. C -d, -u, -c, H 8c 8. C 8d, 8c. 9. A-d, B-c, E 15c, I 10c 10. "The Solar-Wood Connection," *Solar Age*, ed. *Solar Retrofitting* 1981 12. 80% private, 20% government

Robert G. Flower
342 Franklin Street
Alburtis, PA 18011
1-215-966-4545

5. ABCDEF,H 6. A 5c, E 2c, G 1c 7. A 10c, B 5c, C 2c 8. A 3u, 1c, B 1u 9. A 2c, C 1u, D 6c, G 10c, I 4c 10. Contributor, *The Solar Greenhouse Book, Rodales Solar Growing Frame* 11. 6 years 12. 100% private

John E. Flynn
Department of Architectural Engineering
Pennsylvania State University
University Park, PA 16802
1-814-863-2085

5. D,FG 6. E -u, H -u 7. A -u, B -u 10. *State-of-Art for Daylighting Buildings* (1979), *Architectural Interior Systems* (1970), *Experiments in Architectural Form* (1977, 1979) 12. Private and Government 13. Architect (registered)

D. B. Pennington
494 Madison Avenue
York, PA 17404
1-717-843-5112

5. H

Christopher Shipp
R. D. #1, Box 331D
West Grove, PA 19390
1-215-869-8323

5. AB 6. A 2c, D 1d, E 13c 7. A 3c 8. A -d 9. I 1c 11. 8 years 12 Private 13. PA Home Builder 14. Variable

## TENNESSEE

Bill Barth
Barth & Ramsbottom
5006 Whitaker Street
Knoxville, TN 37919
1-615-588-9983

5. ABC,F,H 6. A -d, -c; B -d, -c; C -d, -u, -c; D -d, -c; H -d, -c 7. A -c, B -c 9. A -c, B -c, C -c, D -c, E -c 10. Wauford Analysis - Proceedings 3rd Passive Solar Conference. *Solar House Design Report* - TVA. *Passive Residential/Commercial Solar Strategy Report* - TVA. Winterization Guidelines for the Highlander Center, New Market, TN 11. 6 years 12. 95% private, 5% government 13. Architectural Registration in Tennessee and Licensed General Contractor, Tennessee

## TEXAS

J. W. Griffith
3921 Caruth Blvd.
Dallas, TX 75225
1-214-368-3753

5. A,FG 6. E -d, 100's c 7. A -c, B -c, C -d 10. *Lifecycle Cost Benefit Analysis* (Book). "Benefits of Daylighting," *ASHRAE*, Jan. 1978. *Resource Optimization and Economic Planning*, LD&A, Sept. 1973 11. 30 years 12. 50% private, 50% government 13. Texas 14. $100

Paul W. Lodholz, AIA
Tackett Way Lodholz, Inc.
3121 Buffalo Speedway, #400
Houston, TX 77098
1-713-622-3130

5. AB,DEFG 6. A 5c, B 1u, E 3c, H 1c 7. A 2u, 10c; B 3c 9. B 10c, C 8c, E 3c, I 1u 10. Solar Realities, 1979. New Horizons in Solar Heating and Cooling 11. 3 years 12.95% private, 5% government 13. Architect, State of Texas #7025

Frank Moreland
904 Boland
Fort Worth, Tx 76107
1-817-335-2883

5. ABCDEF 6. AD 7. A -c, B -c

## VERMONT

Green Mountain Homes
Royalton, VT 05068
802-763-8384

5. H 6. A -c, H -c 7. A -c 100% private

Northern Owner Builder
RFT 1
Plainfield, VT 05667
1-802-454-7808

5. AB,F,H 6. A 9d, 1u, 2c; C 1d, D 2d, 1c; E 7d, 1u, 4c; G 1d, 1u 7. A 13d, 5u, 5c 9. I 1d, K (Classes & Workshops) 11. 3 years 12. 100% private 14. Design/draft: 35¢/ft²

## VIRGINIA

Dr. C. James Bier
Rt 2, Box 35
Ferrum, VA 24088
1-703-365-2230

5. AB,DEF 6. A 2u, C 2c, E 1c, H 1c 7. A 1u, 2c; B 2u 9. A 1c, C -c, D -c, E 1c 11. 3 years 12. 80% private, 20% government 13. Ph.D. Chemistry 14. $0 - $20

S. Robert Hastings
11571 Links Drive
Reston, VA 22090
1-703-437-3450

5. AB,DEF 6. A 2d, 1c; E 4d, 2u, 1c 7. A 4d, 2u, 1c; B -d 10. Window Design Strategies to Conserve Energy. Three Typical House Designs for Energy Conservation Research. Quantified Occupant Use Factors for Energy Analysis. Daylighting—Constraints to Architectural Solutions 12. 25% private, 75% government 13. Architectural License, Massachusetts, Virginia 14. $20.00

## WASHINGTON

Ecotope Group
2332 E. Madison
Seattle, WA 98112
1-206-322-3753

5. ABCDEFG 6. A -c, B -u, C -d, E -u, G -c 7. A 10d, 10u, 16c; B -c, C -c 8. B -c 9. C 1u, 1c; E 1d, 1u 11. 6 years 12. 20% private, 80% government 14. $25 - $35

Marietta S. Millet
610 23rd Avenue, East
Seattle, WA 98112
1-206-324-2693

5. A,FG 7. A -d, -u, -c; B -d, -u, -c, C -d 10. "Designing for Daylight: A New Prediction Technique," 3rd National Passive Solar Conference. "Slice of the Sun," Plenary address, 4th Passive Solar Conference. "Lighting and Building Codes for Energy Consultation," *Research and Innovation in the Building Regulator Process*, NBS Special Pub. 552, July, 1979 11. 7 years

Alternative Energy Systems
PO 295
Winthrop, WA 98862
509-996-2490

5. ABCDEFGH Cost Effective Solar Heating Project Cost Effectiveness of Weatherization in Three Counties Commercial Solar Greenhouse Project Performance of Two Level Commercial Solar Greenhouse — In Progress — 6. A 1d 1u 1c, B 2d 5u 1c, C 1u 1c, D 2c, E 1u 4c, G 5c 7. A 3d 5u 15c, B 3u 8. A 1u, B 1c, C 2c 7 yrs 30% private $25 hr.

## WISCONSIN

Willis & Lillian Leenhouts, FAIA
Architects
3332 N. Dousman Street
Milwaukee, WI 53212
1-414-964-1204

5. AB,GH 6. A 3c, C 1u, E 5c, G 2c, H 2c 7. A 125d, 7u, 95c; B 3c 10c 8. A 3d 9. B 95c, C 2u, 1c; G 1c, I 2c, J -u 10. *Regional Guidelines for Building Passive Energy Conserving Homes*, Sec. Climate 1-B Assistance by Lillian Leenhouts 11. 34 years 12. 95% private, 5% government 13. Architects, Wisconsin, Illinois

Don Schramm
PRADO
Box 1128
Madison, WI 53701
1-608-256-4908

5. ABC,FG 6. A 4d, -c, B 3d, D 3d, E 6d, G -d 7. A 10d, -u, -c; B 2d 8. A -d, B -d 9. A -d, -u, B -d, -u, C -d, D -d, -u, -c 10. "Low Cost Performance Testing Procedures for Passive Solar Retrofit." "Passive Solar Remodeling in Northern Climates." "A Method of Estimating Monthly Average Radiation of Shaded Surfaces." "Analysis of Building Components Related to Direct Solar Heating of Buildings." 11. 3 years 12. Private 13. Wisconsin

W.T. Brown
317 Sixth Avenue
Stevens Point, Wisc. 54481
715-341-8672

5. AB

# CANADA

## ALBERTA

Vinto Engineering
10950 119th Street
Edmonton, Alberta T5J 2N2
452-1800

5. C,EF 7. B -u, -c; C -u, -c 9. A -u, B -u, C -u, D -u 12. Private and government

## BRITISH COLUMBIA

Island Solar Supply Ltd.
2915 Prior Street
Victoria, BC V8T 3Y4
1-604-384-0106

5. ABCD,FGH 6. A 12d, 4u, 8c; E 18d, 4u, 16c; G 6d, 2u, 4c 7. A 10d, 4u, 8c; B 2d, 1c 8. B 10d, 10c; C 2d, 2c 9. C 20d, 5u, 15c; D 20d, 5u, 15c; E 3d, 3c; I 3d, 2u, 3c 11. 1 year 12. 100% private 13. B.C. business license 14. $15.00

Solar Applications & Research Ltd.
3683 West 4th Avenue
Vancouver, BC V6R 1P2
1-604-733-5631

5. ABCD,FGH 6. A 1u, 4c; B 1c; C 8c; E 3d, 1u, 9c; G 1u 7. A 4d 2u, 14c; B 1u 8. B 4d, 2u, 42c 9. G 1d, I 1u, 2c 10. "Solar Systems in B.C., A Step by Step Guide to Help You Build Your Own Solar Water Heater." "Basic Guidelines for Energy Conservation in Community Planning and Building Design." "Design, Construction and Initial Operating Experience with a Passively Solar Heated Townhouse." 11. 4 years 12. 50% private, 50% government 13. Professional Engineer 14. $45

## NOVA SCOTIA

Larry Bogan (Alternate Energy Consultants)
RR #2 Cambridge Station
Nova Scotia, Canada B0P 1G0
1-902-678-9258

5. AB,EF 6. E 1c 7. A 1c 9. A 1u 11. 2 years 12. Private

## ONTARIO

llen Drerup White
34 King Street, East, Studio 505
oronto, Ontario, Canada M5A 1L7
-416-368-2821

. ABCDEFGH 6. A -c, C -c, E -c, G -c, H -c 7. A -c, B -c, C -c 8. -c, B -c, C -c 9. A -c, B -c, C -c, D -c, E -c, I -c, J -c 10. larrowsmith Articles (several) 11. 10 years 12. 60% private, 40% overnment

codomus Developments Ltd.
355 Wellington Street/Suite 203
Ottawa, Ontario, Canada K1Y 3C2

. ABCD,F 6. A -c, D -d, E -c, G -u 7. A -u, -c; B -d 9. A -d, C -d, -u 11. 3 years 12. Private 13. Board of Trade, Better Business ureau, HUDAC, Assoc. Prof. Eng.

ohn Hix Architect
07 Queens Quay West
oronto, Ontario, Canada M5J 1A7
-416-861-1245

. ABCDEF,H 6. A -c, D -c, E -c, G -d, H -d, -u, -c 7. A -d, -u, -c; -d 9. A -c, B -c, C -c, D -c, E -c, G -c, I -c, J -c 10. *The Glass louse*, MIT Press. Canadian Architect. 11. 7 years 12. 50% rivate, 50% government 13. MRAIC, OAA, NCARB

lorvath and Associates
ohn A. Horvath
50 Consumers Road, Suite 701
Villowdale, Ontario, Canada M2J 4V6
-416-493-6900

. ABCD,H 6. H -d 7. A -d, -u 9. C -d, I -d, -u, J -d, -u 11. 2 years 2. 40% private, 60% government 13. P. Eng. 14. $20 to $30 depending on size of project, Minimum $50

Moffat Engineering Ltd.
5 Eglinton Avenue East
oronto, Ontario, Canada M4P 1G8
-416-487-1135

. ABC,EF 6. A -c, D -c, E -c 7. A -d, B -c, C -c 8. B -c, I -u 11. 25 years 12. 80% private, 20% government

Ernest H. Jenkins
R.R. #2, Peterborough
Ontario, Canada K9J 6X3
-705-745-9480

. AB,D,FH 6. A 3d, 1c; E 3d, 1u, 2c; G 2d, 1u, H 1d, 1u 7. A 7d, 3u, 3c; B 2d, 1u; C 1d 8. B 3d, 1u, 1c 9. C 3d, 1u; I 4d, 1u, 1c 10 'Layman's Solar Energy Notes." 11. 4 years 12. 100% private 13. M.A.A.T.O. - Member Associate of Architectural Technologists of Ontario 14. $15

Joseph C. Somfay
442 Union Street
Salem, Ontario, Canada
-519-846-0251

. AB,D,FGH 6. A -d, -u, -c; C -d, D -d, E -d, -u, -c; G -d, -u, -c; H -d, -u, -c 7. A 10d, 6u, 10c 9. A -d, -u, B -d, -u, -c; C -d, -u, -c; E -d, -u, -c; I -d, -u, -c 11. 12 years 12. 100% private 13. RAIC, OAA

## PRINCE EDWARD ISLAND

Miller Solsearch Ltd.
P.O. Box 2320, 49 Pownal Street
Charlottetown, P.E.I., Canada C1D 1H9
1-902-892-9898

5. ABCDEFGH 6. A 1d, 2u, 10c; B 4c, E 10c, G -d, 6c; H -d, -u, 10c 7. A 10c, B 10b, C 10c 8. 2u, 2c; C 10c 9. A 3c, B 10c, C 10c, E 10c, G 2c, I 10c, J 10c 10. Paper, SESCI Conference 1979. Paper—New Alchemy Institute, Woods Hole 11. 5 years 12. 50% private, 50% government 13. MRAIL, AAPEI, AIA, BSA, P.E.

Island Energy Associates
Box 2277
Charlottetown, P.E.I.
1-902-892-4067

5. ABCD,F 6. A -d, -c 8. C -c 9. E -d, -u, -c

## QUEBEC

Nick Nicholson
Walker Road
Ayer's Cliff, Quebec, Canada J0B 1C0
1-819-838-4871

5. B,D 6. A -c, D -c, E -c, H -c 7. A -c 8. C -c 9. A -c, D -c, E -c, F -c, I -c 10. Nicholson Solar Energy Bldg. Manual 300 pp. Prototype Canada (standard solar system plan). Harvest the Sun - 224 pp. The Autonomous House Report. 11. 20 years 12. Private and government

Charles Ross
304 E Street, St. Germain
Rimouski, Quebec, Canada G5L 1B9
1-418-722-7655

5. AB,F,H 6. A -d, -c; C -d, -c; D -d, -c; E -d, -c; H -d, -c 7. A -d, -u, -c, B -d, -u; I -d, -c 11. 3 years 12. 60% private, 40% government 13. Order of Architects of Province of Quebec 14. $30

Witold Rybczynski, Director
Minimum Cost Housing Group;
School of Architecture, McGill University
3480 University Street
Montreal, PQ, Canada H3A 2A7
1-514-392-8021

5. B,D 6. A -c, C -d, E -d 7. A -d, -u, -c 8. A -c 9. A -u, I -d 10. *The Ecol Operation*, MCHG. *Stop the 5 Gallon Flush*, MCHG. *Water Conservation and the Mist Experience*, MCHG. *Rooftop Wastelands*, MCHG 11. 7 years 12. 25% private, 75% government 13. Registered Architect

## SASKATCHEWAN

Peter J. Catania
Faculty of Engineering
University of Regina
Regina, Saskatchewan, Canada S4S 0A2
1-306-584-4702

5. A,D,F,H 6. C -d, D -d, H -d 7. A -d, B -u, C -u 9. I -u, C -c, D -c, E -c, J -c 11. 4 years 12. 90% private, 10% government

Concept Construction
2234 Hanselman Avenue
Saskatoon, Saskatchewan
1-306-652-2233

5. ABCD,F,H 6. A-c, C-c, E-c 7. A-c, B-c 8. B-c 9. A-c, C-c, D-c, E-c, J-c 11. 4 years 12. 90% private, 10% government

Robert S. Dumont
1333 2nd Avenue N
Saskatoon, Saskatchewan, Canada
1-306-665-8156

5. A,C 6. E -d, -u, -c 7. A -d, -u, -c 8. B -d, -u, -c 9. A -d, -u, -c; C -d, -u, -c; J -d, -u, -c 10. Low Energy Passive Solar Housing Extension Division, Univ. of Saskatchewan. 11. 4 years 12. Private 13. Ph.D. Mechanical Engr. 14. Based on need

ENERCON CONSULTANTS LIMITED
2073 Cornwall Street
Regina, Saskatchewan S4P 2K6
1-306-585-0022

5. ABCDEFGH 6. A 5d, 2c; I (Energy Conserving/Passive Solar) 12d, 19u, 10c 7. A 12d, 19u, 10c 9. A 3d, 3u, 3c; B 12d, 19u, 10c; C 12d, 19u, 10c; D 12d, 19u, 10c 11. 2 years 12. 75% private, 25% government 13. Professional Engineer

Solar Heating Ltd.
2114 14th Avenue
Regina, Saksatchewan S4P 0X5
1-306-527-9822

5. AB,D,F 6. E -d, -u, -c 7. A -d, -u, -c, B -d 9. D -d, -u, -c; G -d, I -d, -u, -c 11. 4 years 12. 100% private

Sal R. Sottile, Architect
2014 14th Avenue
Regina, Saskatchewan S4P 0X5
1-306-522-8333

5. AB,F,H 6. E -d, -u, -c; H -d 7. A -d, -u, -c; B -d 8. D (solar assisted) -d 9. B -d, -c; D -d, -u, -c; E -d, G -d, I -d, -u, -c 11. 3 years 12. Private 13. Architect

Passive Solar Products Ltd.
2234 Hanselman Ave
Saskatoon, Sask.
306-652-2233

5. ABCD,F,H 6. A -c, H -c 7. A -c, B -c 8. B -c 9. A -c, C -c, D -c 100% private

## FRANCE

Jacques Jean-Paul Michel
14, Rue des Poissonniers - B.P.32
92204 Neuilly-sur-Seine Cedex France
722-07-94, 747-01-71

5. ABCD,FGH 6. E 1c, C 26c; H 1u, 1c 7. A 5u, 20c; B 2d; C 1c 9. A -u, B -u, C -c, E -c, F -c, G -c, H -c, I -c 10. Solar Energy Application in Buildings, Academic Press. F.N.P.C. (joint) 11. 19 years 12. 80% private, 20% government 13. Ordre des Architectes, Francais No. 12013 14. Bareme de l'Ordre des Architecte Francais

# Chapter 10: Further Information

For further information on various types of passive systems, I would recommend the following:

First: Join the American Section of the International Solar Energy Society. Full membership is only $25 and brings you **Solar Age** magazine (in my opinion the best solar magazine) as well as discount prices on the proceedings of the National Passive Solar Conferences, order them all.

**Second Passive Solar Conference** (Philadelphia, PA). Editor: Don Prowler, Contains 156 papers in a three-book set, total 1000 pages. ISES Member Price: $14, Non-Member Price: $50.

**Third Passive Solar Conference** (San Jose, CA). Editors: David Richards, Harry Miller and Michael Riordan, Contains 130 papers in one book, total 925 pages. ISES Member Price: $19, Non-Member Price: $65.

**Fourth Passive Solar Conference** (Kansas City, MO) Editor: Gregory Franta, Contains 200 papers in one book total 800 pages, ISES Member Price: $17, Non-Member Price: $65.

In addition, you will probably want to order the **First Passive Solar Conference Proceedings** $10.50 ($13 foreign) from NTIS, U.S. Department of Commerce, Springfield, VA 22161 and the **First Passive Solar Catalog** $7.50 PSI, P.O. Box 722, Davis, CA 95616. These collections will provide a wealth of information as well as a look at most of the types of passive systems that have been tried and good references to further work.

You may also wish to join the Solar Energy Society of Canada, Inc., P.O. Box 1353, Winnipeg, Manitoba R3C 2Z1. You will receive the SESCI publication Sol as part of your membership.

You should also make contact with your local AS-ISES or SESCI Chapter. I have included a list of U.S. contacts for your convenience. Having people with local experience can be very helpful.

## American Section, I.S.E.S.
## A.T.U. P.O. Box 1416
## Kileen, Texas 76541

**ORGANIZED CHAPTER CONTACTS**

CHAPTER COORDINATOR
G. Lee Salmon
P.O. Box 238
Loveland,Colorado 80537
(303) 669-4811

ALABAMA SOLAR ENERGY ASSOCIATION
Dr. Don Wallace, Chairman
University of Alabama
P.O. Box 1247
Huntsville, Alabama 35807
(205) 895-6323

Bernard Levine, Vice-Chairman
7205 Hickory Hill Lane
Huntsville, Alabama 35802
(205) 881-9174

ARIZONA SOLAR ENERGY ASSOCIATION
Don Osborn, Chairman
Arizona Solar Energy Commission
Capital Tower Room 502
1700 W. Washington
Phoenix, Arizona 85007
(602) 965-6210

COLORADO SOLAR ENERGY ASSOCIATION
Rachel Snyder, President
P.O. Box 5272-TA
Denver, Colorado 80217
(303) 231-1192

Tom Feliu, Office Manager
CSEA Book Store
2239 E. Colfax Avenue
Denver, Colorado 80206
(303) 321-1645

EASTERN NEW YORK SOLAR ENERGY ASSN.
Ronald Stewart
P.O. Box 5181
Albany, New York 12205
(518) 457-7584

GEORGIA SOLAR ENERGY ASSOCIATION
Tom McGowan, Chairman
Engineering Experiment Station
Energy Research Lab
Georgia Institute of Technology
Atlanta, Georgia 30332
(404) 894-3448

HOOSIER SOLAR ENERGY ASSOCIATION, INC.
James O. Johnson, Chairman
Johnson Richhart & Associates, Inc.
227 West 11th Street
Anderson, Indiana 46016
(317) 649-2258

Monte Hoover, Sec.-Treasurer
Box 4448

Monte Hoover, Sec. - Treasurer
Box 44448
Indianapolis, Indiana 46204

KANSAS SOLAR ENERGY ASSOCIATION
Donald R. Stewart
1202 South Washington
Wichita, Kansas 67211
(316) 262-7427

METROPOLITAN NEW YORK SOLAR ENERGY ASSN.
Kurk J. Wasserman
P.O. Box Z
Port Jervis, New York 12771
(916) 663-7799

MICHIGAN SOLAR ENERGY ASSOCIATION
Mr. Edward J. Kelly, Jr.
201 East Liberty Street, Suite 2
Ann Arbor, Michigan 48104
(313)663-7799

Karen Kanniainen, Coordinator

MID-ATLANTIC SOLAR ENERGY ASSOCIATION
Linda Knapp, Director
2233 Grays Ferry Avenue
Philadelphia, Pennsylvania 19146
(215) 963-0880

Bob Bennett, Chairman
6 Snowden Road
Bala Cynwyd, Pennsylvania 19004
(215) 667-7365

MINNESOTA SOLAR ENERGY ASSOCIATION, INC.
Peter Pfister
P.O. Box 171
Minneapolis, Minnesota 55440
(612) 871-5703

MISSISSIPPI SOLAR ENERGY ASSOC. (MISSEA)
Dr. Pablo Okhuysen
225 West Lampkin Road
Starkville, Mississippi 39759
(601) 323-7246

MO-ARK SOLAR ENERGY ASSOCIATION
Box 1643
Jefferson City, Missouri 65102

NEBRASKA SOLAR ENERGY ASSOCIATION
Dr. Bing Chen
University of Nebraska
Department of Electrical Technology
60th & Dodge Streets
Omaha, Nebraska 68182
(402) 554-2769
(402) 553-3104 (Home)

NEW ENGLAND SOLAR ENERGY ASSOCIATION
Tom Minnon, Executive Director
P.O. Box 541
Brattleboro, Vermont 05301
(802) 254-2386

Drew Gillett, Chairman
Northeast Solar Energy Center
70 Memorial Drive
Cambridge, Massachusetts 02142
(617) 661-3500

NEW MEXICO SOLAR ENERGY ASSOCIATION
David Miller, Executive Director
P.O. Box 2004
Santa Fe, New Mexico 87501
(505) 983-1006

Mary Beth Bliss, Assistant
(505) 983-1006

NORTH CAROLINA SOLAR ENERGY ASSOCIATION
Leon Neal
P.O. box 12235
Research Triangle Park, North Carolina 27709
(916) 549-0671

NORTHERN CALIFORNIA SOLAR ENERGY ASSN.
Dr. Marshal F. Merriam, President
Dept. of Materials Science
Hearst Mining Building
University of California
Berkeley, California 94720
(415) 642-3664
(415) 845-0272 (Home)

Harry Miller, Vice-President
808 Cerrito Street
Albany, California 94706
(415) 526-4594

NORTHERN ILLINOIS SOLAR ENERGY ASSN.
James A. Hartley, Chairman
P.O. Box 1592
Aurora, Illinois 60507
(312) 892-4705

Jerry Wurm, Vice-President
(312) 567-3839

OHIO SOLAR ENERGY ASSOCIATION
Dr. Bruce T. Austin, Chairman
Wright State University
Dayton, Ohio 45435
(513) 873-2169 or
(513) 767-7324, ext. 78
(513) 767-5341 (Home)

OKLAHOMA SOLAR ENERGY ASSOCIATION
Dr. Bruce V. Ketcham
Solar Energy Laboratory
University of Tulsa
Tulsa, Oklahoma 74104
(918) 939-6351

PACIFIC NORTHWEST SOLAR ENERGY ASSN.
2332 E. Madison
Seattle, Washington 98112

Dr. Douglas R. Boleyn, Chairman
17610 Spring Hill Place
Gladstone, Oregon 97027
(503) 226-8478

Evan Brown, Secretary
(206) 322-3753

SOUTH CENTRAL ILLINOIS SOLAR ENERGY ASSN.
Earl G. Powell, Chairman
637 Eccles
Hillsboro, Illinois 62049
(217) 532-2195

SOUTHERN CALIFORNIA SOLAR ENERGY ASSOCIATI
Ildiko Demeter, Executive Director
San Diego City Hall
202 C Street
San Diego, California 92101
(714) 749-3083

Howard Kraye, Chairman
(213) 246-8408

TENNESSEE SOLAR ENERGY ASSOCIATION
Mr. Joe Hultquist, Executive Director
P.O. Box 127
Kodak, Tennessee 37764

TEXAS SOLAR ENERGY SOCIETY
Russell E. Smith, Executive Director
1007 South Congress, Suite 348
Austin, Texas 78704
(512) 443-2528

Dr. Gary Vliet, Vice-Chairman
(512) 471-7571

VIRGINIA SOLAR ENERGY ASSOCIATION
Lawrence Ç. Perry
P.O. Box 12365
Roanoke, Virginia 24025
(703) 342-1816

Craig Crawford, Secretary
P.O. Box 11084
Richmond, Virginia 23230
(804) 358-9111

WISCONSIN SOLAR ENERGY ASSOCIATION
Ernest Rogers, Chairman
Windworks
Rt. 3, Box 44A
Mukwonago, Wisconsin 53149
(608) 831-4446 (Home)
(414) 363-4088

Allan Anderson, Vice-Chairman
(608) 274-6997

# Solar Energy Society of Canada, Inc.
# P.O. Box 13531
# Winnipeg, Manitoba R3C 2Z1

**ORGANIZED CHAPTER CONTACTS**

SESCI Prince Edward Island
P.O. Box 1041
CHARLOTTETOWN, P.E.I.
C1A 7M4

SESCI New Brunswick Chapter
P.O. Box 1114, Station A
FREDERICTON, N.B.

SESCI Section Quebecoise
Dept. Genie Mecanique
Ecole Polytechnique
C.P. 6079, Succ. A
MONTREAL, Que.
H3C 3A7

SESCI Eastern Ontario
P.O. ox 162
ALEXANDRIA, Ont.
KOC 1AO

SESCI Kawartha Chapter
c/o Judy Shaw
250 Aberdeen Ave.
PETERBOROUGH, Ont.
K9H 2W9

SESCI Kent Chapter
c/o Wilson Kerr
#807 - 150 Mary Street
CHATHAM, Ont.
N7L 4V2

SESCI Kingston Chapter
P.O. Box 153, Station A
KINGSTON, Ont.
K7M 6R1

SESCI Mississauga-Oakville Chapter
c/o T.Gruner
2222 Belfast Crescent
MISSISSAUGA, Ont.
L5K 1N9

SESCI Ottawa Chapter
c/o Algonquin College
Physics Dept.
Lees Ave.
OTTAWA, Ont.
K1S OC5

SESCI Sarnia Chapter
c/o W. Himmelman
1034 Lombardy Drive
SARNIA, Ont.
N7S 2E2

SESCI South Central Ont. Chapter
16 Crescent Street
KITCHENER, Ont.
N2H 1H8

SESCI Simcoe County Chapter
c/o The Technology
Georgian College
401 Duckworth Street
BARRIE, Ont.

SESCI Southwest Ontario
P.O. Box 2220, Station A
LONDON, Ont.
N6A 4E3

SESCI Toronto Chapter
P.O. Box 396, Station D
TORONTO, Ont.
M6P 3J9

SESCI Thunder Bay Chapter
c/o E.J. Tymura
Hilldale Road
THUNDER BAY, Ont.
P7B 5N1

SESCI Windsor-Essex Chapter
Chemical Engineering Dept.
University of Windsor
WINDSOR, Ont.
N9B 3P4

SESCI Manitoba Chapter
P.O. Box 164 Station F
187 Henderson Hwy.
WINNIPEG, Man.
R2L 2A5

SESCI Northern Saskatchewan Chap.
c/o Rob Dumont
Dept. Mechanical Engineering
U. of Saskatchewan
SASKATOON, Sask
S7N OWO

SESCI Regina Chapter
P.O. Box 3959
REGINA, Sask.
S4P 3R9

SESCI Swift Current Chapter
416 Hayes Drive
SWIFT CURRENT, Sask.
S9H 4H9

SESCI Northern Alberta Chapter
P.O. Box 11062
EDMONTON, Alta.
T5J 3K4

SESCI Peach Region Chapter
c/o Box 204
GRANDE PRAIRIE, Alta.
T8V 3A4

SESCI British Columbia Chapter
c/o Richard Kadulski
504 Davie Street
VANCOUVER, B.C.
V6B 2G4

SESCI Vancouver Island Chapter
c/o Environmental Centre
101 - 645 Fort Street
VICTORIA, B.C.
V8W 1G2

Second: I would recommend subscriptions to the following magazines in order of my preference:

**Alternative Sources of Energy,** 107 S. Central Ave., Milaca, MN 56353, $15: U.S., $18: Canada, $32: Foreign.

**Bulletin of the New Mexico Solar Energy Association,** P.O. Box 2004, Sante Fe, NM 87501, $10: U.S., $15: Foreign.

**Canada Renewable Energy News,** P.O. Box 4869, Station E, Ottawa, Ont. K1S 5B4, $10: U.S. & Canada, $25: Foreign.

**Solar Living and Solar Greenhouse Digest,** P.O. Box 2626, Flagstaff, AZ 86003, $10: U.S., $13: Canada, $18: Foreign.

**Solar Engineering,** 8435 N. Stemmons Freeway, Ste 880, Dallas, TX 75247, $15: U.S., $22: Foreign.

**Solar Times,** 3 Old Post Road, Madision, CT 06443, $12.50: U.S. and Canada, $17.50: Foreign.

**Solar Heating and Cooling,** P.O. Box 2106, Morristown, NJ 07960, $15: U.S., $23: Foreign.

**Solar Energy Digest,** P.O. Box 1776, San Diego, CA 92117 $28 year.

Third: You might also enjoy the following, which include many subjects other than solar design but do occasionally include an excellent passive solar or conservation article:

**New Shelter,** 33 East Minor, Emmanus, PA 18049, $8.97 U.S.

**Mother Earth News,** P.O. Box 70, Hendersonville, NC 28739, $15: U.S., $17: Foreign.

**Organic Gardening,** Organic Park, Emmaus, PA 18099, $6.97 U.S.

**Harrowsmith,** Camden East, Ontario, Canada KOK IJO, $10: U.S. and Canada, $14: Foreign, $25: air mail.

**CoEvolution Quarterly,** P.O. Box 428, Sausalito, CA 94965, $12: U.S.

**TRANET,** P.O. Box 567, Rangeley, ME 04970, $15: U.S. and Foreign.

**Countryside,** 312 Highway 19E, Waterloo, WI 53594.

**Natural Life,** P.O. Box 640, Jarvis, Ont. $8.

**The Country Journal,** 139 Main Street, Brattleboro, VT 05301 $12.

**Rain,** 2270 N.W. Irving, Portland, OR 97210, $12.50 year.

**Popular Science,** P.O. Box 2871, Boulder, CO 80302, $9.94.

**Popular Mechanics,** P.O. Box 10064, Des Moines, Iowa 50350, $7.97.

**Sunset,** Menlo Park, CA 94025, $10 U.S. (Western States), $14: East, $18.25: Foreign.

**Hudson Home Guide,** 289 South San Antonio Rd., Los Altos, CA 94022, $15: U.S., $17: Canada, $20: Foreign.

**Family Handyman,** P.O. Box 2897, Boulder, CO 80302. $8.95.

**Professional Builder,** 5 S. Wabash, Chicago, IL 60603, $30: U.S., $40: Canada, $45: Foreign.

**Not Man Apart,** F.O.E. 124 Spear, San Francisco, CA 94105, $10.

**Adobe Today,** P.O. Box 1178, Belen, NM 87002, $9.50 U.S.

**Wood Burning Quarterly and Home Energy Digest,** 8009 34th Ave., Minneapolis, MN 55420.

**Wind Power Digest**, 54469 CR 31, Bristol, IN 46507.

**Conserver Society News,** P.O. Box 2097, Station C., Downsview, Ontario, M3N 2S8 $5. year.

**Alternative Magazine,** Trent University, Petersborough, Ontario, $15.00 year

Fourth: For more general questions about solar systems, you can call: 1-800-523-2929 or Write:

National Solar Heating and Cooling
Information Center
P.O. Box 1607
Rockville, MD 20850

You may also wish to contact your State Solar Office or Energy Office.

YES—I want to help Solar Lobby in its efforts to advance solar energy. Please keep me informed about new developments in solar and other energy issues.

______ Enclosed are my membership dues of $15.

______ I want to help with an additional contribution. The total of my check to Solar Lobby is for:

______ $25 ______ $50 ______ $100 ______ $500 Other $______

______ I am already a Solar Lobby member, here's a special contribution.

NOTE: If you would like to support Solar Lobby's educational programs and to deduct your contribution from taxes, please make your check payable to Center for Renewable Resources.

**$2.75 of your Solar Lobby dues is allocated to the cost of the *Sun Times*. For a copy of the full financial report, write to Solar Lobby.**

**Solar Lobby**
**1001 Connecticut Avenue, N.W., 5th fl.**
**Washington, D.C. 20036**

Fifth: For access to current work, you may be able to benefit from the following computerized energy research networks:

**In the U.S.**

**RECON -** Energy Data Base, Projects and Reports
Department of Energy, Tech. Info.
P.O. Box 62
Oakridge, TN 37830

**In Canada:**

Research in Progress:
Cam McNeil
Renewable Energy Resources Branch
Energy, Mines and Resources
580 Booth St.
Ottawa, Canada K1A OE4

Canadian Literature (in progress, not finished yet, will be on **CANOLAY** network)
David Holmes
Energy Library
CISTI, NRC
Building M-55
Ottawa, Ontario K1A 052

General Wind and Solar:
Dr. Terry Rachuk, SWERP
Solar Energy Research Program
Alberta Research Council
11315 87th Avenue
Edmonton, Alberta T6G 202

**In Australia:**

**ARER Index**
The Manager
Central Information Service
CSIRO
P.O. Box 89
East Melbourne, Victoria, Australia 3002

Your local utility may also be a resource for information on energy conservation and solar design. The national Residential Conservation Service program is getting even the more reluctant utilities involved.

# Chapter 11: WOMEN IN PASSIVE SOLAR

Women have played a major role in passive solar developments. This essay has been included to encourage even greater involvement of women in the many aspects of passive solar development for although women have been in the forefront of passive solar development they account for considerably less than 10% of the registered architects and less than 1% of the builders in the U.S. The achievements of the olar pioneers have clearly demonstrated that we would have better buildings if women were more heavily involved.

Special tribute is due for the women who have been so successful in passive solar developments including:

- Karen Terry, builder, Sun-Up Construction
- Lynn Nelson, builder-designer, Habitat Center
- Virginia Thigpen, builder-designer, Bent Nail Construction
- Judy Corbett, developer, Village Homes
- Susan Nicholls, design and development, Communico
- Sarah Balcomb, Passive Solar Associates
- Lola Redford, President, Consumer Action Now
- Susan Alusil, promotion, Adirondack Energy Group
- Peggy Wrenn, Director, Colorado Renewable Fuels Program
- Jan Hamrin, Solar Office Manager, California Energy Commission
- Lillian Leehouts, architect
- Polly Cooper, designer
- Vivian Loftness, design, formerly Manager AIA Research Corporation
- Penelope Niland, writer, developer
- Karen Crowther, engineer, Living Systems
- Valerie Anderson, instructor, Sacramento State College
- Elizabeth Chase, designer, Tennessee Valley Authority
- Holly Antolini, writer/editor, Sunset Magazine

And the many others, particularly the owner-builders. The following statements were submitted in response to a PSI letter sent to women working in passive solar design and construction.

## Virginia Thigpen, Contractor-Designer, owner: Bent Nail Construction

*Bent Nail Construction is an equal opportunity, small design and construction firm. Bent Nail was established in 1977 for the purpose of designing and building low-to-moderate cost, low-technology, effective, custom passive solar homes.*

*Bent Nail tries hard to achieve some lofty, idealistic goals through various types of appropriate technology:*

**TRIES TO**

*Increase owner involvement thus satisfaction Create an awareness in people* regarding our need to be self-sufficient yet together and responsible.*

**homeowners as well as Bent Nail employees*

*Eliminate the need to use nuclear energy and other non-renewable energy supplies.*

**BY**

*Involving owners in many aspects of design and construction . . . As much as they will stand for!*

*Giving them the skills and tools to be more capable of controlling their own lives - construction lessons, solar design, gardening, etc.*

*Developing passive solar designs and solar components which are mass producible*

*Each Bent Nail Home (there have been about ten or twelve so far) comes complete with all the equipment needed to be 80% plus energy-self-sufficient; assorted construction lessons in painting, electrical, roofing, you-name-it, and four to ninety mini-lectures on living the good life.*

Bent Nail Construction

## Valerie Anderson Owner-designer-builder, Water Wall Passive Solar House, Davis.

*The six months from starting out to moving in were some of the best in my life. I was physically active, feeling strong and capable; learning something everyday; feeling confidence level go up and up; watch a house grow; seeing my own design translated into a real structure.*

*All skills and techniques were within our capabilities—access to information was most important—also having good "teachers" -- Virginia, Marshall, Reed and you helped immeasurably.*

*Any time one woman might not feel she has the muscle to do a job - she can be assured all it takes is a friend or two, not a muscle building course or a defeat.*

*As I wrote up the story of the building process for an article I could see the main theme is "We can do it," "I can do it" - "Women can do it."*

**Positive Factors**

1. *Chance to work on real project with others - camaraderie, friends.*
2. *Sense of achievement - accomplishment*
3. *Sense of capability - I can learn and can do.*
4. *Seeing tangible results.*

*Important to be involved in the design process-- you know best, your own needs, wants, and dreams for your shelter.*

**Energy Use Comparison**

**Duplex, West Davis**
**Single Story**

860 sq. ft.

| 1978 | Therms | KWH |
|---|---|---|
| March | 64 | 373 |
| April | 63 | 335 |
| May | 32 | 272 |
| June | 25 | 349 |
| July | 21 | 509 |
| Aug. | 21 | 515 |
| Spet. | 24 | 313 |
| Oct. | 28 | 464 |
| Nov. | 56 | 441 |
| Dec. | 112 | 482 |
| Total | 446 | 4053 |

**Solar home, West Davis**
**Ground floor + 2 lofts**

1200 sq. ft.

| 1979 | Therms | KWH |
|---|---|---|
| March | 20 | 391 |
| April | 19 | 275 |
| May | 14 | 260 |
| June | 8 | 214 |
| July | 8 | 266 |
| Aug. | 8 | 275 |
| Sept. | 14 | 261 |
| Oct. | 18 | 303 |
| Nov. | 25 | 394 |
| Dec. | 22 | 349 |
| Total | 156 | 2988 |

1. Occupants same both homes; mother, teen-age daughter.
2. Work/school schedules same both years.

**Energy Saving** Gas consumption down 66%
Electricity use down 27%

Valerie Anderson

## Lillian Leenhouts, Architect

*The year I finished art school was the same year that Frank Lloyd Wright published AN AUTOBIOGRAPHY and the Chicago World's Fair exhibited the round glass house by the pioneer solar architect, George Fred Keck. The book and the fair had much to do with my return to architectural school. I have been working in architecture ever since and like it very much as is probably evidenced by the fact that my husband and I were the first couple to be made Fellows of the American Institute of Architects at the same time in Atlanta in 1975.*

*Most of the people who graduated with me from Michigan in 1936 did not find jobs in architecture because of the depression. I was fortunate. I worked for a Milwaukee architect for six years until all materials were frozen except for use in World War II. I then became a draftsman for powder plants and air craft. Finally my husband and I returned to Milwaukee to open our own office in 1945. Since then, we have designed all our buildings, unless it were physically impossible, using passive solar, ground temperature and prevailing breezes.*

Elderly project - Lillian Leenhouts

## Lola Redford, President of Consumer Action Now

*The women of this country represent an important resource that has yet to be tapped in helping the country make this transition to clean renewable energy sources.*

*Women are responsible for 80 percent of all consumer purchases, and more women than ever before are becoming heads of households; yet, most women are not aware of the overwhelming impact that energy has on family budgets and the environment. With the proper consumer education, women can learn that oil prices affect many more areas of their lives than simply gasoline for automobiles and electricity for homes.*

*Today, more and more women are entering the work force. Their economic and social aspirations are greater now than at any previous time in our history. In a world which utilizes clean, safe, renewable resources, women will find many new, exciting and rewarding possibilities.*

*However, it is not easy to break down age-old stereotyping and sex roles and, in order to demystify energy for women and encourage them to realize their full potential to affect change, Consumer Action Now (CAN) has embarked on a national energy education program for women.*

*The goals of CAN's Energy Project for Women are:*

- *To interest women in energy decision making by showing them how these decisions have economic impact on their daily lives.*
- *To encourage women to assess energy needs and problems at the local level.*
- *To provide them with the tools to impact state and national energy policy.*
- *To encourage more women of all ages to pursue jobs and business opportunities in the energy field.*
- *To convince women that they should promote methods of energy production which are renewable and less damaging to the environment than those now prevalent.*

## Cynthia Easton, Architect

*There is no particular reason for women to be less involved than men in Solar Architecture and Construction. Every architect, designer and builder should understand the implications of, and design considerations for, solar energy. At the present time, an architect must either consult with structural and mechanical engineers or do their own engineering. From now on, the architect will also need to consult with, or be his or her own solar engineer.*

*Since women are only now entering the field of architecture in any number, they are less prejudiced toward mechanical system solutions to the problems. Thus, they can be more sensitive to the environment; how to use it, work with it and accept it, rather than how to ignore it.*

Water wall house - Cynthia Easton

## Polly Cooper, designer, San Luis Obispo Solar Group

*I have had an interest in a holistic approach to design and building for a long time. With a varied academic background, my professional experience moved from conventional teaching and conventional architecture to work in environmental planning and passive solar architectural design.*

*I was turned off by the misuse of cheap power and the consequent callous approach to the world and felt our design work as well as our culture needed reorientation. A clearer understanding of ecological processes made the forgotten processes of the past and appropriate technologies of today again compatible.*

*For me, designing passive solar architecture is a step in this direction. And if a building encourages conservation in the users and allows them to be more in touch with the nature of materials, with natural lighting, and with climate in general, we have contributed more than just another efficient structure with passive solar design. We have produced an enviroment which helps to reconnect people to the natural systems upon which we all depend.*

Water wall house - Polly Cooper

## Sarah Balcomb, Passive Solar Associates

*Consider these points: *almost one third of the energy used in this country is for heating and cooling buildings and water; *virtually all new buildings and many of the existing ones could - with existing knowledge - be constructed or modified to use one-third to two-thirds less energy and still be comfortable; *even though the full time housewife is an endangered species, it is still women who are the primary purchasers of homes and who spend more time in them, and who are responsible for the internal energy management in most schools and small commercial structures.*

*It doesn't take much imagination to put those statements together and realize how very important the role of women is in just that one aspect of solar energy. (Similar sets of conditions could be stated for solar fields other than the heating and cooling of buildings). The opportunities are vast and varied. More important - for me - the responsibility and the challenge are great and the notion that I might be able to help others realize their own potential for saving energy, for using our resources more wisely, for simply living more rationally and realistically, is very exciting.*

Water wall house - Bent Nail Construction

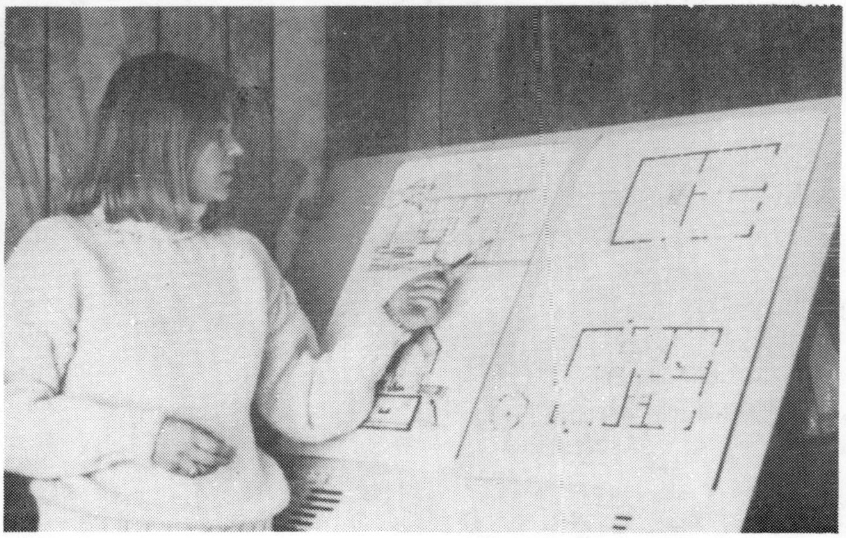

# Chapter 12: Low Energy Use Materials

The energy cost of building materials is an important factor in the pattern of energy use in the U.S. In 1967, the last year for which data on energy use for the building sector have been analyzed, over 8% of all energy use was for building construction and maintenance. When compared to an estimated 25-30% for space heating and cooling this may not sound terribly important, yet it still represents the equivalent of 38 billion gallons of gasoline or the yearly output of 25 one thousand megawatt generating plants. And, almost as easily as passive solar design can reduce energy use for space heating and cooling, wise design can reduce the energy cost of materials.

The energy cost of building has not been well studied, energy cost having been so heavily subsidized for many years, but some preliminary work is now available. These studies suggest the following energy cost for typical construction.

| Type | Energy Intensiveness BTU $ | Total Energy Required, U.S. 1967 | BTU/Sq. Ft. |
|---|---|---|---|
| Single family housing | 55,500 | 781 Trillion BTU | 700,000 |
| Remodels | 51,600 | 262 Trillion BTU | n.a. |
| Highrise apartments | 60,000 | 118 Trillion BTU | 740,000 |

It is instructive to compare this with the energy cost of constructing a passive solar adobe house. If we use energy conserving design the house will have an oiled mud floor, vigas (round rather than sawn beams), wood bond beams, mud mortared stone foundation, and a willow roof deck. Almost the entire structure can be built using materials available on site or at least locally. The exceptions, are the glass, insulation for between doubled walls (mineral wool) and roll roofing. The vigas may have to be hauled in as well, but require no other energy. The energy cost of this structure then will be:

| Component | Energy | Energy Cost/BTU |
|---|---|---|
| Foundation | Labor | 4,000 cal/day |
| Floor | Labor | 4,000 cal/day |
| | Oil, 6 gallons | 200,000 BTU |
| Walls | Labor | 4,000 cal/day |
| | Insulation, R-11 | 624,000 BTU |
| | Glass, 200 sq. ft. | 2,687,400 BTU |
| Roof | Roll roofing, mineral coated | 11,700,700 BTU |
| | Labor | 4,000 BTU |
| Total Labor | 8 man months | 3,870,000 BTU |
| TOTAL ENERGY | | 19,082,100 BTU |
| RENEWABLE | | 3,870,000 BTU |
| TOTAL NONRENEWABLE | | 15,212,100 BTU |
| ENERGY COST, BTU/SQ. FT. | | 15,212 |
| % TRADITIONAL CONSTRUCTION ENERGY COST | | 2 |

We may not all be able to do this well of course, but we can do much better than we do now. Consider lifetime energy cost and environmental impact as well as the energy savings possible when choosing building materials. The initial energy cost of a good passive structure built with energy intensive materials may equal the energy use for space heating and cooling over 80 years while using energy efficient materials in a passive solar house can reduce this to about 2 years.

There are many ways for the builder to conserve energy in the use of building materials. These include, using materials that are durable, require little maintenance, and where possible, have a low initial energy cost. The energy savings usually translate to life-cycle labor and financial savings as well, so it is worth careful consideration even if the initial cost is higher.

Wood is an excellent material and can become a renewable resource if we manage our nation's forest land better. It is often best left unpainted, because it will usually last longer, look better, and not require the adverse energy, dollar and environmental impacts associated with repainting. In many cases it is also possible to mill your own lumber, some of the better mills are referenced at the end of this chapter.

Earth and stone are also almost ideal materials, widely distributed, durable, and low in energy cost and environmental impact. They are also excellent for thermal storage in passive design. Both, however, are labor intensive and are currently discriminated against by building codes.

For windows, glass is best in almost every case. It is more durable than plastics, uses less energy, and is more easily recycled. Only where impact is a consideration should the plastics be considered.

The energy conscious builder can utilize these durable, low energy cost, materials effectively in passive solar design. Two of the best and least widely used candidates are adobe and rammed earth. Building with adobe is fairly widely recognized (though little used), but rammed earth is still almost unknown. The two are complementary, for in areas with sandier soil where adobe is impractical, rammed earth may be perfect.

The First Passive Solar Catalog featured a passive solar house built over 25 years ago with rammed earth. As the owner-builder David Miller said during our interview, "This type of building made good sense in the resource short years after World War II. It makes even more sense today and for tomorrow!" The references at the end of this chapter will help you build either a rammed earth or adobe house.

Four other methods of construction utilize low energy cost materials. These are: 1) Building with recycled materials, 2) building with stone, 3) building a cordwood house, and 4) building a log house. Each has its advantages and disadvantages and may be just the thing for your next house.

In many respects the house built of recycled aterials may be best. Rather than letting energy ‹pensive materials (5,600 gallons of gas equivalent er house) be consigned to the sanitary (sic) land fill e should recycle them whenever feasible. My arents' current house is built primarily from re- ycled materials saved from an existing ramshackle ouse on their homestead. We took it apart very arefully and managed to save probably 90% of the ood, all windows, fixtures, and wire, and most of the ails which were straightened in the evening and olished in the cement mixer.

By cleaning boards carefully before stacking by ze and length it was possible to minimize handling nd increase efficiency. Using this method pieces ould often be used without cutting during construc- on. The quality of wood used in older houses was mazing, many of the full dimension 2" x 4" 's had o knots. And 1" sheathing up to 24" wide was sually clear or close to it. This material should ever be bulldozed and carried to the dump.

The tools for recycling are generally inexpen- ve, but do require some skill, which is quickly arned. A basic set would include: gloves, a small nd large crowbar, a wonder bar (flat-stock crow ar), slide hammer nail puller (Crescent seems best), edge, hammers (steel handle), linemens pliers, ppers, a couple of saws (a circular saw 7-1/4 inch is elpful), and a wire brush. With mixed dimension mber, a table saw for ripping is desirable, get a ood 10" saw with at least 1 hp (expect to pay $600 r a good new one).

A stone house may also be a good choice, if the stones are on site and you don't have an aversion to hard work. The slip form system seems to be the best for beginners and is the one I would suggest. The stone house is typically built with insulation on the inside, but itwouldbe better to include insulation on the outside or between double walls. In temperate climates with plenty of sun insulation may not be necessary because of the high mass of the walls.

If wood is readily available I would be tempted to build a cordwood house rather than a log house. Using 14" or 16" cordwood walls or double walls with insulation between can make a reasonable energy efficient house. The skill required is less than that for a log house and the short lengths are also more readily available.

Only very rarely is a log house desirable because it is hard to make a log house an energy efficient house. The exceptionis where the logs available on site are of excellent quality and will allow you to build a log wall averaging 12-16" in thickness. I would be tempted to mill the top, bottom and inside of the logs if I were to build with logs and might just mill lumber to build a post and beam house instead.

The following table may help you chose mater- ials more wisely.

Hopefully this brief introduction to low energy cost materials will get your interest up and get you started toward more energy efficient construction for your passive solar buildings.
The following books will help you understand the use of low energy use materials.


**Energy Cost of Materials:**

Richard Stein (1978) *Architecture and Energy*, Anchor Books, N.Y. $6.95.

Bruce Hannon et al. (1978) "Energy and Labor in the Construction Sector," *Science*, V. 202, Nov. 24.

William Lumpkins, in (1972) *Life Support Technics*, Biotechnic Press, Albuquerque.

David MacKillop (1972) "Low Cost Housng," *The Ecologist*, v. 2, n. 12 December.

Oregon Office of Energy (1974) *Transition*, Salem, Oregon.

**Recycling**

"Wrecking and Salvage" in (1973) *Shelter*, Shelter Publications. Not much - but I don't know of any more complete works (have to write one, I guess!)

**Cordwood**

NHC *Stackwall: How To Build It*, Univ. Manitoba, 417 Engineering Building, Winnipeg, Mantiboa, $12.50.

Jack Henstridge *Building the Cordwood House*, Plain Deal Pauper Press, 824 Charlotte St., Fredericton, NB $6.00.

Jack Henstridge and Francois Tanguay *La Maison de Bois Corde*, Aurore, RR 1, Inverness Quebec, Goslko, $6.95.

**Log**

B. Allan Mackie *Building with Logs*, Log House Publishing, Box 1205 Prince George, B.C. V2L 4Ue, $10.00.

Christian Bruyere and Robert Inwood (1975) *In Harmony with Nature*, Drake Pub., 800 2nd Ave., NY, NY 10017, $6.95.

**Post and Beam**

Steward Elliott and Eugenie Wallas (1977) *The Timber Framing Book*, Housesmiths Press, P.O. Box 157, Kittery, ME 93905, $10.95.

Steward Elliott with Janes Brooks (1978) *The Timber Frame Raising Book*, Dovetail Press, P.O. Box 1496, Boulder, CO 80306, $9.95.

Stewart Elliott (1979) *The Timber Frame Planning Book*, Dovetail Press, P.O. Box 1496, Boulder, CO 80306, $16.00.


## ENERGY COST OF BUILDING

| Material | Unit | Energy Cost BTU/Unit Before To Job Site | After Delivery |
|---|---|---|---|
| Rough softwood lumber | Board foot | 5,299 | 7,661 |
| Finished softwood lumber | Board foot | 5,399 | 7,859 |
| Softwood exterior plywood | Square foot | 4,393 | 5,779 |
| Shakes and shingles | Square foot | 4,682 | 7,315 |
| Roll roofing, mineral surface | Square foot | 10,673 | 11,012 |
| Asphalt strip shingles | Square foot | 24,553 | 25,334 |
| Galvanized steel sheet | Square foot (poured) | 26,458 | 27,836 |
| Aluminum sheet | Square foot | 31,532 | 31,981 |
| | Pound | 94,596 | 95,943 |
| Paint, latex exterior | Gallon | 413,519 | 489,063 |
| | Square foot | 1,181 | 1,397 |
| Glass, double strength | Square foot | 13,437 | 15,436 |
| Ready mix concrete | Cubic foot | 95,738 | 96,086 |
| Rebar | Pound | 14,888 | 15,664 |
| Mineral Wool Insulation | 3-1/2" batt, square foot | 6,112 | 6,860 |
| Fiberglass insulation | 3-1/2" batt, square foot | | 9,000 (est.) |
| Cellulose insulation | 3-1/2" deep, | | 1,000 (est.) |

**Mill your own**

Lumber Maker - Haddon Tool, 4719 West Rte 120, McHenry, Il. 60050.

Alaskan Mill - Granberg Industries, 200 South Garrard Blvd., Richmond, CA 94804.

Oregon Mill - Mobile Mfg., Co., Box 258, Troutdale, Or.

Oneman Mill - Belsaw, 3679 Field Bldg., Kansas City, MO 64111

**Adobe**

Archer *Dirt Cheap*, Compedium (Australia)

Paul McHenry (1973) *Adobe, Build it Yourself*, University of Arizona Press, $7.95

Marcia Southwick (1974) *Build your Own Adobe*, Swallow.

Ken Kern (1975) *The Owner Built Home*, Owner Builder Pub' $6.95.

Eugene Boudreau (1971) *Making the Adobe Brick*, Random House, $2.95.

Karl Schultz (1972) *Adobe Craft*, from Adobe Craft, 198322 Carlwyn Drive, Castro Valley, CA $5.50.

Paul Garrison (1979) *How to Build Adobe Hosues . . . etc.*, TAB Books, $7.95.

Rex Hopson (1979) *Adobe: A Comprehensive Bibliography*, Lightning Tree Press $6.95.

**Rammed Earth**

Middleton *Build Your House of Earth*, Compendium (Australia).

Cytryn *Soil Construction*, Weismann Science Press (Israel).

Lyle Wolfskill *Handbook for Building Home of Earth*, NTIS, Springfield, VA 22151, $3.00

Dancy *A Manual of Building Construction*, ITG., (England).

Lyle Wolfskill, Wayne Dunlap, Bob Gallaway (1962) *Earthen Home Construction*, from AERO 435 Stapleton Blvd., Billings, MT 59101

AERO (1979) "Rammed Earth Construction," Billings, MT 59101

USDA (1963) *Rammed Earth Walls for Buildings*, Farmers Bulletin #1500, from AERO, $2.50

Ken Kern (1975) *The Owner Built Home*, Scribner.

**Stone**

Helen and Scott Nearing (1952) *Living the Goodlife*, Schocken Books.

Karl and Sue Schwenke (1975) *Build Your Own Stone House*, $5.95.

Lewis and Sharon Watson (1974) *How to Build a Low Cost House of Stone*, Stone House Public., Box 4002, Sweet Idaho, ID 83670, $5.95.

Ken Kern, Steve Magers, Lou Penfield (1976) *Stone Masonry*, Owner Builder Public., Box 550, Oakhurst, CA 93644, $6. ppd.

To encourage discussion of the value of saving older buildings from the bulldozer in your town, you might submit a copy of the following ordinance to your city council. I would prefer not to see anymore government intervension in private affairs, yet it is one of the few avenues for combatting the governments incredible energy subsidies. If we paid real cost for our energy (including indirect costs) there is no question that all buildings would be recycled rather than demolished.

Ordinance No. ____________ Series____________

AN ORDINANCE PROHIBITING BUILDING DEMOLITION IN____________________________

**Section1: Findings**

A. Buildings contain valuable materials that have an energy value of as much as 7 trillion BTU for a 1,000 square foot house. (1)

B. When buildings are demolished (using more energy), these materials are trucked to the solid waste disposal site (using more energy) and buried, removing valuable resources from circulation and burying the equivalent of 5,600 gallons of gasoline in terms of energy value of materials.

C. Buildings can be successfully recycled without excessive energy input with almost complete recovery of materials for reuse. (2)

D. Building recycling is labor intensive.

**Section2: Ordinance**

Now therefore, the City Council of the City of ____________________________ does hereby add section ____________________________ to the City Code of the City of ______________

"No building may be destructively demolished and the materials buried in the solid waste site unless special conditions may be shown which prohibit "recycling" the building through careful deconstruction."

Passed and adopted by the City Council of the City of ____________ ____________________.

____________________________________ 1980.

Attest: City Clerk


1. Bruce Hannon et al. (1978) *"Energy and Labor in the Construction Sector," Science*, v. 202, Nov. 24.
2. David Bainbridge (1980) *The 2nd Passive Solar Catalog*, PSI, Davis.

# Chapter 13: Living the Good Life

*"Like multitudes of people all over the world we are seeking a good life — a simple, balanced, satisfying life style. Like them, our aim is to lend a hand in shaping the planet into a homelike living place for successive generations of human beings and for the many other life forms domiciled in and on Mother Earth, her lands and waters."*
*Helen and Scott Nearing (1979)*

Village Homes: A 200 Unit Solar Subdivision with Community Farm

Living the good life is what the Nearings call it, and their good health and satisfying lives are a testament to the value of their approach. Scott is now 97, Helen over 75, yet both are still very active on their homestead in Maine. The simple living they first sought and discovered in the 1930's on their first homestead in Vermont is equally valuable for all of us today. As the subtitle of their fine book, *Living the Good Life,* tells us, "how to live simply and sanely in a troubled world."

What is the good life? Well, it certainly depends on who you are, not all of us would be comfortable with the Nearings method; but it should incorporate many of the things described here.

Probably the most important element of any "good life" is what is now becoming known as voluntary simplicity. In many ways this is counter to current practice in America but it has many rewards. These include: the need for less money to buy and support the conveniences, a chance to become reacquainted with your family and friends, a feeling of independence (if you have simple things they are usually easy to fix or don't break), and more time to relax and play.

The backpacker, bicycle tourist, or canoeist is perhaps best able to realize how little we need for a comfortable and satisfying life. A shelter from the storm, simple yet tasty and filling food, a book to read, and a good companion are really all we need. For a month in wild country our necessities might total 50 pounds, 20 for food and 30 for shelter, entertainment, cooking, etc. Many backpackers consider their times out on the loose the only time they are really alive, all the other time is just waiting.

And in many cases they are right. They fail to heed the lessons learned along the way and return to a spacious house in the suburbs. To support it and the energy guzzling car they use to get to work, they labor for weeks and months and years at a desk in a poorly lit, overheated and isolated office in a smoggy city—getting exercise only rarely and eating richly and too often. Hardly living...perhaps if they but listened to their bodies they would heed the lessons learned backpacking, and try simple living all the time.

A small passive solar house is the ideal home for living the good life. It is easy to build, inexpensive, durable, healthy, comfortable, and requires little energy. It does, however, need attention to the weather, opening and shutting the drapes, shutters and windows, when necessary for example. This provides a good excuse for getting outside often to keep track of the developments in the weather.

After building your passive house, furnishing it elegantly (but simply), preferably with furniture you've made yourself, you can devote more attention to developing your homestead. Even a small lot in the city can provide most of your food if it is wisely planned and carefully tended (only two or three hours work per week). The choice of fruit trees is particularly important, and should probably be made even before the house is built. They will provide a lifetime of benefits for very little work. The garden can provide all of your vegetables for the summer, and most for the winter and spring as well. With a solar green house you can probably meet almost all of your needs.

All the food you grow at home represents energy saved, particularly if you use the French-intensive biodynamic deep bed method. Rather than investing 8 calories in non-renewable energy for each calorie you eat (as conventional agriculture does) you will spend only from 0.1 - .5 calories of renewable energy for each calorie you eat. And you will have food that is tastier and more nutritious.

The choice of animals to keep, if any, is also important. The Nearings keep none—and present some good reasons for it. But I think most folks would be happier with at least chickens (and eggs) and possibly a pig or two and some rabbits. All are relatively easy to care for, subsist in large part on garden scraps (rather than purchased food) and are easy to process. For a larger place, an acre or more rather than a 1/4 acre, you may want to keep a draft horse, to help with the work, and a cow or two for milk and/or meat. Beware, however, as care is more critical and upkeep can be more expensive.

After a good place to live and good food to eat comes the necessity for a good way to make a living. This should be healthful, productive, and contribute to a better future. With a simple, yet fulfilling, lifestyle the need for money is considerably less pressing, allowing for choice of more satisfying, if less lucrative employment. As the Nearings have discovered over the years, only 1/2 of their work is devoted to life support, and this includes work on the garden and homestead where there is no financial reward—except money saved. Certainly cash income in the poverty level can be more than sufficient.

A number of options are available, including many of the service industries such as nursing, doctoring, food serving (the Van Dresser's another pioneering pair advocating the good life, supported themselves by operating a small organic restaurant in New Mexico from the 1940's to 1970's), home repair, and similar ventures. Manufacturing is also open and can be very satisfying for both the craftsperson weaving rugs, the machinist making pelton wheels for hydropower installations, the sawyer making hardwood lumber, or the gardener growing vegetables for sale.

Passive solar design offers many opportunities for business people interested in the good life. Some of the best openings are for: 1) House Doctors - specializing in energy audits and analysis, telling people what to do and how, 2) Remodelers - to implement the house doctors recommendations, plugging leaks, adding windows and mass, etc. 3) Installers - to add mass, passive solar water heaters, shutters, etc., to new houses, 4) Manufacturers - for passive solar water heaters, insulated drapes, insulated shutters, thermal mass, etc., 5) Distributors - for all the above, as well as energy conservation products and materials, 6) Teachers - to help communities and homeowners reduce their energy use, and many more.

And finally, the good life includes close family and friends and a healthy community. The good life has a way of bringing the family together and keeping them together. The less hectic pace, the fact that people are away less, and the lack of conveniences contribute to a closeness most people have never known. You'll probably find that working together is even better than recreating together—because it has meaning. Children are included too, because work has real meaning for them too.

Co-operating with other families and the community is also desirable but can be more difficult to achieve. It is the one area where the Nearing's felt disappointed. It is in some respects easier now than it was in the 1930's in Vermont, but it can still be hard unless you can find a community with a good framework for co-operation already established.

And finally, the good life is a life with little adverse affect on the environment. This is made easier with a passive solar house and simple life, yet it still requires thought and effort to buy wisely, use carefully, and recycle completely. It is both possible and economically feasible to do so. For example, should you buy a wool carpet or nylon carpet? Well, the wool will usually last longer, be more comfortable, and can be recycled by composting when at long last it is worn out. The nylon carpet cannot. Or should you buy a homeowners model table saw for $250 or the industrial grade model for $700. Well, in industrial grade model will allow you to do better work more safely and will still be working well when your children teach their children how to use it, the homeowners model won't. It is easier to be a responsible consumer than you might expect—until you've tried it.

Another aspect of responsible consumership is supporting responsible businesses. Do you buy from the chain store or local independent business, offering the same quality product at a higher price, yet with better service? I'd say it's better to spend a dollar on your local business because it stays in the community, helping someone else maintain their independence and helping keep the community strong. Co-operatives are also a very attractive alternative, as the profits help reduce everyone's cost as well as bringing people together.

Bartering is also a very valuable tool for the consumer because it can eliminate the middleman, and sometimes the intervening hand of government, trying so hard to increase its size and hold on our lives. It's surprising how effective barter can be, whether it is service for service, goods for service, or goods for goods. A strong back and capable hands can be bartered for virtually anything, even these days. Try it.

This, then, is the good life. A meaningful, healthy, low impact lifestyle that contributes very directly to a livable future for all of us. It is very different from how many of us live today but there is little doubt that most of us will find it better. It is perhaps the only answer to America's energy, environmental, and social problems.

The following books will help you explore the Good Life further:


Helen and Scott Nearing (1952) *Living the Good Life*, Schocken

Helen and Scott Nearing (1979) *Continuing the Good Life*, Schoken

John Seymour (1978) *The Self-Sufficient Gardener*, Dolphin Books

Peter Van Dresser (1976) *Design for a Human Landscape*, Lightning Tree Press

David Bainbridge (1979) *The First Passive Solar Catalog*, P.S.I.

David Bainbridge et al (1979) *Village Home's Solar House Designs*, Rodale Press

Steward Brand, ed. (1974) *The Whole Earth Catalog and Epilog*, Penguin Books

Robert Argue et al (1978) *The Sun Builders*, Renewable Energy in Canada, Toronto

James Lawrence, ed. (1979) *The Harrowsmith Source Book*, Camden House

Stewart Brand and Jay Baldwin (1979) *Soft Tech*, Penguin Books

Godfrey Boyle and Peter Harper, eds. (1976) *Radical Technology*, Random House

Richard Congdon, ed. (1977) *Appropriate Technology*, Rondale Press

*Organic Gardening Magazine*

*Harrowsmith Magazine*

*Mother Earth News Magazine*

# Chapter 14: The Future is Passive Solar

After reading through this book you should have a reasonably good idea of what passive solar is—and what it can do. This final chapter examines the role passive solar must play in America's future. It begins by examining America's current energy position, precarious at best; then describes the vital role passive solar can play in the future; and finally explores the paths to this survivable future.

## America's Energy Position: 1980

*"America, relatively, is weak, is vulnerable—because America must import huge quantities of oil to keep its economy afloat . . . Until America resolves its energy dependence, it is a shackled giant . . ."*
— Manchester Guardian

America's position is probably worse now than it has been since the bleak winter of despair, 1977-78, when George Washington and has ragtag army suffered at Valley Forge. Low on money, low on supplies, suffering from the cold, and discouraged by a series of defeats on the battlefield they still retain independent sources of supply (meagre though they were).

*"Washington's veterans wintered at Valley Forge like animals in burrows, the tide of war fell to its lowest ebb. Not only did cabals, real or fancied, split the army and the Congress, but also many civilians began to give up the cause for lost. Paper money, that barometer of public faith, issued by both the Congress and the states, was practically worthless."*
—Rebels and Redcoats, G.F. Scheer and H.F. Rankin (1957) The World Publishing Company

Today by contrast, America is shackled by her dependence on imported oil. This dependence on imported oil (more than 60% in 1979), like other additions, has a number of serious side effects other than the most serious loss of security and freedom. These include: increasing inflation, weakening the dollar against foreign exchange, isallocation of resources (particularly capital), and increasing the national debt. If we hope to remain a free and independent country (at least insofar as possible with current world market inter-dependence) we have no choice but to end this addiction.

The question is whether we have the will to do it. The colonials struggle for independence was triggered by the Tea Tax which would have provided only 3001 per year to King George by contrast the price (an energy tax if you will) we pay to OPEC was $60 billion in 1979 and expected to reach $85 billion in 1980. Yet, we still haven't made any real effort to cut energy use. "Pay more, sure, why not - it's easiest," seems to be the current philosophy.

We could eliminate muh of this demand for foreign oil within a few months if we try. This has been demonstrated in a number of studies, including the Energy Project at the Harvard Business School, and more practically by a number of energy programs in industry (Dow for example, has reduced energy use by over 40% in 10 years) and the efforts of families who have made the effort to reduce their own reliance on non-renewable funds. Yet, as a nation, we seem to lack the will and the leadership necessary to voluntarily take this step. In fact, it now looks as if we will wait until our oil supplies are cut off before undertaking this type of program. This will certainly be more traumatic than if we did it voluntarily.

Why might our supplies be curtailed? Well, events in Iran and Afghanistan are hardly encouraging when you realize most of the oil we import passes through the narrow Straights of Hormuz. The sinking of a single supertanker, mining, or missile attacks could easily shut the Straits—even if we remain on good terms with the Gulf Oil States. Then we will have no choice but to begin the transition to energy independence.

*"The nearer any disease approaches to a crisis, the nearer it is to a cure. Danger and deliverance make their advances together; and it is only in the last push that one or the other takes the lead."*

Thomas Paine 1937-1809

The key to energy independence will be the transition to an economy based on decentralized renewable fuels. The first step in this process will involve a concerted effort to reduce energy use. How much can we conserve? The answer varies with who you believe, and what the assumptions you make are. However, there now seems little doubt that a savings of 50% would be possible with no change in the standard of living. The main effort here is in plugging the leaks and improving the efficiency of our machines, transportation, buildings, and industrial processes. It will also involve a much greater effort at using energy more wisely through cogeneration and othe multiple uses, such as the district heating widely practiced in Sweden.

This is all that would be necessary—if eliminating dependence on foreign oil were our only goal. Yet, this would still leave us dependent on sources of energy that are exhaustible, particularly oil and natural gas. Prices are rising rapidly for domestic sources also—and they will rise even faster when our imported oil is cut off. The rapidly rising prices and uncertain supplies of these non-renewable fuels are also a threat to our security and we should begin immediately to wean ourselves from them.

Can it be done? Once again there is no qustion that it can be done—if we have the will. The box left includes several reasons why we think so. The available renewable resources, the sun, wind, hydropower, and biofuels could easily supply half of the space heating, cooling, lighting and hot water for residential uses by 1985. A major component of this saving will be from the passive solar space heating and cooling, hot water heating and daylighting.

A concerted biofuels effort (particularly ethanol from crop waste and residue) could probably replace 15% of the demand for motor fuel by 1985. If Brazil's success with their alcohol programs is any indication, we might do considerably better.

Industry poses a more difficult challenge because many processes require high quality energy for processing and mechanical operations. Here the major savings will come from preheating, using solar energy, and replacement of non-renewable feed stocks with renewable ones. A concerted effort may reduce use of non-renewable energy by 25% by 1985.

What then, are the total potential savings possible from a concerted effort to conserve energy and replace non-renewable fuels wiht renewable fuels. Well, they are as follows: By conservation

50%, and by replacement of fuels 33%, particularly for heating and cooling, of the remaining 50% for a total reduction of non-renewable energy use of 65% by 1985—with little or no change in standard of living.

If we are willing to make changes in our standard of living, many of which we might find desirable after they are made, we could do even better. This would involve such things as relying more on our own gardens, using the biodynamic French intensive methods of agriculture (and generally eating less energy costly foods!), working at home more, traveling less but enjoying it more, accepting wider temperature swings int he house, buying fewer more durable goods, and intensively recycling. Making these changes in addition to the savings from conservation and renewable fuels could lead to a total saving of 80% of current non-renewable energy consumption. Over the longer term these changes will probably occur anyway as the price of non-renewable energy rises but it will be much easier on us if we undertake a more rapid transition.

The chart below shows the changing pattern of future energy consumption discussed here. It may be useful to begin thinking more about our childrens' future, or their childrens' childrens future, rather than our own immediate demand to have it all NOW—Right Away!

# How do we get there from Here

A complete energy program for America is beyond the scope of this book, but it may help to suggest some of the methods and bring about the change necessary for America's survival. These include: Committment, Ending subsidies, reducing government involvement in energy, and increased investment in conservation and renewable fuels.

## Commitment

The first step is often the hardest, making the commitment. This will have to come from the grass roots—because there is no national leader or organization capable of doing it. Of the contenders for president, none, with the possible exception of Barry Commoner and Jerry Brown, seem qualified to lead this effort. So make up your own mind, and begin each in your own way.

You'll probably find it very helpful to calculate how much energy you used in 1979. Add up your energy bills for your house, your car, and traveling you did by plane, food, and wastes. (We hope to have a booklet out soon to help you do this). Then make the commitment to reduce your total energy use by 1/3 in 1980.

You'll probably find its surprisingly easy unless you are already using energy very wisely. For 1981 resolve to drop it 30% again, thi may require more careful calculation and the investment of some money—but the returns (both financial and emotional) will be well worth it.

Entering 1982, you will be using 1/2 the energy you use today. After that, you may wish to try for only 10% savings per year. This may be hard, but it will probably be feasible and economically sound to do so.

## End Subsidies

As we suggested earlier, we probably would still be using passive solar systems and other renewable sources of energy if the government subsidies for non-renewable fuels were not so high. Therefore, the first (probably futile) effort should be to reduce or eliminate the many subsidies for non-renewable fuels. These include such things as: fuel tax credits, special depreciation allowances, the depletion allowance, credit for foreign taxes paid, etc. In 1978 these subsidies were calculated to be over $14 billion for oil and $20 billion for electricity. Between 1960-1977 over $24 billion dollars in subsidies were provided for nuclear energy. It is surprising that solar energy is as cost competitive as it is . . .

These figures mentioned above do not include the more indirect subsidies paid by society in covering the indirect costs of using non-renewable fuels. These include such things as: air pollution and its costly impact on health, crop production, and natural ecosystem (10's of billion per year); oil spills and resulting damage, particularly to the ocean biosystems (estimated spill total, 4 million metric tons in 1980); virtually irrevocable damage to land productivity caused by strip mining (2 million acres unreclaimed by 1970 in the U.S. alone), and the list goes on and on. I haven't seen an estimate of these indirect costs - but I would hazard a guess that it is on the order of 50 billion dollars a year.

If we were to eliminate these subsidies for non-renewable fuels we would have to do nothing else, the market would quickly provide the alternatives, which are in most cses competitive today. Unfortunately, the special interests are adept at protecting themselves and there is little chance of doing more than putting a dent in these subsidies.

Our best option, then, may be to raise the prices by increasing taxes on non-renewable energy. Although this is less desirable than simply letting the market rule, it can still be very effective. I would suggest the following: elimination of all property tax on principal residence and land and replacement of that value with increased energy cost. This does two things, first it enables the farmer or homeowner to operate on a much lower income, encouraging better use of the land and time, and second it puts a real incentive on energy conservation and the transition to renewable fuels.

## Reduce Government

At the same time it would be very helpful to reduce government involvement in energy management and research and development. This is important for several reasons:

First, the government is generally very poor at allocating resources among different alternatives (the spending in solar is illustrative, as of 1977, 400 million had been spent, of this only 100 million on active solar heating and cooling (primarily on almost ridiculous, overly technical and uneconomic systems), and less than $2 million had been spent on passive solar (this might be why passive solar is doing so well!).

Second, the government is ineffective and inefficient at conducting research and development, as are the larger corporations they favor with their largesse. Anyone who has worked in the bureaucracy or large company can see why—there is little or no reward for brilliance or hard work. Instead the incentives are for maintaining a low profile, not rocking the boat, and retreating home at night to the "real" world where bills have to be paid on time, commitments met, and decisions made. A recent study credited small business with over 1/2 of the industrial innovations between 1953-1973, yet only during the same period 4% of all Federal payouts went to small firms and within the Department of Energy, only 2.6% went to small firms.

Third, and perhaps even more critical is the loss of talented people who might otherwise make valuable contributions in energy conservation and alternative energy development. Although many

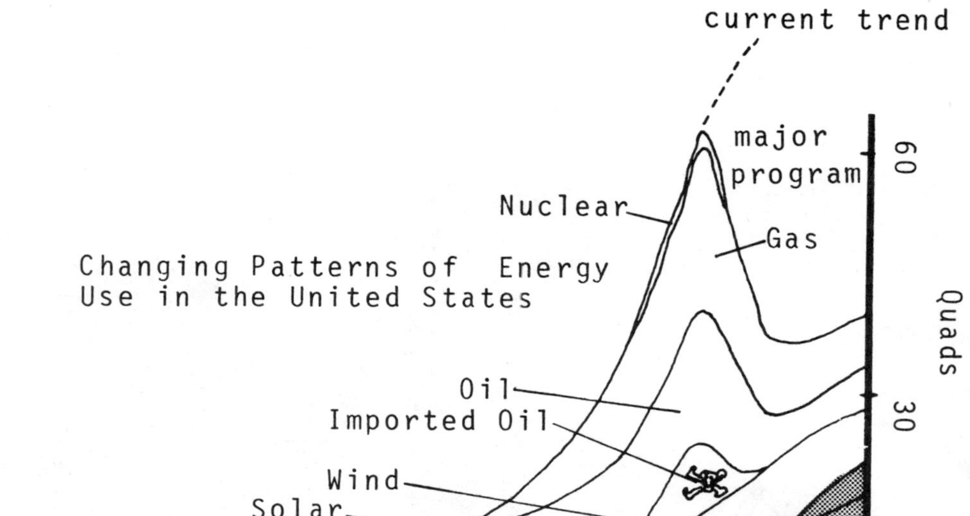

of the best prospects last only a few months in the bureaucracy, countless others are seduced by the high pay (often twice what they might make in business), short hours, and lack of responsibility, no bills to pay or eal schedules to meet. The Department of Energy and its related labs, the big energy corporations, and various state energy bureaucracies probably account for over 95% of the people involved in energy R & D. If we could reduce this drag on development even slightly who knows what we might achieve.

## Increased Investment

We will also have to increase our investment for conservation, renewable fuels, and particularly passive solar systems. This will mean increased spending of both time and money by the individual or family, the builder, business and industry, the utilities, and governments. The potential for these various sectors is briefly described in this section.

Personal investment in passive solar design is already the major source of funding for development and installation of passive systems. People are wiser than the economists believe and are making the investment because it will save money and provide increased security. Passive solar is one of the best investments in an inflationary economy and as fuel costs soar, the low fixed cost of solar-heated homes will grow increasingly attractive. Returns even today are often more than 30% per year. With a reduction in subsidies this could probably reach 60-70% a year, considerably better than stocks and bonds and these savings increase with inflation.

The houses of America are predominately designed and built by small builders. It is an area of American business where independence and hard work are still the rule—rather than the exception and it is a business that is beginning to move rapidly toward passive design. The California Building Industry Association was one of the first groups to really understand the implications of the continuing reliance on foreign oil and nonrenewable fuels. In a resolution passed in August, 1978, they concluded: "the California Building Industry Association encourages home builders in California to incorporate energy conservation and passive solar design in houses . . .".

The building industry can, and I believe will, make almost all of the new housing built in 1982 passive solar. In 1980, several large scale passive developers will be begun, including a 2,900 unit development near Los Angeles.

Business and industry can also move fairly quickly to more widespread use of passive solar design. Commercial buildings are particularly attractive because even in cold areas internal heat gain and fairly limited solar glazing can meet the full heating demand. The potential savings in energy now used for cooling from good solar control programs, with shadescreen and roof coatings, are staggering and should lead to some fast and high profitable growth in those industries.

The Passive Solar Institute has and will continue to devote much of its effort to stimulating activity in Passive Solar market development. The demand is there but in many cases the products aren't. Business and industry have fallen down in this regard and the implications of this failure are not good. Research and development has suffered from corporate complacency and indolence. Companies should be investing much more both in their own research and also in supporting independent research. Particular growth potential in the 1980's is in manufacturing passive soalr components: thermal shutters and drapes, thermal mass tanks, greenhouse module, passive solar water heater, and others. Growth will probably be as rapid as that experienced in the wood stove industry. In 1970, there were about 10 companies involved in wood stove manufacturing—today there are over 200.

The utilities also have an important role to play. Being somewhere between an industry and a governmenttal body, regulated and protected, they suffer from the worst effects of both. Profits are often guaranteed by regulating agencies, based on investments. Hence, it is reasonable to invest in as many bit power plants as possible because they cost more.

If they weren't so oddly constitued they would have long ago invested in energy conservation, passive solar design, and renewable fuels because producing more power using nonrenewable fuels is almost always the most expensive alternative. Utilities are slowly but steadily making progress however. The most notable success is the Tennesee Valley Authorities' passive solar program. Others, including virtually all of California's utilities have developed passive solar and conservation programs. The Sacramento Municipal Utilities District has been particularly active, and sponsored a Passive Solar Institute seminar on energy efficient design in 1979. They also issued a Passive Solar Study Report in 1979 which should be read by every utility manager in the U.S. It showed that the eight passive systems studied could meet an average of 68% of the heating demand, 74% of the cooling, and 53% of the hot water. Energy savings of up to $200/year are projected, on a house that with Federal and State tax credits should cost at worst $1,000 more than a conventional house, and could more commonly be built for the same price.

*"A national crusade for energy independence would end our bondage to imported oil. It would go a long way toward controlling runaway inflation. It would give us a common purpose, direction and goals in which every American could become intimately involved—qualities that have been sadly lacking in American life lately. And it would help us learn to live on a limited planet. Using energy efficiently could set a pattern for developing efficiency in the use of all of the planet's limited resources—minerals, water, air, soil."*

Harold Gilliam (1979) This World,
The San Francisco Chronicle

The Federal, State, and local governments can also do much to promote passive solar design. Rather than mandating solar, like San Diego County, or investing in ever more energy studies and plans they should remove the subsidies for renewable fuels, remove regulatory barriers, set a good example by retrofitting their buildings, and let the market work.

And rather than overregulating and swamping the market with rule changes and conflicting rules governments should set long term standards which protect the public (if they set any at all) and stick to them. Better yet, they should set specifications instead and buy a few hundred million dollars of passive solar products a year for retrofitting their own buildings. Those that can meet the specifications will prosper and build even better products.

## How Much Would it Cost

The revenue now flowing to the Mid-East would be enough to implement all of the changes suggested earlier—50% saving by conservation and transition to 33% renewable fuels by 1986. The $360 billion dollars that would otherwise go to OPEC would allow us to spend: $1500 per house in America (we could probably save more than 50% and over 33% here); another 100 billion for industrial retrofit, redesign, and particularly operations management (we might make 50% conservation but 33% renewable fuel would be unlikely); and we still would have over 80 billion dollars for development of renewable fuels.

If we split this equally between solar, hydropower, wind, and biofuels this would probably provide the following amounts of energy, based on best available estimates. Twenty billion dollars for solar, split 7 billion for passive (process heat, etc.), 7 billion for active (a wide variety of uses), 6 billion for photovotaics (electricity), and 1 billion for soalr thermal would provide approximately 1/2 of the process wtaer used in industry and the equivalent of 800 megawatts of peak electricity generating capacity.

Our 20 billion dollars for hydro-power, predominately for small scale x 10 kw plants, would provide an additional 10,000 megawatts of electrical capacity. Even wind, which is more expensive, might provide an additional 8,000 megawatts of capacity.

The 18,800 megawatts of capacity from direct thermal, protovotaics, wind, and hydro would at best produce only about 2 quadrillion BTU?year, or about as much as nuclear power plants did in 1975. But they will easily do much more by 1996.

The impact of the biofuel program is a little more difficult to estimate, however, based on estimates from Canada and the U.S. an investment of 20 billion dollars should provide alcohol capacity equivalent to 370 million barrels of oil per year in 1985, barring development delays. This represents about 15% of U.S. motor gasoline demand. This combined with the existing use of wood for direct combustion primarily in homes and also in power plants will easily meet the goal established earlier.

It will admittedly take a great deal of work, and a considerable effort, and will cost a great deal. Yet the cost of not doing it will be much higher.

PLEASE TYPE USING A CARBON RIBBON (10 pt. Elite preferred)

**PASSIVE SOLAR INSTITUTE**
**P.O. BOX 722**
**DAVIS, CA 95616**

*We are also looking for good retrofit and commercial projects to include in the text. Please send us a brief description and photo.*

---

## CONSULTANT LISTING — PASSIVE SOLAR CATALOG

1. Name ______________________________
2. Address ______________________________
   ______________________________
3. Phone ______________________________

**5.** TYPE (Check Those That Apply)

| | |
|---|---|
| A. Consulting | ☐ |
| B. Design | ☐ |
| C. Engineering | ☐ |
| D. Research & Development | ☐ |
| E. Energy Audit | ☐ |
| F. Energy Conservation | ☐ |
| G. Daylighting | ☐ |
| H. Other Passive | ☐ |

**6.** System Used (Indicate Number)

| | Designed | Underway | Completed |
|---|---|---|---|
| A. Green House/Solarium | ☐ | ☐ | ☐ |
| B. Water Wall | ☐ | ☐ | ☐ |
| C. Trombe Wall | ☐ | ☐ | ☐ |
| D. Underground | ☐ | ☐ | ☐ |
| E. Simple Passive | ☐ | ☐ | ☐ |
| F. Roof Pond | ☐ | ☐ | ☐ |
| G. Hybrid | ☐ | ☐ | ☐ |
| H. Other Passive | ☐ | ☐ | ☐ |

**7.** Emphasis (Indicate Number)

| | Designed | Underway | Completed |
|---|---|---|---|
| A. Residential | ☐ | ☐ | ☐ |
| B. Commercial | ☐ | ☐ | ☐ |
| C. Industrial | ☐ | ☐ | ☐ |

**8.** Passive Water Heating System Used

| | Designed | Underway | Completed |
|---|---|---|---|
| A. Bread Box | ☐ | ☐ | ☐ |
| B. Thermosiphon | ☐ | ☐ | ☐ |
| C. Wood Assisted | ☐ | ☐ | ☐ |

**9.** Other Specialties (Indicate Number)

| | Designed | Underway | Completed |
|---|---|---|---|
| A. Movable Insulation | ☐ | ☐ | ☐ |
| B. Shading | ☐ | ☐ | ☐ |
| C. Ventilation | ☐ | ☐ | ☐ |
| D. Weatherization | ☐ | ☐ | ☐ |
| E. Wood Heating | ☐ | ☐ | ☐ |
| F. Photo Voltaics | ☐ | ☐ | ☐ |
| G. Wind | ☐ | ☐ | ☐ |
| H. Hydropower | ☐ | ☐ | ☐ |
| I. Active Solar | ☐ | ☐ | ☐ |
| J. Back Up System | ☐ | ☐ | ☐ |

**10.** Articles & Books (4 Most Important)

A.

B.

C.

D.

**11.** Years in Business

**12.** % Work Private ☐ Government ☐

**13.** Licenses & Registration

**14.** Hourly Fee (opt.)

# SOLAR HOUSE PLANS $25

**Performance of the Solar Clerestory Water Wall House**

| Area | Solar Heating | Natural Cooling |
|---|---|---|
| Los Angeles, CA | 100% | 100% |
| Sacramento, CA | 94% | 100% |
| Medford, OR | 72% | 100% |
| Phoenix, AZ | 100% | 50% |
| Denver, CO | 82% | 100% |
| Ft. Worth, TX | 100% | 30% |
| Little Rock, AK | 89% | 37% |
| Knoxville, TN | 75% | 100% |
| Raleigh, NC | 99% | 43% |
| Miami, FL | 100% | 57% |

**Even better performance and greater savings could be achieved in a more sophisticated passive design. The flat roof on the south of the clerestory was included so that a more sophisticated cooling system could be installed without complete revision of the plan. A Skytherm" system, "Energy Roof", flooded roof, or sod roof could all provide additional cooling in the hotter areas. The added mass of these systems would require revisions to the framing and ceiling and should be worked out with the system designer or a structural engineer.**

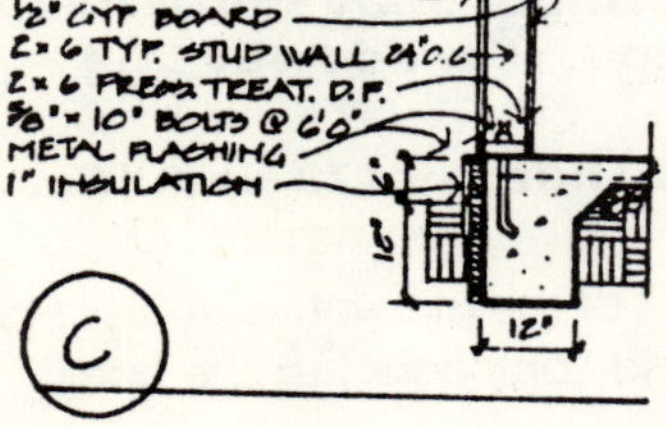

*photo by Chad Ankele*

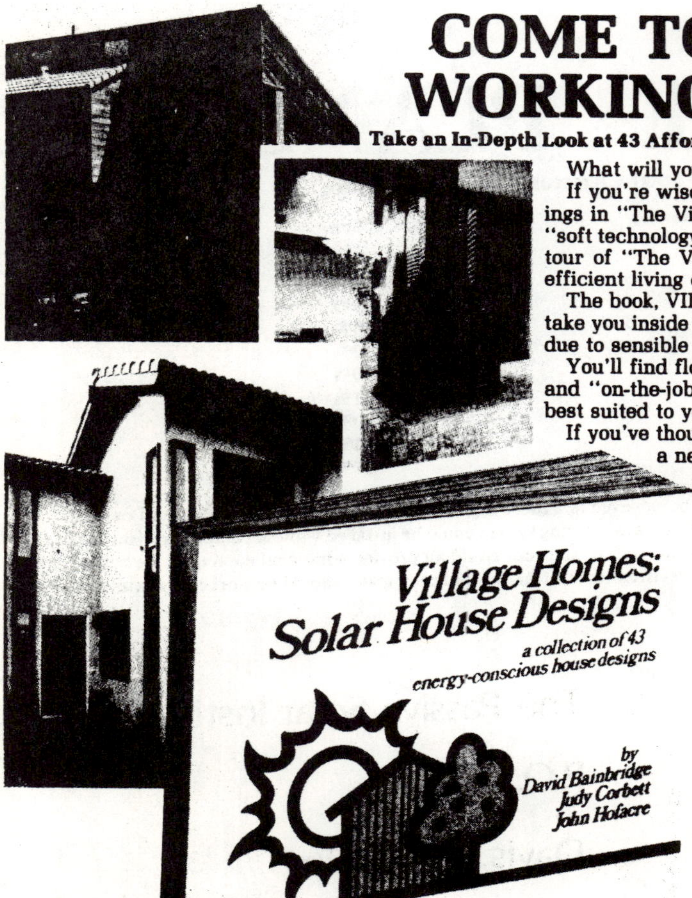


# COME TOUR AMERICA'S FIRST WORKING SOLAR SUBDIVISION

**Take an In-Depth Look at 43 Affordable New Age Homes Designed Especially for Middle-Income Families**

What will your next home look like?

If you're wise and energy-conscious, it will probably resemble the "new age" dwellings in "The Village," a solar subdivision in Davis, California, that is making the new "soft technology" of the future work *today.* And now there is a book that will take you on tour of "The Village" homes so you can see just how practical and feasible energy-efficient living can be.

The book, VILLAGE HOMES, has over 90 photographs and over 100 illustrations that take you inside 43 *working,* moderately-priced homes that use at least 50% less energy due to sensible design and the use of solar features.

You'll find floor plans, specific construction steps integral to each design, and price and "on-the-job" performance reports so you can choose the design and components best suited to your needs.

If you've thought about solar energy for your house, or if you're considering building a new solar home, or if you just want to know more about affordable, good-looking homes with solar features, then VILLAGE HOMES is for you.

Send for your copy soon and get a glimpse at tomorrow's housing . . . at work today!

**Village Homes: Solar House Designs, 1978 David A. Bainbridge, Judy Corbett, John Hofacre $7.95 ppd. (Ca. add 40¢ tax)**

**Village Homes Slide Set $20 ppd. (Ca. add $1.20 tax)**

Twenty color slides of the Innovative Village Homes Subdivision. A good companion to the book.

## About the Author:

David Bainbridge was born in 1948 in Schenectady, N.Y. At the time his parents lived in a log cabin nestled in the woods by a beautiful mountain stream and the mountains and wilderness have continued to play an important role in his life. Family moves and travel have introduced him to the North Cascades, Canadian and Colorado Rockies, and the Sierra Nevada.

After graduating from U.C. San Diego with a major in earth sciences and a minor in surfing he worked for the San Diego Environmental Development Agency in a successful effort to save the coastal lagoons from imminent development. This project led to a Masters degree from U.C. Davis in Ecology with emphasis on ecology, planning, and environmental geology.

While going to school in Davis he worked with the state Joint Committee on Environmental Quality and formed an environmental consulting firm which specialized in Bikeway Planning, Environmental Impact Analysis, and Environmental Planning and Management.

He retired after two hectic and rewarding years to help his parents build a homestead in Cortez, Colorado. This involved completely dismantling and reccycling a dilapidated house, including the nails, into an energy conserving house.

On his return to California he worked for the Association of Bay Area Governments on an innovative Land Capability study. At the same time he began the restoration of an 1892 Victorian house in a small town near Sacramento that was to take two years. After completing the ABAG report he began what was to become two years of very interesting and productive work with a solar design group in Winters, California.

These two years with Living Systems included research, writing, photography, and design of passive solar and energy conservation projects. David was in charge of the Planning for Energy Conservation studies in Davis, Indio, and Sacramento County. He also was involved in natural lighting studies for the State Office Building, the Energy Conservation Building Code Workbooks for Indio and Sacramento, and the first drafts of a report on the challenge of planning for Solar Access.

He retired to devote more time to completing his passive solar home in Davis and to his writing. Unfortunately, retirement lasted only a month before an urgent call put him to work on the development of the Passive Section of California's Solar Tax credit at the Solar Office of the California Energy Commission. Other reports and studies followed at regular intervals.

He resigned from the Energy Commission to form the Passive Solar Institute and promote the widespread and rapid transition to passive solar utilization. He has many reports underway in addition to the Passive Solar Catalog.

After finishing his house in Village Homes in January 1978 he became more involved in the affairs of this remarkable 70 acre solar subdivision. He has served on the Board of Directors, the Small Farms review board, and has written a book about the development which will be published by Rodale Press early in 1979.

In addition to his work on the Passive Solar Institute he does freelance writing, photography, energy consulting, lecturing, and concrete and carpentry work for friends building in Village Homes. On his time off, which is all too rare, he enjoys backpacking, skiing, surfing, canoeing, playing guitar, and writing poetry.

# The Transition: Further Reading: A Sampler


### The United States

Robert Stobaugh and Daniel Yergin, Eds. (1979) Energy Future: Report of the Energy Project at the Harvard Business School, Random House (A good, though conservative look at conservation look at conservation and renewable energy. Recognition at last by the establishment).

Council on Environmental Quality (1978) Solar Energy: Prospects and Promises, 722 Jackson Place, N.W. Washington, D.C. 20006 (Projects solar meeting 25% of U.S. energy demand in 2000, doesn't include strong enough conservation program).

Robert Williams (1978) "Industrial CoGeneration" Center for Environmental Studies, Princeton University, Report #66, May. To be reprinted in Annual Reviews of Energy, v. 3, 1978. (Est. 208 gw of cogeneration possible by 2000).

Jon Hammond, Bill Kopper, David Bainbridge, et al. (1979) The Davis Energy Conservation Report, from Living Systems, Rte. 1, Box 170, Winters, (The Davis project, Building Code, Planning Policy, and demonstration houses).

Dennis Hayes, Gary DeLoss, Jon Lash, et al. (1979) Blue Print For A Solar America, Solar Lobby, 1028 Connecticut Ave., N.W., Washington, D.C. 20036. $2.
(A fairly conservative look at the transition, overly enthusiastic about government involvement.

John Huetter, et al. (1978) Status, Potential, and Problems of Small Hydro-electric Power Development in the U.S., Energy Research and Applications for C.E.Q.
(Estimates 200 gw capacity in U.S. untapped, v/s 68 gw now in use.)

Paul Craig, et al. (1979) Distributed Energy Systems in California's Future, D.O. E., O.T.I., Washington, D.C.
(A conservative and slightly off target project. Greatly underestimates solar, particularly passive solar contribution. Also misses the boat on biomass but still worth a look.)

Marc Ross and Robert Williams (1979) "Drilling for Oil and Gas in Our Buildings" The Center for Energy and Enviromental Studies, Princeton.
(Calculate savings equivalent to 912 million barrels/year at half the cost of imported oil by conservation retrofit).

Joe Szostak (1979) "$60 million U.S. program helps accelerate wind power," CREN, March. (American Wind Energy Association estimates wind potential equal 100 million barrels of oil/year by 1985 with a modest investment).

### CANADA

Intergroup (1978) Liquid Fuels from Renewable Resources, Government of Canada, Interdepartmental Steering Committee on Canadian Liquid Fuels, from Intergroup, 704-283 Portage Ave., Winnipeg, Manitoba.
(Near term estimate 3 billion gallons alcohol each year but up to 10 billion by late 80's, and 16 billion by year 2000).

### ENGLAND

Gerald Leach, et al. (1979) A Low Energy Strategy for the United Kingdom, Science Reviews and IIED, 10 Percy Street, London, WIP ODR.

### FRANCE

Tout Solaire, Amis de la Terre, 14 bis, Rue de l'Arbalete, 7005, Paris.

Project Alter, Le Group de Bellevue, 85 Boulevarde Port Royal, 75031, Paris 8F.

### GERMANY

Florentin Krause (1979) "A Soft Energy Path for Germany, "Soft Energy Notes V8n2 Dec.

### DENMARK

Jorgen Norgard (1979) Households and Energy, DEMO Project, Technical Univ. of Denmark.

### BRAZIL

National Alcohol Program, Funda cao de Technologia Industrial, Attn: Dr. Marchado, Ave. Venekzuela, 827 Andar, 20081 Rio de Janeiro, Brazil.

### GENERAL

Ivan Illich (1974) Energy and Equity, Calders and Boyer, London.

Amory B. Lovins (1977) Soft Energy Paths, Ballinger/FOE.

David Bainbridge (1979) Patwin: An Energy Self Sufficient New Town: Preliminary Report, PSI P.O. Box 722, Davis, CA 95616. $10.

Helen and Scott Nearing (1979) Continuing the Good Life, Schocken

Warren Johnson (1978) Muddling Towards Frugality, SC. Books, San Francisco.

Peter Van Dresser (1976) A Landscape for Humans Lighting Tree Press, Sante Fe.

Wendel Berry (1976) The Unsettling of America, S.C Books, San Francisco.

Richard Stein (1978) Architecture and Energy Anchor Books.


## Average of all Eight SMUD Passive House Designs as a Percent of Calculated Energy Budgets used in Comparison with Existing Housing

| Space Heating | | Space Cooling | |
|---|---|---|---|
| Cal. Code Title 24 | Conventional 1960's Home | Cal. Code Title 24 | Conventional 1960's Home |
| 37.7% | 17.8% | 20.2% | 11.2% |

## SMUD Passive Solar Design #8A, 1280 Sq. ft.

### Solar Contributions
### Heating BTU x 10 6/Mo.

| | Jan. | Feb. | Mar. | Apr. | May | Jun. |
|---|---|---|---|---|---|---|
| Load: | 5.73 | 3.98 | 3.49 | 2.12 | 0.85 | 0.18 |
| Solar | 3.59 | 3.43 | 3.49 | 2.12 | 0.85 | 0.18 |
| Per Cent | 62.6% | 86.2% | 100.0% | 100.0% | 100.0% | 100.0% |

| | July | Aug. | Sept. | Oct. | Nov. | Dec. |
|---|---|---|---|---|---|---|
| Load: | 0.00 | 0.00 | 0.08 | 0.90 | 3.27 | 5.44 |
| Solar | 0.00 | 0.00 | 0.08 | 0.90 | 3.27 | 3.53 |
| Per Cent | 100.0% | 100.0% | 100.0% | 100.0% | 100.0% | 64.9% |

**Annual Load: 26.04 Solar: 21.44 82.3%**

### Cooling BTU x 10 6/Mo.

| | Jan. | Feb. | Mar. | Apr. | May | June |
|---|---|---|---|---|---|---|
| Load: | 0.00 | 0.00 | 0.00 | 0.17 | 0.47 | 1.23 |
| Natural: | 0.00 | 0.00 | 0.00 | 0.17 | 0.47 | 1.23 |
| Per Cent Natural | 100.0% | 100.0% | 100.0% | 100.0% | 100.0% | 100.0% |

| | July | Aug. | Spet. | Oct. | Nov. | Dec. |
|---|---|---|---|---|---|---|
| Load: | 1.97 | 1.68 | 1.25 | 0.25 | 0.00 | 0.00 |
| Natural: | 1.47 | 1.68 | 1.25 | 0.25 | 0.00 | 0.00 |
| Per Cent Natural | 100.0% | 100.0% | 100.0% | 100.0% | 100.0% | 100.0% |

**Annual Load: 7.01 Natural: 7.01 100.0%**

## Estimated Annual Energy Costs for Heating and Cooling

| | Specified Design | State House | Older House |
|---|---|---|---|
| **Heating** | | | |
| Kilobtus: | 4603 | 35405 | 68871 |
| Natural Gas: (at $.250 Therm) | $ 19.28 | $147.52 | $286.96 |
| Electricity: (at $.018/KHW) | ($ 24.95) | ($191.91) | ($ 373.31) |
| **Cooling** | | | |
| Kilobtus: | 0 | 15555 | 28254 |
| Electricity: (at $.018/KHW & EER 7) | 0.00 | $ 41.11 | $ 74.67 |
| **Total Kilobtus:** | **4603** | **50,960** | **97,125** |
| **Total Costs:** | **$19.18** | **$ 380.54** | **$723.26** |

## Corrections p = page c = column p = paragraph

pb c1 p5 already
p2 c1 p1 than it has received
p2 c3 p2 delete extra "Assuming the estimate of desired investment in solar systems."
p2 chart is California experience
p2 c3 p4 indiscriminately
p3 c3 p2 increasing
p5 Graph is my estimate, DHW & Space Conditioning
p6 c2 p2 It can go up
p7 c1 rule 4 close paren.
p7 c2 p2 oriented so a
p7 c2 p3 significant
p7 c3 p3 thermal mass
p8 c1 p2 graphics changed, lids can slide as shown here or lift.
p8 c3 p2 However, it is
p8 c3 p3 incentive to build
p8 c3 bottom PSWH book will be available March 1981, cost $10.95 from PSI.
p10 c1 p4 for an established one
p10 c2 add insulated shutters and shades (should be near the top of this list)
p10 c3 p5 sealant
p11 c1 p3 only when
p11 c1 p5 casement windows
p11 c2 p3 acrylic
p11 c3 p6 sealant
p13 c1 p6 delete paren. around 1/8"
p13 c2 p1 you have
p13 c2 p7 you might
p14 c1 p5 vapor barrier; multiple layers are used; embarrassing
p14 c2 p1 In colder; ice damming
p14 c2 p4 joists
p15 c1 p2 espaliered
p15 c2 p1 window
p16 c3 p5 for the ventilator
p17 c1 p5 tilted down
p19 c1 p4 paren. on 4)
p19 c3 p1 occasionally
p19 c3 p5 your building's
p20 Table 1 water specific heat is 1
p21 c3 p2 A plastic bag liner is advisable for most 5-gallon cans.
p22 c1 p2 considerable
p22 c1 p5 will be determined
p23 c3 p2 shear and overturning
p23 c3 p5 lateral acceleration
p25 c1 p2 delete paren. 60-70%
p25 c2 p2 flow restrictors
p25 c3 p6 preceding; If it is properly
p26 c2 p7 delete large type "If we bought commercial insu"
p27 c3 p3 investment
p28 The Handbook can be ordered for $7.42 from CEC, 1111 Howe Ave., Sacramento, CA 95825
p28 c1 p2 Berkeley
p44 c1 p3 convected
p48 p1 This retrofit
p48 p2 in the addition
p50 c1 p2 recapture
p52 p2 building was operated
p55 c1 p2 noise
p57 c1 p3 absorption
p58 c1 p3 recommendation
p59 c2 Andersen Windowalls
p85 c1 p1 local
p85 c1 p2 was as complete, however
p85 c1 p3 recommendation; work is done.
p85 c2 p3 review with
p85 c2 bottom letter d stands
p92 William Shurcliff's book on Super Insulated and Double Envelope Homes is recommended reading. $10.95
p95 c1 p1 solar pioneers
p98 c3 Solar Energy Digest should be Box 17776, New Shelter should read Emmaus, PA
p99 c2 p2 reasonably
p102 c1 p1 Well, the industrial
p103 c1 p2 1777-78; his ragtag army
p103 c1 p3 like other addictions; misallocation
p103 c1 p4 300 pounds
p103 c1 p5 much of this
p103 c2 p2 Straits of Hormuz
p103 c3 p1 no question
p103 c3 p2 The chart on page 104 includes several reasons why we think so.
p103 c3 p5 with renewable
p104 c2 p3 this may
p104 c3 p3 cases competitive
p104 c3 p6 yet during
p105 c1 p1 real
p105 c1 p4 predominantly
p105 c2 p2 solar components
p105 c2 p3 governmental; big power plants
p105 c2 p4 oddly constituted
p105 c3 p4 solar thermal
p105 c3 p6 photovoltaics; BTU/year
p106 page should be reversed with 110
p110 c1 p1 delete extra "look at conservation"

## Additions: Products

**1.4** Pella Sunroom
Rollscreen Co.
Pella, Iowa 50219
Wood frame window components engineered for use as a sunroom or solar greenhouse.

**7.3** In-Sol Sliders
Energy Industries
2010 N. Redwood Drive, Rte. 1
Independence, MO 64050
816-257-1919
Double walled rigid polypropylene plastic panels for interior storm windows. 65¢ per sq. ft. Includes instructions.

**8.2** Climatite, Inc.
Box 269
Berlin, NY 12022
Jordan Holt
518-658-2860
Tight, well insulated manufactured homes with timber frame and insulated panel skin. House kits include AT drapes, Andersen windows, etc. Less than $30 per sq. ft.

**9.1** The Integral Passive Solar Water Heater Book, $10.95. Available March 1, 1981 from PSI, P.O. 722, Davis, CA 95616. The first comprehensive look at IPSWHs. Theory, design, many examples, plans, how to get started in business and much more.

**11.** Helios 1 Solar Daylight Heliostat
American Solar
2116 Center Street
Garden City, KS 67846
316-275-7488
Automated mirror system for enhanced daylighting.

**12.2** The Solar Photovoltaic Monitor
Monegon Ltd.
4 Professional Drive, Suite 130
Gaithersburg, MD 20760
Bimonthly of latest work in photovoltaics. $350/year.

**12.3** Jacobs Wind Electric
2180 West First Street, Suite 410
Fort Myers, Fla. 33901
The master builder, Marcellus Jacobs, is back in business with an 8-10 kv machine. $12,000.

**13.** Aeolian Kinetics
P.O. Box 100
Providence, RI 02901
Instruments for solar and wind evaluation. PDL-24 Data Logger has 24 channels, cassette storage, microcomputer, interface, etc. $4,000

**17.1** AIA Energy Notebook
1735 New York Ave. N.W.
Washington, D.C. 20006
Two volume practical references with updates. Passive section is weak but more to come. AIA members $90, others $120.

**17.2** Solar Energy Resources
2330 Old Middlefield Way
Mountain View, CA 94043
Distributor of solar/energy conservation products.

## Additions: Consultants

**Florida**

Nelson Hellmuth, P.E.
Solar Designs
1738 Kingsley Ave.
Orange Park, FL 32073

5. A 6. E

**New York**

Steven Robinson
2114 Broadway
New York, NY 10023

5. ABEFG 6. E4c 7.A 8.C 9. ABCDEFGI 10. *The Energy Efficient Home* (1978) New American Library

**Pennsylvania**

Charles Klein
525 Main Street
Bethlehem, PA 18018

5. ABH 6.A 5d 5c C 3d 2u 1c D 1d 1c E 20d 4u 16c G 1d 1u H 3d 1u 2c 7. a 24d 3u 17c B 3d 2u 9. A 12d 2u 10c B 10d 2u 8c C 3d 1u 2c D 4d 1u 3c E 3d 2u 3c I 1d 1c 11. 8 yrs 12. 100% private 13. PA License, National Certification

**Rhode Island**

Beckman, Blydenburgh, and Associates
P.O. Box 100
Annex Station
Providence, RI 02901
401-274-3690

5. ABCDEFGH 6. AEH 7. AB 9. ABCDG 13. RI arch.

# INDEX

Xerox page 85 if you wish for ease in reaching the right consultant. Don't do anymore, though, without permission.

## characteristics

COST:
ABOUT \$400.00 FOR MATERIAL ON A 3-TANK, 90 GAL. SYSTEM.

WILL HEAT TO 160°F IN MOST AREAS.

NEEDS FREEZE PROTECTION IF IN TEMPS. BELOW 32°F FOR DAYS.

## variations

IN A SKYLIGHT...

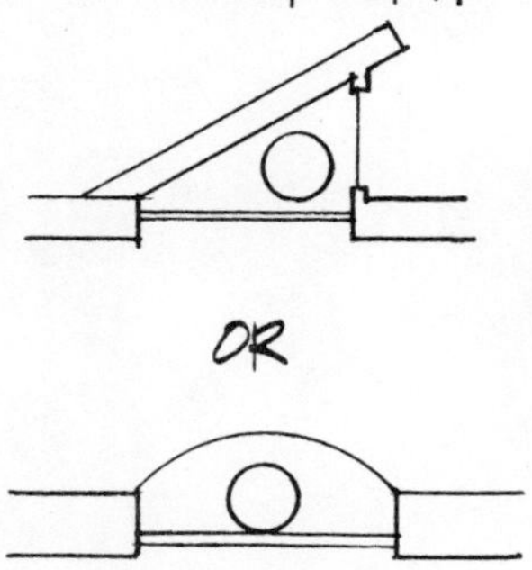


IN A GREENHOUSE

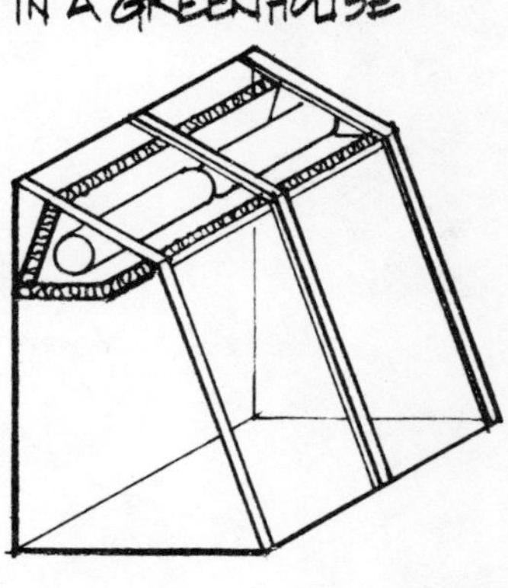

IN SOUTH WALL

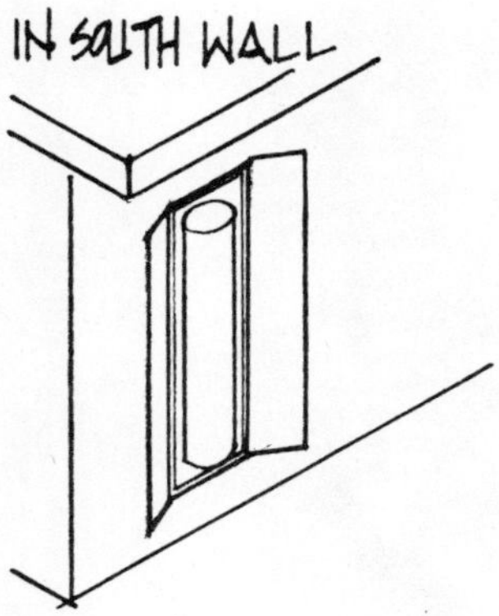

## 3 TANK PASSIVE SOLAR WATER HEATER

### CONSTRUCTION

1. BUILD ON SITE. WT. 400 LBS. EMPTY, 1200 LBS. FULL. BRACE ROOF AS NECESSARY
2. BUILD FRAME AND TANK DECK.
3. EXT. PLYWOOD
4. INSULATE
5. INTERIOR PLYWOOD OR SHEETROCK.
6. INSTALL TANKS AND PLUMB.
7. PRESSURE TEST
8. INSTALL GLAZING
9. TRIM
10. INSULATE PIPES AND DOPE IF ARMAFLEX.

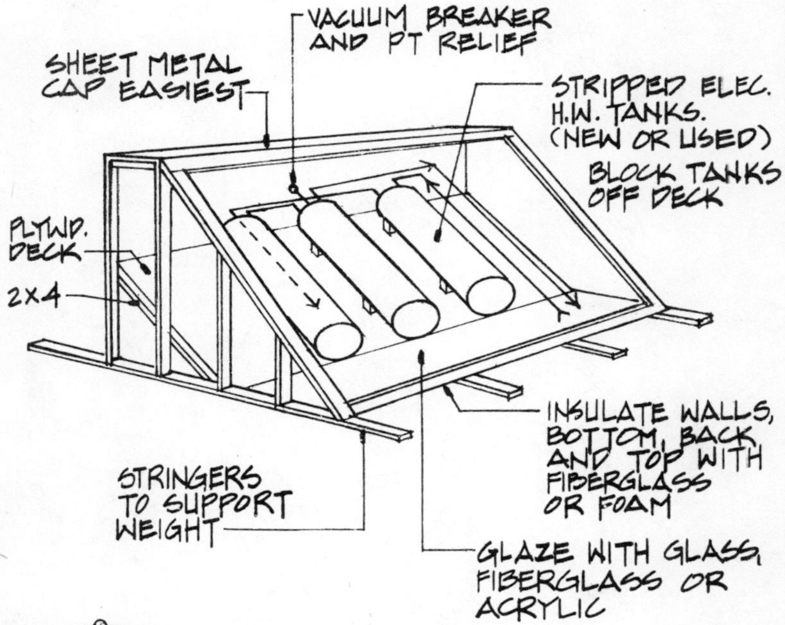


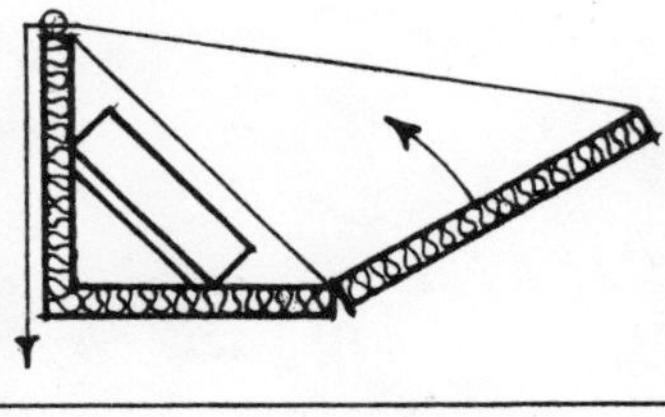

AN INSULATED COVER CAN BE VERY USEFUL. HINGE AT BOTTOM WITH PULLY AND ROPE TO GROUND IF NECESSARY.

## relation to other components

CAN BE USED FOR SPACE HEATING WITH A "CHILL CHASER".

MOVABLE INSULATION IS DESIRABLE TO REDUCE NIGHT HEAT LOSS:

CAN BE INTEGRATED WITH WOOD HEATER BY ADDING COIL OR TANK TO FIREBOX OR STACK.

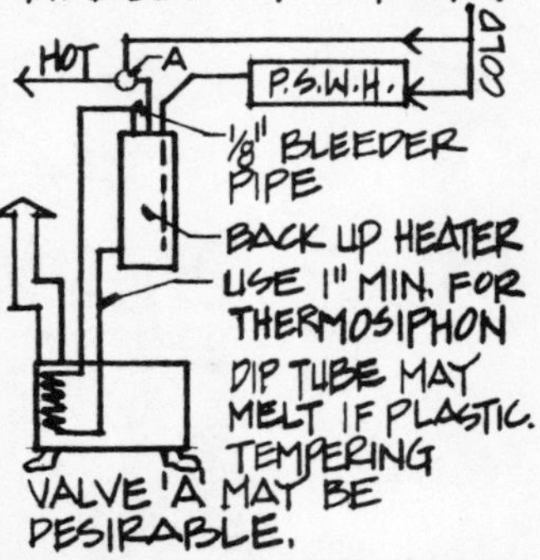


### VALVE POSITIONS

FULL SOLAR:
B, C CLOSED
A, B, D OPEN

PREHEAT:
A, B, E OPEN
C, D CLOSED

BYPASS:
A, B, D CLOSED
E, C OPEN

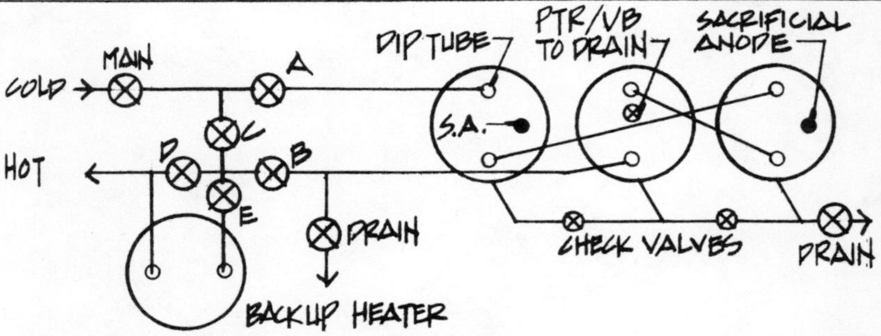


**applicability:** BEST IN TEMPERATE AREAS BUT CAN BE EFFECTIVE IN COLDER AREAS IF BUILT IN GREENHOUSE OR WITH MOVABLE INSULATION.